# Graduate Texts in Mathematics 224

**Springer**

*New York*
*Berlin*
*Heidelberg*
*Hong Kong*
*London*
*Milan*
*Paris*
*Tokyo*

# Graduate Texts in Mathematics

*(continued after index)*

Gerard Walschap

# Metric Structures in Differential Geometry

With 15 Figures

Springer

Gerard Walschap
Department of Mathematics
University of Oklahoma
Norman, OK 73019-0315
USA
gerard@math.ou.edu

Mathematics Subject Classification (2000): 53-xx, 58Axx, 57Rxx

Library of Congress Cataloging-in-Publication Data
Walschap, Gerard, 1954–
    Metric structures in differential geometry/Gerard Walschap.
      p. cm.
    Includes bibliographical references and index.
    ISBN 0-387-20430-X
  1. Geometry, Differential.  I. Title.
  QA641.W327 2004
  516.3′6—dc22                                          2003066219

ISBN 0-387-20430-X          Printed on acid-free paper.

Printed in the United States of America.     (EB)

9 8 7 6 5 4 3 2 1          SPIN 10958674

Springer-Verlag is a part of *Springer Science+Business Media*

*springeronline.com*

# Preface

This text is an elementary introduction to differential geometry. Although it was written for a graduate-level audience, the only requisite is a solid background in calculus, linear algebra, and basic point-set topology.

The first chapter covers the fundamentals of differentiable manifolds that are the bread and butter of differential geometry. All the usual topics are covered, culminating in Stokes' theorem together with some applications. The students' first contact with the subject can be overwhelming because of the wealth of abstract definitions involved, so examples have been stressed throughout. One concept, for instance, that students often find confusing is the definition of tangent vectors. They are first told that these are derivations on certain equivalence classes of functions, but later that the tangent space of $\mathbb{R}^n$ is "the same" as $\mathbb{R}^n$. We have tried to keep these spaces separate and to carefully explain how a vector space $E$ is canonically isomorphic to its tangent space at a point. This subtle distinction becomes essential when later discussing the vertical bundle of a given vector bundle.

The following two chapters are devoted to fiber bundles and homotopy theory of fibrations. Vector bundles have been emphasized, although principal bundles are also discussed in detail. Special attention has been given to bundles over spheres because the sphere is the simplest base space for nontrivial bundles, and the latter can be explicitly classified. The tangent bundle of the sphere, in particular, provides a clear and concrete illustration of the relation between the principal frame bundle and the associated vector bundle, and a short section has been specifically devoted to it.

Chapter 4 studies bundles from the point of view of differential geometry, by introducing connections, holonomy, and curvature. Here again, the emphasis is on vector bundles. The last section discusses connections on principal bundles, and examines the relation between a connection on the frame bundle and that on the associated vector bundle.

Chapter 5 introduces Euclidean bundles and Riemannian connections, and then embarks on a brief excursion into the realm of Riemannian geometry. The basic tools, such as Levi-Civita connections, isometric immersions, Riemannian submersions, the Hopf-Rinow theorem, etc., are introduced, and should prepare the reader for more advanced texts on the subject. The relation between curvature and topology is illustrated by the classical theorems of Hadamard-Cartan and Bonnet-Myers.

Chapter 6 concludes with Chern-Weil theory, introducing the Pontrjagin, Euler, and Chern characteristic classes of a vector bundle. In order to illustrate these concepts, vector bundles over spheres of dimension $\leq 4$ are reinterpreted in terms of their characteristic classes. The generalized Gauss-Bonnet theorem is also discussed here.

This book grew out of a series of graduate courses taught over the years at the University of Oklahoma. Although there were many outstanding texts available that collectively contained the sequence of topics I wished to present, none did this on its own, with the possible exception of Spivak's monumental treatise. In the end, I often found myself during a course following one author on a particular topic, another on a second one, and so on. As a result, the approach here at times closely parallels that of other texts, most notably Gromoll-Klingenberg-Meyer [15], Poor [32], Steenrod [35], Spivak [34], and Warner [36].

There are several options for using the material as the textbook for a course, depending on the instructor's inclination and the pace she/he wants to set. A leisurely paced one-semester course on manifolds could cover the first chapter. Similarly, a one-semester course on bundles could be based on Chapters 2 and 3, assuming the students are already familiar with the concept of manifolds. I have also used Chapter 1, parts of Chapter 4, and Chapter 5 for a two-semester course in differential geometry.

I would like to thank Yelin Ou for reading parts of the manuscript and making valuable suggestions, and Gary Gray for offering his considerable LaTeX-pertise.

Gerard Walschap

# Contents

# Differentiable Manifolds

In differential geometry, $n$-dimensional Euclidean space is replaced by a differentiable manifold. In essence, this is a set $M$ constructed by gluing together pieces that are homeomorphic to $\mathbb{R}^n$, so that $M$ looks locally, if not globally, like Euclidean space. The idea is that all *local* concepts, such as the derivative of a function $f : \mathbb{R}^n \to \mathbb{R}$ at a point, can be carried over to $M$ by means of these identifications. A simple, yet useful example to keep in mind is that of the two-dimensional unit sphere $S^2$, where for any point $p \in S^2$, the neighborhood $S^2 \setminus \{-p\}$ of $p$ is homeomorphic to $\mathbb{R}^2$.

## 1. Basic Definitions

Recall that the vector space $\mathbb{R}^n$ is the set $\{(p_1, \ldots, p_n) \mid p_i \in \mathbb{R}\}$, together with coordinate-wise addition and scalar multiplication. The $i$-th projection is the map $u^i : \mathbb{R}^n \to \mathbb{R}$ given by $u^i(p_1, \ldots, p_n) = p_i$, and the $j$-th standard basis vector $\mathbf{e}_j$ is defined by $u^i(\mathbf{e}_j) = \delta_{ij}$.

Let $U$ be a subset of $\mathbb{R}^n$. Given a function $f : U \to \mathbb{R}$, $p \in U$, the *i-th partial derivative* of $f$ at $p$ is

$$D_i f(p) = \lim_{t \to 0} \frac{f(p + t\mathbf{e}_i) - f(p)}{t} = (f \circ c)'(0),$$

where $c$ is the line $c(t) = p + t\mathbf{e}_i$ through $p$ in direction $\mathbf{e}_i$. $f$ is said to be *smooth* or *differentiable* on $U$ if it has continuous partial derivatives of any order on $U$.

A map $f : U \to \mathbb{R}^k$ is said to be smooth if all the component functions $f^i := u^i \circ f : U \to \mathbb{R}$ of $f$ are smooth. In this case, the *Jacobian matrix* of $f$ at $p$ is the $k \times n$ matrix $Df(p)$ whose $(i,j)$-th entry is $D_j f^i(p)$. The Jacobian will often be identified with the linear transformation $\mathbb{R}^n \to \mathbb{R}^k$ it determines.

DEFINITION 1.1. A second countable Hausdorff topological space $M$ is said to be a *topological n-dimensional manifold* if it is locally homeomorphic to $\mathbb{R}^n$; ·i.e., if for any $p \in M$ there exists a homeomorphism $x$ of some neighborhood $U$ of $p$ with some open set in $\mathbb{R}^n$. $(U, x)$ is called a *chart*, or *coordinate system*, and $x$ a *coordinate map*.

DEFINITION 1.2. A *differentiable atlas* on a topological $n$-dimensional manifold $M$ is a collection $\mathcal{A}$ of charts of $M$ such that

(1) the domains of the charts cover $M$, and
(2) if $(U, x)$ and $(V, y) \in \mathcal{A}$, then $y \circ x^{-1} : x(U \cap V) \to \mathbb{R}^n$ is smooth.

The map $y \circ x^{-1}$ is often referred to as the *transition map* from the chart $(U, x)$ to $(V, y)$.

If $\mathcal{A}$ is an atlas on $M$, a chart $(U, x)$ is said to be *compatible* with $\mathcal{A}$ if $\{(U, x)\} \cup \mathcal{A}$ is again an atlas on $M$. A *differentiable structure* on $M$ is a maximal differentiable atlas $\mathcal{A}$: Any chart compatible with $\mathcal{A}$ belongs to the atlas. Alternatively—for those uncomfortable with the term "maximal"—given two atlases $\mathcal{A}$ and $\mathcal{A}'$, define $\mathcal{A} \sim \mathcal{A}'$ if for any charts $(U, x) \in \mathcal{A}$ and $(V, y) \in \mathcal{A}'$, $y \circ x^{-1}$ and $x \circ y^{-1}$ are differentiable. A differentiable structure is then an equivalence class of the relation $\sim$ defined above.

DEFINITION 1.3. A *differentiable $n$-dimensional manifold* is a topological $n$-dimensional manifold together with a differentiable structure.

From now on, the term *manifold* will always denote a differentiable manifold.

EXAMPLES AND REMARKS 1.1. (i) In order to specify a differentiable structure, it suffices to provide some atlas $\mathcal{A}$: This atlas then determines a differentiable structure $\mathcal{A}'$ which consists of all charts $(U, x)$ such that $x \circ y^{-1}$ and $y \circ x^{-1}$ are smooth for any coordinate map $y$ of $\mathcal{A}$.

(ii) The *standard differentiable structure* on $\mathbb{R}^n$ is the one determined (as in (i)) by the atlas consisting of the single chart $(\mathbb{R}^n, 1_{\mathbb{R}^n})$, where $1_{\mathbb{R}^n}$ denotes the identity map.

(iii) Let $V$ denote an $n$-dimensional real vector space. The *standard differentiable structure* on $V$ is the one induced by the atlas $\{(V, L)\}$, where $L : V \to \mathbb{R}^n$ is some isomorphism. Why is this structure independent of the choice of $L$?

(iv) Any open subset $U$ of a manifold $M$ inherits a natural differentiable structure (of the same dimension) from that of $M$: An atlas $\{(U_\alpha; x_\alpha)\}_{\alpha \in A}$ of $M$ induces an atlas $\{(U \cap U_\alpha, x_\alpha|_{U \cap U_\alpha})\}_{\alpha \in A}$ of $U$. For example, the set $GL(n) \subset M_{n,n} \cong \mathbb{R}^{n^2}$ of all invertible $n \times n$ real matrices is an $n^2$-dimensional manifold.

(v) Let $r > 0$. The *$n$-sphere $S_r^n$ of radius $r$* is the compact topological subspace of $\mathbb{R}^{n+1}$ consisting of all points at distance $r$ from the origin. Let $p_N = (0, \ldots, 0, r)$ and $p_S = (0, \ldots, 0, -r)$ denote the north and south poles, respectively, and set $U_N = S_r^n \setminus \{p_N\}, U_S = S_r^n \setminus \{p_S\}$. Then the collection $\{(U_N, x_N), (U_S, x_S)\}$ is a differentiable atlas on the sphere, where $x_N$ and $x_S$ are the "stereographic projections"

$$x_N(p_1, \ldots, p_{n+1}) = \frac{r}{r - p_{n+1}}(p_1, \ldots, p_n),$$

$$x_S(p_1, \ldots, p_{n+1}) = \frac{r}{r + p_{n+1}}(p_1, \ldots, p_n).$$

In fact, the transition map is given by

$$x_N \circ x_S^{-1} = x_S \circ x_N^{-1} = \frac{r^2}{|1_{\mathbb{R}^n}|^2} 1_{\mathbb{R}^n} : \mathbb{R}^n \setminus \{0\} \to \mathbb{R}^n \setminus \{0\}$$

and is clearly differentiable.

The sphere is thus described by two charts, and can therefore be considered to be the simplest nontrivial example of a manifold.

(vi) Let $(M_i^{n_i}, \mathcal{A}_i)$ be manifolds of dimension $n_i$, $i = 1, 2$. The collection

$$\mathcal{A}_1 \times \mathcal{A}_2 := \{(U \times V, x \times y) \mid (U, x) \in \mathcal{A}_1, (V, y) \in \mathcal{A}_2\}$$

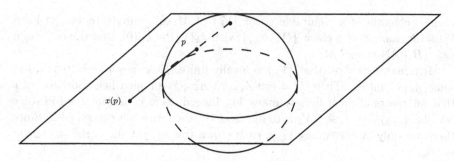

FIGURE 1. Stereographic projection from the north pole.

is an atlas on $M_1 \times M_2$. Here, $(x \times y)(p, q) = (x(p), y(q))$. The induced differentiable structure is called the *product manifold* $M_1 \times M_2$.

DEFINITION 1.4. A function $f : M \to \mathbb{R}$ is said to be *smooth* if $f \circ x^{-1} : x(U) \to \mathbb{R}$ is smooth for any chart $(U, x)$ of $M$.

DEFINITION 1.5. A *partition of unity* on $M$ is a collection $\{\phi_\alpha\}_{\alpha \in A}$ of smooth nonnegative functions $\phi_\alpha$ on $M$ such that

(1) $\{\operatorname{supp} \phi_\alpha\}_{\alpha \in A}$ is a locally finite cover of $M$. Recall that the support of a function is the closure of the set on which the function is nonzero. A collection of sets is locally finite if any point has a neighborhood that intersects at most finitely many of the sets.

(2) $\sum_\alpha \phi_\alpha \equiv 1$. (Why does this possibly infinite sum make sense?)

THEOREM 1.1. *Any open cover $\{U_\alpha\}_{\alpha \in A}$ of a manifold $M$ admits a countable subordinate partition of unity $\{\phi_k\}_{k \in \mathbb{N}}$; i.e., for any integer $k$, there exists an $\alpha \in A$ such that $\operatorname{supp} \phi_k \subset U_\alpha$.*

There are several steps involved in the proof of Theorem 1.1. Given $\epsilon > 0$, $q \in \mathbb{R}^n$, $B_\epsilon(q)$ will denote the set of points at distance less than than $\epsilon$ from $q$.

THEOREM 1.2. *If $\{U_\alpha\}$ is an open cover of $M$, then there is a countable differentiable atlas $\{(V_k, x_k)\}$ of $M$ such that*

(1) *$\{V_k\}$ is a locally finite refinement of $\{U_\alpha\}$;*
(2) *$x_k(V_k) = B_3(0)$;*
(3) *the collection $\{W_k\}$, where $W_k = x_k^{-1}(B_1(0))$, is a cover of $M$.*

PROOF OF THEOREM 1.2. Since $M$ is locally compact (i.e., every point has a neighborhood with compact closure), Hausdorff, and second countable, there exists a countable basis $\{Z_k\}$ for $M$ with $\bar{Z}_k$ compact. Let $A_1 = \bar{Z}_1$. Given $A_i$ compact, let $j$ denote the smallest integer such that $A_i \subset Z_1 \cup \cdots \cup Z_j$; define $A_{i+1} = \bar{Z}_1 \cup \cdots \cup \bar{Z}_j \cup \bar{Z}_{i+1}$. Then $\{A_k\}$ is a sequence of compact sets with $A_k \subset \operatorname{int} A_{k+1}$, and $\cup_k A_k = M$. Define $A_0$ to be the empty set. Since $M = \cup_{i \geq 0}(A_{i+1} \setminus \operatorname{int} A_i)$, we may assume that for each $p \in M$, there exists a chart $(V_p, x_p)$ sending $p$ to 0, such that

$$x_p(V_p) = B_3(0), \quad V_p \subset U_\alpha \text{ for some } \alpha, \quad \text{and} \quad V_p \subset (\operatorname{int} A_{i+2}) \setminus A_{i-1} \text{ for some } i.$$

Then $\{x_p^{-1}(B_1(0))\}_{p \in A_{i+1} \setminus \operatorname{int} A_i}$ is an open cover of the compact $A_{i+1} \setminus \operatorname{int} A_i$, and contains a finite subcover which we denote $P_i$. If $P = P_0 \cup P_1 \cup \cdots$, then $P$ consists of a countable cover $\{V_k\}$ of $M$ subordinate to $\{U_\alpha\}$. Each $V_k$ is the domain of a chart $\{(V_k, x_k)\}$ with $x_k(V_k) = B_3(0)$, and the collection $\{x_k^{-1}(B_1(0))\}$ covers $M$.

It remains to show that $\{V_k\}$ is locally finite. Now, any $p \in M$ belongs to some $A_{i+1} \setminus \text{int } A_i$. Then $W = (\text{int } A_{i+2}) \setminus A_{i-1}$ is an open neighborhood of $p$ that intersects at most finitely many $V_k$: Indeed, each $V_k$ is contained in some set $(\text{int } A_{j+2}) \setminus A_{j-1}$, so if $V_k$ is to intersect $W$, then $j$ cannot exceed $i+2$. Since there are only finitely many $V_k$ in each crown $(\text{int } A_{j+2}) \setminus A_{j-1}$, the statement follows.                                                                                $\square$

Given $\epsilon > 0$, denote by $C_\epsilon(0)$ the open cube $(-\epsilon, \epsilon)^n$ in $\mathbb{R}^n$.

LEMMA 1.1. *There exists a differentiable function* $\phi : \mathbb{R}^n \to \mathbb{R}$ *satisfying*

(1) $\phi \equiv 1$ *on* $\bar{C}_1(0)$,
(2) $0 < \phi < 1$ *on* $C_2(0) \setminus \bar{C}_1(0)$, *and*
(3) $\phi \equiv 0$ *on* $\mathbb{R}^n \setminus C_2(0)$.

PROOF OF LEMMA 1.1. Let $h : \mathbb{R} \to \mathbb{R}$ be given by

$$h(x) = \begin{cases} e^{-1/x}, & \text{if } x > 0, \\ 0, & \text{otherwise}, \end{cases}$$

and define

$$f(x) = \frac{h(2+x)h(2-x)}{h(2+x)h(2-x) + h(x-1) + h(-x-1)}.$$

This expression makes sense because $h(x-1) + h(-x-1)$ is nonnegative, and equals 0 only when $|x| \leq 1$, in which case $h(2+x)h(2-x) > 0$. Furthermore, $f(x) = 1$ if $|x| \leq 1$, $0 < f(x) < 1$ if $1 < |x| < 2$, and $f(x) = 0$ if $|x| \geq 2$. Now let $\phi(a_1, \ldots, a_n) = \Pi_{i=1}^n f(a_i)$.                                                      $\square$

PROOF OF THEOREM 1.1. Let $\{(V_k, x_k)\}$ be a differentiable atlas as in Theorem 1.2, and $\phi$ the function from Lemma 1.1, where $n$ equals the dimension of $M$. For each $k$ define a function $\theta_k : M \to \mathbb{R}$ by

$$\theta_k(p) = \begin{cases} \phi \circ x_k(p), & \text{if } p \in V_k, \\ 0, & \text{otherwise}. \end{cases}$$

$\theta_k$ is differentiable on $M$, since it is differentiable on $V_k$, and is identically zero on the open neighborhood $M \setminus x_k^{-1}(\bar{C}_2(0))$ of $M \setminus V_k$. Any $p \in M$ belongs to $x_j^{-1}(B_1(0))$ for some $j$, so that $\theta_j(p) > 0$. Since $\{V_k\}$ is locally finite and $\text{supp } \theta_k \subset V_k$, the collection $\{\text{supp } \theta_k\}$ is a locally finite cover of $M$. This means that $\sum_k \theta_k(p)$ is finite for every $p \in M$; now set $\phi_k := \theta_k / (\sum_i \theta_i)$.          $\square$

EXERCISE 1. Show that the transition maps for the atlas in Examples and Remarks 1.1(iv) are given by

$$x_N \circ x_S^{-1} = x_S \circ x_N^{-1} = \frac{r^2}{|1_{\mathbb{R}^n}|^2} 1_{\mathbb{R}^n} : \mathbb{R}^n \setminus \{0\} \to \mathbb{R}^n \setminus \{0\},$$

and deduce that $\{(U_N, x_N), (U_S, x_S)\}$ is indeed a differentiable atlas on the sphere.

(Notation: Given a manifold $M$, $1_M : M \to M$ denotes the identity map of $M$.)

EXERCISE 2. Let $U$ be an open subset of $M$, $V$ a set whose closure is contained in $U$. Show that there exists a smooth nonnegative $\phi : M \to \mathbb{R}$ that is identically 1 on the closure of $V$, and the support of which is contained in $U$.

## 2. Differentiable Maps

The superscript in the symbol $M^n$ will refer to the dimension of the manifold $M$.

DEFINITION 2.1. Let $M^n, N^k$ denote manifolds, and suppose $U$ is open in $M$. A map $f : U \to N$ is said to be *differentiable* or *smooth* if $y \circ f \circ x^{-1}$ is smooth as a map from $\mathbb{R}^n$ to $\mathbb{R}^k$ for any coordinate maps $x$ of $M$ and $y$ of $N$.

If $A$ is an arbitrary subset of $M$, $f : A \to N$ is said to be smooth if it can be extended to a smooth map $\bar{f} : U \to N$ for some open set $U$ containing $A$.

Observe that the composition of differentiable maps is differentiable. $f : M \to N$ is said to be a *diffeomorphism* if it is bijective and both $f$ and its inverse $f^{-1}$ are smooth. The collection $\mathrm{Diff}(M)$ of all diffeomorphisms of $M$ with itself is clearly a group under composition.

EXAMPLES AND REMARKS 2.1. (i) For a function $f : M \to \mathbb{R}$, the Definition 2.1 coincides with 1.4.

(ii) If $(U, x)$ is a chart, then $x : U \to x(U) \subset \mathbb{R}^n$ is a diffeomorphism.

(iii) It is known that any two differentiable manifolds of dimension no larger than 3 which are homeomorphic are actually diffeomorphic. On the other hand, there exist "exotic" $\mathbb{R}^4$'s; i.e., manifolds that are homeomorphic but not diffeomorphic to $\mathbb{R}^4$ with the standard differentiable structure.

Given a subset $A$ of $M$, let $\mathcal{F}(A)$ denote the set of all smooth functions $f : A \to \mathbb{R}$. $\mathcal{F}(A)$ is a real algebra (and in particular, both a ring and a vector space) under the operations

$$(f + g)(p) = f(p) + g(p), \quad (f \cdot g)(p) = f(p)g(p), \quad (\alpha f)(p) = \alpha f(p), \quad \alpha \in \mathbb{R}.$$

For example, if $(U, x)$ is a chart, then $x^i \in \mathcal{F}(U)$, where $x^i := u^i \circ x$, $1 \leq i \leq \dim M$.

DEFINITION 2.2. Let $U$ be an open subset of $M$, $p \in U$, and set $\mathcal{F}_p^0(U) = \{f \in \mathcal{F}(U) \mid f \equiv 0 \text{ in a neighborhood of } p\}$. $\mathcal{F}_p^0(U)$ is an ideal in $\mathcal{F}(U)$, and the quotient algebra $\mathcal{F}_p = \mathcal{F}(U)/\mathcal{F}_p^0(U)$ is called the algebra of *germs of functions* at $p$.

Thus, a germ is an equivalence class of functions, with two functions being equivalent iff they agree on a neighborhood of the point. The reason we omitted $U$ in the terminology for $\mathcal{F}_p = \mathcal{F}_p(U)$ is due to the fact that the map $\mathcal{F}(M) \to \mathcal{F}(U)$ given by $f \mapsto f \circ \imath$, where $\imath : U \to M$ denotes inclusion, induces an isomorphism $\mathcal{F}_p(M) \cong \mathcal{F}_p(U)$: This map is clearly injective; to see that it's surjective, let $f \in \mathcal{F}(U)$, and consider an open set $V$ whose closure is contained in $U$. Let $\phi$ be the function from Exercise 2, and define a smooth function $g$ on $M$ by setting it equal to $\phi f$ on $U$ and 0 outside $U$. Since $f$ and $g$ coincide on $V$, the germ of $g$ at $p$ is mapped to the germ of $f$ at $p$.

EXERCISE 3. Consider $\mathbb{R}$ with the two atlases $\{1_\mathbb{R}\}$ and $\{\phi\}$, where $\phi(t) = t^3$.

(a) Show that these atlases are not compatible; i.e., they determine different differentiable structures on $\mathbb{R}$.

(b) Show that the two differentiable manifolds from (a) are diffeomorphic.

EXERCISE 4. (a) Show that $f : S^n_r \to \mathbb{R}$, where $f(p_1, \ldots, p_{n+1}) = \sum_i p_i$, is smooth.

(b) Show that $f : S^n_1 \to S^n_r$, where $f(p) = -rp$, is a diffeomorphism.

## 3. Tangent Vectors

A vector $v$ in $\mathbb{R}^n$ acts on differentiable functions in a natural way, by assigning to $f : \mathbb{R}^n \to \mathbb{R}$ the derivative $D_v f(p) := Df(p) \cdot v$ of $f$ in direction $v$. This assignment depends of course on the point $p$ at which the derivative is evaluated; furthermore, it is linear, and satisfies the product rule $D_v(fg)(p) = f(p)D_v(g)(p) + g(p)D_v(f)(p)$. This is essentially the motivation behind the following:

DEFINITION 3.1. Let $p \in M$. A *tangent vector* $v$ at $p$ is a map $v : \mathcal{F}_p(M) \to \mathbb{R}$ satisfying

(1) $v(\alpha f + \beta g) = \alpha v(f) + \beta v(g)$; and
(2) $v(fg) = f(p)v(g) + g(p)v(f)$

for $\alpha, \beta \in \mathbb{R}$, $f, g \in \mathcal{F}_p(M)$.

In the above definition, we have used the same letter to denote both a germ and a function belonging to that germ: If $U$ is a neighborhood of $p$, then a tangent vector $v$ at $p$ induces a map $\mathcal{F}(U) \to \mathbb{R}$ given by $v(f) := v([f])$. The point $p$ is called the *footpoint* of $v$, and the set $M_p$ of all tangent vectors at $p$ is called the *tangent space* of $M$ at $p$. It is a real vector space under the operations $(v + w)(f) = v(f) + w(f)$, $(\alpha v)(f) = \alpha v(f)$.

In the familiar context of Euclidean space, one can think of a tangent vector at $p$ as simply being a vector $v$ whose origin has been translated to $p$, denoted $(p, v)$. Then $(p, v)(f) = D_v f(p)$. Notice that one recovers $v$ from the way $(p, v)$ acts on functions: $v = ((p, v)(u^1), \ldots, (p, v)(u^n))$.

The first condition in Definition 3.1 says that a tangent vector is a linear operator on (germs of) functions, and the second that it is a *derivation*.

Let $x$ be a coordinate map around $p$ (that is, $p$ belongs to the domain of $x$), and as usual, let $x^i = u^i \circ x$. The *coordinate vector fields* at $p$ are the tangent vectors $\partial/\partial x^i(p) \in M_p$ given by

$$(3.1) \qquad \frac{\partial}{\partial x^i}(p)(f) := D_i(f \circ x^{-1})(x(p)), \qquad f \in \mathcal{F}(M), \quad 1 \le i \le n.$$

One often denotes the left side of (3.1) by $\partial f/\partial x^i(p)$. For example, in $\mathbb{R}^n$, the standard coordinate vector fields at $p$ are $\partial/\partial u^i(p)$, where $\partial f/\partial u^i(p) = D_i f(p)$. We will often denote them simply by $D_i$. When $n = 1$, we write $D$ instead of $\partial/\partial u$, so that $Df(a) = f'(a)$.

The coordinate vector fields actually form a basis for the tangent space at a point. In order to show this, we need the following:

LEMMA 3.1. *Let $U$ denote a star-shaped neighborhood of $0 \in \mathbb{R}^n$ — that is, the line segment connecting the origin to any point of $U$ is also contained inside $U$. Given $f \in \mathcal{F}U$, there exist $n$ functions $\psi_i \in \mathcal{F}U$, with $\psi_i(0) = D_i f(0)$, such that*

$$f = f(0) + \sum_i u^i \psi_i.$$

PROOF. For any fixed $p \in U$, consider the line segment $c(t) = tp$, and set $\phi = f \circ c$. $\phi$ is a differentiable function on $[0,1]$, and $\phi'(t) = \sum_i p_i D_i f(tp)$. Thus,

$$f(p) - f(0) = \phi(1) - \phi(0) = \int_0^1 \phi' = \sum_i p_i \int_0^1 D_i f(tp)\, dt.$$

The claim then follows by setting $\psi_i(p) := \int_0^1 D_i f(tp)\, dt$.     □

PROPOSITION 3.1. *Let $(U, x)$ be a chart around $p$. Then any tangent vector $v \in M_p$ can be uniquely written as a linear combination $v = \sum_i \alpha_i \partial/\partial x^i(p)$. In fact, $\alpha_i = v(x^i)$.*

*Thus, $M_p^n$ is an $n$-dimensional vector space with basis $\{\partial/\partial x^i(p)\}_{1 \le i \le n}$.*

PROOF. We may assume without loss of generality that $x(p) = 0$, and that $x(U)$ is star-shaped. By Lemma 3.1, any $f \in \mathcal{F}M$ satisfies $f \circ x^{-1} = f(p) + \sum u^i \psi_i$, with $\psi_i(0) = \partial/\partial x^i(p)(f)$. Thus, $f|_U = f(p) + \sum_i x^i(\psi_i \circ x)|_U$, and

$$v(f) = v(f(p)) + \sum_i [v(x^i) \cdot \psi_i(0) + x^i(p) \cdot v(\psi_i \circ x)] = \sum_i v(x^i) \frac{\partial}{\partial x^i}(p)(f),$$

where we have used the result of Exercise 5 below. It remains to show that the $\partial/\partial x^i(p)$ are linearly independent; observe that

$$\frac{\partial}{\partial x^i}(p)(x^j) = D_i(x^j \circ x^{-1})(0) = D_i(u^j)(0) = \delta_{ij}.$$

Thus, if $\sum \alpha_i \partial/\partial x^i(p) = 0$, then $0 = \sum \alpha_i \partial/\partial x^i(p)(x^j) = \alpha_j$.     □

Notice that if $x$ and $y$ are two coordinate systems at $p$, then taking $v = \partial/\partial y^i(p)$ in Proposition 3.1 yields

$$(3.2) \qquad \frac{\partial}{\partial y^i}(p) = \sum_{j=1}^n \frac{\partial x^j}{\partial y^i}(p) \frac{\partial}{\partial x^j}(p) = \sum_{j=1}^n D_i(u^j \circ x \circ y^{-1})(y(p)) \frac{\partial}{\partial x^j}(p)$$

for $1 \le i \le n$. This means that the transition matrix from the basis $\{\partial/\partial x^i(p)\}$ to the basis $\{\partial/\partial y^i(p)\}$ is the Jacobian matrix of $x \circ y^{-1}$ at $y(p)$.

EXERCISE 5. Let $c \in \mathbb{R}$. Show that if $c \in \mathcal{F}M$ denotes the constant function $c(p) :\equiv c$ for all $p \in M$, then $v(c) = 0$ for any tangent vector $v$ at any point of $M$.

EXERCISE 6. Write down (3.2) explicitly for the $n$-sphere of radius $r$, if $x$ and $y$ denote stereographic projections.

## 4. The Derivative

In calculus, one usually thinks of the Jacobian $Df(p)$ of $f : \mathbb{R}^n \to \mathbb{R}^k$ as the derivative of $f$ at $p$. It is therefore natural, when seeking a meaningful generalization of this concept for a map $f : M \to N$ between manifolds $M$ and $N$, to look for a linear transformation. In view of the previous section, where we defined vector spaces at each point of a manifold, this suggests a linear transformation $f_{*p} : M_p \to N_{f(p)}$ between the respective tangent spaces. We would of course like $f_{*p}$ to correspond to $Df(p)$ when $M = \mathbb{R}^n$ and $N = \mathbb{R}^k$, if $\mathbb{R}^n_p$ is identified with the set of pairs $(p, v)$, $v \in \mathbb{R}^n$; i.e, we require that $f_{*p}(p, v) = (f(p), Df(p)v)$ for all $v \in \mathbb{R}^n$. Now, if $\phi : \mathbb{R}^k \to \mathbb{R}$ is differentiable, then by the Chain rule,

$$f_{*p}(p, v)(\phi) = (f(p), Df(p)v)(\phi) = D_{Df(p)v}\phi(f(p)) = D\phi(f(p))\, Df(p)v$$
$$= D_v(\phi \circ f)(p) = (p, v)(\phi \circ f).$$

This motivates the following:

DEFINITION 4.1. Let $M$ and $N$ denote differentiable manifolds of dimensions $n$ and $k$ respectively, $f : U \to N$ a differentiable map, where $U$ is open in $M$, and $p \in U$. The *derivative of $f$ at $p$* is the map $f_{*p} : M_p \to N_{f(p)}$ given by

$$(f_{*p}v)(\phi) := v(\phi \circ f), \qquad \phi \in \mathcal{F}(N), \quad v \in M_p.$$

It is clear from the definition that $f_{*p}$ is a linear transformation.

PROPOSITION 4.1. *With notation as in Definition 4.1, let $x$ be a coordinate map around $p \in U$, $y$ a coordinate map around $f(p) \in N$. Then the matrix of $f_{*p}$ with respect to the bases $\{\partial/\partial x^i(p)\}$ and $\{\partial/\partial y^j((f(p))\}$ is the Jacobian matrix of $y \circ f \circ x^{-1}$ at $x(p)$.*

PROOF.

$$f_{*p}\frac{\partial}{\partial x^j}(p) = \sum_i f_{*p}\frac{\partial}{\partial x^j}(p)(y^i)\frac{\partial}{\partial y^i}(f(p)) = \sum_i \frac{\partial}{\partial x^j}(p)(y^i \circ f)\frac{\partial}{\partial y^i}(f(p))$$
$$= \sum_i D_j(u^i \circ (y \circ f \circ x^{-1}))(x(p))\frac{\partial}{\partial y^i}(f(p)).$$

$\square$

EXAMPLES AND REMARKS 4.1. (i) It follows from Definition 4.1 that the identity map $1_M$ of $M$ has as derivative at $p \in M$ the identity map $1_{M_p}$ of $M_p$.

(ii) If $g : N \to Q$ is differentiable, then $g \circ f$ is differentiable, and $(g \circ f)_{*p} = g_{*f(p)} \circ f_{*p}$. In particular, if $f : M \to N$ is a diffeomorphism, then by (i), $f_{*p}$ is an isomorphism with inverse $(f^{-1})_{*f(p)}$. Furthermore, given coordinate maps $x$ and $y$ of $M$ and $N$ respectively, the diagram

$$
\begin{array}{ccc}
M_p & \xrightarrow{\ f_{*p}\ } & N_{f(p)} \\
{\scriptstyle x_{*p}}\big\downarrow & & \big\downarrow{\scriptstyle y_{*f(p)}} \\
\mathbb{R}^n_{x(p)} & \xrightarrow{(y \circ f \circ x^{-1})_{*x(p)}} & \mathbb{R}^k_{(y \circ f)(p)}
\end{array}
$$

commutes. Observe that $x_{*p}\partial/\partial x^i(p) = \partial/\partial u^i(x(p))$, since $x_*\partial/\partial x^i(u^j) = \partial/\partial x^i(u^j \circ x) = \partial/\partial x^i(x^j) = \delta_{ij}$.

(iii) A (smooth) curve in $M$ is a (smooth) map $c : I \to M$, where $I$ is an interval of real numbers. The *tangent vector* to $c$ at $t$ is $\dot{c}(t) := c_{*t}D(t)$. Thus, given $\phi \in \mathcal{F}(M)$,

$$\dot{c}(t)(\phi) = c_{*t}D(t)(\phi) = D(t)(\phi \circ c) = (\phi \circ c)'(t).$$

(iv) Let $E$ be an $n$-dimensional real vector space with its canonical differentiable structure, cf. Examples and Remarks 1.1(iii). For any $v \in E$, $E$ may be naturally identified with its tangent space $E_v$ at $v$ by "parallel translation" $\mathcal{J}_v : E \to E_v$, defined as follows: Given $w \in E$, let $\gamma(t) = v + tw$, and set $\mathcal{J}_v w := \dot{\gamma}(0)$. If $x : E \to \mathbb{R}^n$ is any isomorphism, then

$$\mathcal{J}_v w = \dot{\gamma}(0) = \sum_i \dot{\gamma}(0)(x^i)\frac{\partial}{\partial x^i}(v) = \sum_i D(0)(x^i \circ \gamma)\frac{\partial}{\partial x^i}(v)$$

$$= \sum_i x^i(w)\frac{\partial}{\partial x^i}(v),$$

so that $\mathcal{J}_v$, being linear and one-to-one, is an isomorphism.

Notice that for $E = \mathbb{R}^n$ and $x = 1_{\mathbb{R}^n}$, we obtain $\mathcal{J}_v\mathbf{e}_i = \partial/\partial u^i(v)$. This formalizes our heuristic description of the tangent space of $\mathbb{R}^n$ at $v$ from the previous section, since the map

$$\{v\} \times \mathbb{R}^n \to \mathbb{R}^n_v,$$
$$(v, w) \mapsto \mathcal{J}_v w$$

is an isomorphism that preserves the action on $\mathcal{F}(\mathbb{R}^n)$.

Consider, for example, a linear transformation $L : \mathbb{R}^n \to \mathbb{R}^k$. By Proposition 4.1, the matrix of $L_{*v}$ with respect to the standard coordinate vector fields bases is that of the Jacobian of $L$. But since $L$ is linear,

$$D_i(u^j \circ L)(v) = \lim_{t \to 0} \frac{(u^j \circ L)(v + t\mathbf{e}_i) - (u^j \circ L(v))}{t} = (u^j \circ L)(\mathbf{e}_i),$$

so that the Jacobian matrix of $L$ is just the matrix of $L$ in the standard basis. Thus, the following diagram

$$
\begin{array}{ccc}
\mathbb{R}^n & \xrightarrow{\ L\ } & \mathbb{R}^k \\
\mathcal{J}_v \downarrow & & \downarrow \mathcal{J}_{Lv} \\
\mathbb{R}^n_v & \xrightarrow{\ L_{*v}\ } & \mathbb{R}^k_{Lv}
\end{array}
$$

commutes.

(v) Let $U$ be an open set in $M$, $f \in \mathcal{F}U$, $p \in U$. The *differential* of $f$ at $p$ is the element $df(p)$ of the dual space $M_p^*$ (i.e., $df(p) : M_p \to \mathbb{R}$ is linear) defined by

$$df(p)(v) := v(f), \qquad v \in M_p.$$

Thus, for example, $\{dx^i(p)\}$ is the basis dual to $\{\partial/\partial x^i(p)\}$. Notice also that the diagram

$$
\begin{array}{ccc}
M_p & =\!\!=\!\!= & M_p \\
{\scriptstyle df(p)}\downarrow & & \downarrow{\scriptstyle f_{*p}} \\
\mathbb{R} & \xrightarrow{\ \mathcal{J}_{f(p)}\ } & \mathbb{R}_{f(p)}
\end{array}
$$

commutes:

$$\mathcal{J}_{f(p)}df(p)(v) = \mathcal{J}_{f(p)}v(f) = v(f)D_{f(p)} = f_{*p}v.$$

DEFINITION 4.2. The *tangent bundle* (resp. *cotangent bundle*) of $M$ is the set $TM = \cup_{p \in M} M_p$ (resp. $T^*M = \cup_{p \in M} M_p^*$). The *bundle projections* are the maps $\pi : TM \to M$ and $\tilde{\pi} : T^*M \to M$ which map a tangent or cotangent vector to its footpoint.

PROPOSITION 4.2. *The differentiable structure $\mathcal{D}$ on $M^n$ induces in a natural way $2n$-dimensional differentiable structures on the tangent and cotangent bundles of $M$.*

PROOF. For each chart $(U, x)$ of $M$, define a chart $(\pi^{-1}(U), \bar{x})$ of $TM$, where $\bar{x} : \pi^{-1}(U) \to \mathbb{R}^{2n}$ is given by

$$\bar{x}(v) = (x \circ \pi(v), dx^1(\pi(v))v, \ldots, dx^n(\pi(v))v).$$

Similarly, define $\tilde{x} : \tilde{\pi}^{-1}(U) \to \mathbb{R}^{2n}$ by

$$\tilde{x}(\alpha) = (x \circ \tilde{\pi}(\alpha), \alpha(\partial/\partial x^1(\tilde{\pi}(\alpha))), \ldots, \alpha(\partial/\partial x^n(\tilde{\pi}(\alpha)))).$$

One checks that the collection $\{\bar{x}^{-1}(V) \mid (U, x) \in \mathcal{D}, V \text{ open in } \mathbb{R}^{2n}\}$ forms a basis for a second countable Hausdorff topology on $TM$. A similar argument, using $\tilde{x}$ instead of $\bar{x}$, works for $T^*M$.

Let $\mathcal{A} = \{(\pi^{-1}(U), \bar{x}) \mid (U, x) \in \mathcal{D}\}$. We claim that $\mathcal{A}$ is an atlas for $TM$: clearly, each $\bar{x} : \pi^{-1}(U) \to x(U) \times \mathbb{R}^n$ is a homeomorphism. Furthermore, if $(V, y)$ is another chart of $M$, and $(a, b) \in x(U \cap V) \times \mathbb{R}^n$, then

$$\bar{y} \circ \bar{x}^{-1}(a, b) = (y \circ x^{-1}(a), D(y \circ x^{-1})(a)(b)).$$

To see this, write $b = \sum b_i \mathbf{e}_i$; then

$$\bar{x}^{-1}(a, b) = \sum_i b_i \frac{\partial}{\partial x^i}(x^{-1}(a)) = \sum_{i,j} b_i \frac{\partial y^j}{\partial x^i}(x^{-1}(a)) \frac{\partial}{\partial y^j}(x^{-1}(a)),$$

so that

$$(\bar{y} \circ \bar{x}^{-1})(a, b) = (y \circ x^{-1}(a), \sum_{i,j} b_i \frac{\partial y^j}{\partial x^i}(x^{-1}(a))\mathbf{e}_j)$$

$$= (y \circ x^{-1}(a), D(y \circ x^{-1})(a)(b)).$$

$\square$

For example, the bundle projection $\pi : TM \to M$ is differentiable, since for any pair $(U, x)$, $(\pi^{-1}(U), \bar{x})$ of related charts, $x \circ \pi \circ \bar{x}^{-1} : x(U) \times \mathbb{R}^n \to x(U)$ is the projection onto the first factor.

Any $f : M \to N$ induces a differentiable map $f_* : TM \to TN$, called the *derivative* of $f$: For $v \in M_p$, set $f_* v := f_{*p} v$. Differentiability follows from the easily checked identity:

$$(\bar{y} \circ f_* \circ \bar{x}^{-1})(a, b) = (y \circ f \circ x^{-1}(a), D(y \circ f \circ x^{-1})(a)b).$$

EXERCISE 7. Show that if $M$ is connected, then any two points of $M$ can be joined by a smooth curve.

EXERCISE 8. (a) Prove that $\mathcal{J}_v : \mathbb{R}^n \to (\mathbb{R}^n)_v$ from Examples and Remarks 4.1(iv) satisfies $\mathcal{J}_v w(f) = D_w f(v) = (f \circ c)'(0)$, where $c$ is any curve with $c(0) = v$, $c'(0) = w$.
(b) Show that any $v \in TM$ equals $\dot{c}(0)$ for some curve $c$ in $M$.

EXERCISE 9. For positive $\rho, \sigma$, consider the helix $c : \mathbb{R} \to \mathbb{R}^3$, given by $c(t) = (\rho \cos t, \rho \sin t, \sigma t)$. Express $\dot{c}(t)$ in terms of the standard basis of $\mathbb{R}^3_{c(t)}$.

EXERCISE 10. Let $M$ be connected, $f : M \to N$ a differentiable map. Show that if $f_{*p} = 0$ for all $p$ in $M$, then $f$ is a constant map.

EXERCISE 11. Fill in the details of the argument for the cotangent bundle in the proof of Proposition 4.2.

## 5. The Inverse and Implicit Function Theorems

Let $U$ be an open set in $M$, $f : U \to N$ a differentiable map. The *rank* of $f$ at $p \in U$ is the rank of the linear map $f_{*p} : M_p \to N_{f(p)}$, that is, the dimension of the space $f_*(M_p)$. Recall the following theorem from calculus:

THEOREM 5.1 (Inverse Function Theorem). *Let $U$ be an open set in $\mathbb{R}^n$, $f : U \to \mathbb{R}^n$ a differentiable map. If $f$ has maximal rank $(=n)$ at $p \in U$, then there exists a neighborhood $V$ of $p$ such that the restriction $f : V \to f(V)$ is a diffeomorphism.*

The inverse function theorem immediately generalizes to manifolds:

THEOREM 5.2 (Inverse Function Theorem for Manifolds). *Let $M$ and $N$ be manifolds of dimension $n$, and $f : U \to N$ a smooth map, where $U$ is open in $M$. If $f$ has maximal rank at $p \in U$, then there exists a neighborhood $V$ of $p$ such that the restriction $f : V \to f(V)$ is a diffeomorphism.*

PROOF. Consider coordinate maps $x$ at $p$, $y$ at $f(p)$, and apply Theorem 5.1 to $y \circ f \circ x^{-1}$. Conclude by observing that $x$ and $y$ are diffeomorphisms. $\square$

We now use the inverse function theorem to derive the Euclidean version of one of the essential tools in differential geometry:

THEOREM 5.3 (Implicit Function Theorem). *Let $U$ be a neighborhood of $0$ in $\mathbb{R}^n$, $f : U \to \mathbb{R}^k$ a smooth map with $f(0) = 0$. For $n \leq k$, let $\imath : \mathbb{R}^n \to \mathbb{R}^k$ denote the inclusion $\imath(a_1, \ldots, a_n) = (a_1, \ldots, a_n, 0, \ldots, 0)$, and for $n \geq k$, let $\pi : \mathbb{R}^n \to \mathbb{R}^k$ denote the projection $\pi(a_1, \ldots, a_k, \ldots, a_n) = (a_1, \ldots, a_k)$.*

(1) *If $n \leq k$ and $f$ has maximal rank $(= n)$ at $0$, then there exists a coordinate map $g$ of $\mathbb{R}^k$ around $0$ such that $g \circ f = \imath$ in a neighborhood of $0 \in \mathbb{R}^n$.*

(2) *If $n \geq k$ and $f$ has maximal rank $(= k)$ at $0$, then there exists a coordinate map $h$ of $\mathbb{R}^n$ around $0$ such that $f \circ h = \pi$ in a neighborhood of $0 \in \mathbb{R}^n$.*

PROOF. In order to prove (1), observe that the $k \times n$ matrix $(D_j f^i(0))$ has rank $n$. By rearranging the component functions $f^i$ of $f$ if necessary (which amounts to composing $f$ with an invertible transformation, hence a diffeomorphism of $\mathbb{R}^k$), we may assume that the $n \times n$ submatrix $(D_j f^i(0))_{1 \leq i,j \leq n}$ is invertible. Define $F : U \times \mathbb{R}^{k-n} \to \mathbb{R}^k$ by

$$F(a_1, \ldots, a_n, a_{n+1}, \ldots, a_k) := f(a_1, \ldots, a_n) + (0, \ldots, 0, a_{n+1}, \ldots, a_k).$$

Then $F \circ \imath = f$, and the Jacobian matrix of $F$ at $0$ is

$$\begin{pmatrix} (D_j f^i(0))_{1 \leq i \leq n} & 0 \\ (D_j f^i(0))_{n+1 \leq i \leq k} & 1_{\mathbb{R}^{k-n}} \end{pmatrix},$$

which has nonzero determinant. Consequently, $F$ has a local inverse $g$, and $g \circ f = g \circ F \circ \imath = \imath$. This establishes (1). Similarly, in (2), we may assume that the $k \times k$ submatrix $(D_j f^i(0))_{1 \leq i,j \leq k}$ is invertible. Define $F : U \to \mathbb{R}^n$ by

$$F(a_1, \ldots, a_n) := (f(a_1, \ldots, a_n), a_{k+1}, \ldots, a_n).$$

Then $f = \pi \circ F$, and the Jacobian of $F$ at $0$ is

$$\begin{pmatrix} (D_j f^i(0))_{1 \leq j \leq k} & (D_j f^i(0))_{k+1 \leq j \leq n} \\ 0 & 1_{\mathbb{R}^{n-k}} \end{pmatrix},$$

which is invertible. Thus, $F$ has a local inverse $h$, and $f \circ h = \pi \circ F \circ h = \pi$.  $\square$

## 6. Submanifolds

The implicit function theorem enables us to construct new examples of manifolds. Before doing so, however, there are certain "nice" maps, such as the inclusion $S^n \hookrightarrow \mathbb{R}^{n+1}$, that deserve special recognition:

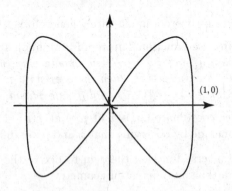

(1,0)

FIGURE 2. The lemniscate $c|_{(0,2\pi)}$.

DEFINITION 6.1. A map $f : M^n \to N^k$ is said to be an *immersion* if for every $p \in M$ the linear map $f_{*p} : M_p \to N_{f(p)}$ is one-to-one (so that $n \leq k$). If in addition $f$ maps $M$ homeomorphically onto $f(M)$ (where $f(M)$ is endowed with the subspace topology), then $f$ is called an *imbedding*.

Notice that if $M$ is compact, then an injective immersion is an imbedding. This is not true in general: For example, the curve $c : \mathbb{R} \to \mathbb{R}^2$ which parametrizes a lemniscate, $c(t) = (\sin t, \sin 2t)$, is an immersion; its restriction to $(0, 2\pi)$ is a one-to-one immersion, but not an imbedding, although $c_{|(0,\pi)}$ is. In fact, an immersion is always *locally* an imbedding:

PROPOSITION 6.1. *If $f : M^n \to N^k$ is an immersion, then for any $p \in M$, there exists a neighborhood $U$ of $p$, and a coordinate map $y$ defined on some neighborhood $V$ of $f(p)$ such that*

(1) *A point $q$ belongs to $f(U) \cap V$ iff $y^{n+1}(q) = \cdots = y^k(q) = 0$, i.e.,*
$$y(f(U) \cap V) = (\mathbb{R}^n \times \{0\}) \cap y(V);$$
(2) *$f|_U$ is an imbedding.*

PROOF. Consider the inclusion $\imath : \mathbb{R}^n \to \mathbb{R}^k$, and let $x$ be a coordinate map around $p$ with $x(p) = 0$, $\tilde{y}$ a coordinate map around $f(p)$ with $(\tilde{y} \circ f)(p) = 0$. Since $\tilde{y} \circ f \circ x^{-1}$ has maximal rank at 0, there exists by the implicit function theorem a chart $g$ of $\mathbb{R}^k$ around 0, and a neighborhood $W$ of $0 \in \mathbb{R}^n$ such that $g \circ \tilde{y} \circ f \circ x^{-1}|_W = \imath|_W$. Set $U = x^{-1}(W)$, $y = g \circ \tilde{y}$; by restricting the domain of $g$ if necessary, (1) clearly holds. (2) follows from the fact that $f|_U = y^{-1} \circ \imath \circ x|_U$ is a composition of imbeddings. $\qquad\square$

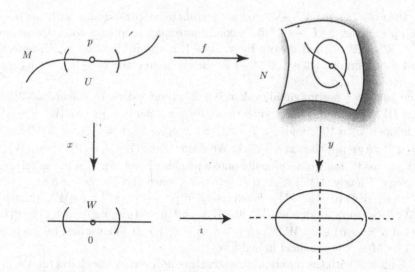

FIGURE 3

REMARK 6.1. When $f$ in Proposition 6.1 is an imbedding, then $f(U)$ equals $f(M) \cap W$ for some open set $W$ in $N$. Thus, in this case, (1) reads

$$f(M) \cap V = \{q \in V \mid y^{n+1}(q) = \cdots = y^k(q) = 0\}.$$

DEFINITION 6.2. Let $M$, $N$ be manifolds with $M \subset N$. $M$ is said to be a *submanifold* of $N$ (respectively an *immersed submanifold* of $N$) if the inclusion map $\imath : M \hookrightarrow N$ is an imbedding (respectively an immersion).

By Remark 6.1, if $M$ is an $n$-dimensional submanifold of $N^k$, then for any $p$ in $M$, there exists a neighborhood $V$ of $p$ in $N$, and a chart $(V, x)$ of $N$ such that

$$M \cap V = \{q \in V \mid x^{n+1}(q) = \cdots = x^k(q) = 0\}.$$

When $f : M \to N$ is a one-to-one immersion (resp. imbedding), then $M$ is diffeomorphic to an immersed submanifold (resp. submanifold) of $N$: namely $f(M)$, where $f(M)$ is endowed with the differentiable structure for which $f : M \to f(M)$ is a diffeomorphism. Clearly, $\imath : f(M) \to N$ is a one-to-one immersion (resp. imbedding). More generally, two immersions $f_1 : M_1 \to N$ and $f_2 : M_2 \to N$ are said to be *equivalent* if there is a diffeomorphism $g : M_1 \to M_2$ such that $f_2 \circ g = f_1$. This defines an equivalence relation where each equivalence class contains a unique immersed submanifold of $N$.

DEFINITION 6.3. Let $f : M^n \to N^k$ be differentiable. A point $p \in M$ is said to be a *regular point* of $f$ if $f_*$ has rank $k$ at $p$; otherwise, $p$ is called a *critical point*. $q \in N$ is said to be a *regular value* of $f$ if its preimage $f^{-1}(q)$ contains no critical points (for example, if $q \notin f(M)$).

THEOREM 6.1. *Let $f : M^n \to N^k$ be a smooth map, with $n \geq k$. If $q \in N$ is a regular value of $f$ and if $A := f^{-1}(q) \neq \emptyset$, then $A$ is a topological manifold of dimension $n - k$. Moreover, there exists a unique differentiable structure for which $A$ becomes a differentiable submanifold of $M$.*

PROOF. Let $y : V \to \mathbb{R}^k$ be a coordinate map around $q$ with $y(q) = 0$; given $p \in A$, let $x : U \to \mathbb{R}^n$ be a coordinate map sending $p$ to 0. Decompose $\mathbb{R}^n = \mathbb{R}^k \times \mathbb{R}^{n-k}$, and denote by $\pi_i$, $i = 1, 2$, the projections of $\mathbb{R}^n$ onto the two factors; finally, let $\imath_2 : \mathbb{R}^{n-k} \to \mathbb{R}^n$ be the map given by $\imath_2(a_1, \ldots, a_{n-k}) = (0, \ldots, 0, a_1, \ldots, a_{n-k})$.
Since $y \circ f \circ x^{-1}$ has maximal rank at $0 \in \mathbb{R}^n$, there exists, by Theorem 5.3(2), a chart $(W, h)$ around 0 in $\mathbb{R}^n$ such that $y \circ f \circ x^{-1} \circ h = \pi_1|_W$. Set $\tilde{W} = \pi_2(W)$. $\tilde{W}$ is open in $\mathbb{R}^{n-k}$, and $y \circ f \circ x^{-1} \circ h \circ \imath_2|_{\tilde{W}} = \pi_1 \circ \imath_2|_{\tilde{W}} = 0$. Thus, if $z := x^{-1} \circ h \circ \imath_2|_{\tilde{W}}$, then $z(\tilde{W}) \subset A$. We claim that $z(\tilde{W}) = A \cap (x^{-1} \circ h)(W)$, so that $z$ maps $\tilde{W}$ homeomorphically onto a neighborhood of $p$ in $A$ in the subspace topology. Clearly, $z(\tilde{W}) \subset A \cap (x^{-1} \circ h)(W)$, since $z(\tilde{W}) = (x^{-1} \circ h \circ \imath_2)(\tilde{W}) = (x^{-1} \circ h)(W \cap (0 \times \mathbb{R}^{n-k}))$. Conversely, if $\tilde{p} \in A \cap (x^{-1} \circ h)(W)$, then $\tilde{p} = (x^{-1} \circ h)(u)$ for a unique $u \in W$, and $0 = y \circ f(\tilde{p}) = (y \circ f \circ x^{-1} \circ h)(u) = \pi_1(u)$, so that $u = (0, a) \in 0 \times \tilde{W}$. Then $\tilde{p} = z(a) \in z(\tilde{W})$. It follows that the inclusion $\imath : A \hookrightarrow M$ is a topological imbedding.
Endow $A$ with the differentiable structure induced by the charts $(z(\tilde{W}), z^{-1})$ as $p$ ranges over $A$. Then $\imath : A \hookrightarrow M$ is smooth, since $x \circ \imath \circ (z^{-1})^{-1} = h \circ \imath_2$. $\square$

EXAMPLES AND REMARKS 6.1. (i) Let $r > 0$, and consider the map $f : \mathbb{R}^{n+1} \to \mathbb{R}$ given by $f(a) = |a|^2 - r^2$. Since $Df(a) = 2(a_1, \ldots, a_{n+1})$, $f$ has maximal rank 1 everywhere except at the origin. Thus, $S_r^n = f^{-1}(0)$ is a differentiable submanifold of $\mathbb{R}^{n+1}$. This differentiable structure coincides with the one introduced in Examples and Remarks 1.1: it is straightforward to check that the inclusion of the sphere into Euclidean space is smooth for the atlas introduced there; i.e., that $\imath \circ x^{-1} : \mathbb{R}^n \to \mathbb{R}^{n+1}$ is differentiable, if $x$ denotes stereographic projection.

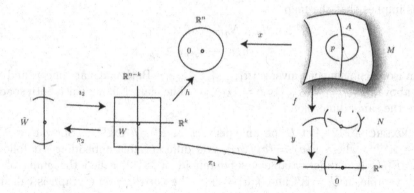

FIGURE 4

(ii) Let $f : M^n \to N^k$ be a differentiable map as in Definition 6.3. A point of $N$ that is not a regular value is called a *critical value of $f$*. Sard proved that if $U$ is an open set in $\mathbb{R}^n$, and $f : U \to \mathbb{R}^k$ is differentiable, then the set of critical values of $f$ has measure zero; i.e., given any $\epsilon > 0$, there exists a sequence of $k$-dimensional cubes containing the set of critical values, whose total volume is less than $\epsilon$. A proof of Sard's theorem can be found in [25]. As a consequence, the set of regular values of a map $f : M \to N$ between manifolds is dense in $N$, since its complement cannot contain an open nonempty set.

(iii) A surjective differentiable map $f : M^n \to N^k$ is said to be a *submersion* if every point of $M$ is a regular point of $f$. In this case, $f$ has no critical values, and each $p \in M$ belongs to the $(n-k)$-dimensional submanifold $f^{-1}(f(p))$.

Let $\imath : A \to M$ be an imbedding. For $p \in A$, $\imath_{*p}$ identifies the tangent space $A_p$ with a subspace of $M_p$.

PROPOSITION 6.2. *Let $q$ be a regular value of $f : M^n \to N^k$, where $n \geq k$, and suppose that $A := f^{-1}(q) \neq \emptyset$. Then for $p \in A$, $\imath_{*p}A_p = \ker f_{*p}$.*

PROOF. Since both subspaces have common dimension $n - k$, it suffices to check that $\imath_{*p}A_p \subset \ker f_{*p}$. Let $v \in A_p$. For $\phi \in \mathcal{F}N$, we have

$$(f_{*p}\imath_{*p}v)(\phi) = (f \circ \imath)_{*p}v(\phi) = v(\phi \circ f \circ \imath) = 0,$$

where the last identity follows from the fact that $f \circ \imath \equiv q$, so that $\phi \circ f \circ \imath$ is a constant function. This establishes the result. $\square$

EXAMPLE 6.1. Given manifolds $M$, $N$ with $p \in M$, $q \in N$, define imbeddings $\imath_q : M \to M \times N$ and $\jmath_p : N \to M \times N$ by $\imath_q(p) = \jmath_p(q) = (p, q)$. If $\pi_1$, $\pi_2$ denote the projections of $M \times N$ onto $M$ and $N$, then

$$\pi_1 \circ \imath_q = 1_M, \quad \pi_2 \circ \jmath_p = 1_N, \quad \pi_1 \circ \jmath_p = p, \quad \pi_2 \circ \imath_q = q,$$

where $p$ is identified with the constant map $M \to M$ sending every point to $p$, and similarly for $q$. Thus,

$$\pi_{1*} \circ \imath_{q*p} = 1_{M_p}, \quad \pi_{2*} \circ \jmath_{p*q} = 1_{N_q}, \quad \pi_{1*} \circ \jmath_{p*q} = 0, \quad \pi_{2*} \circ \imath_{q*p} = 0.$$

This implies that the map

$$L : M_p \times N_q \to (M \times N)_{(p,q)},$$

$$(u, v) \mapsto \imath_{q*p} u + \jmath_{p*q} v$$

is an isomorphism with inverse $(\pi_{1*(p,q)}, \pi_{2*(p,q)})$: Both maps are linear, and by the above, $(\pi_{1*(p,q)}, \pi_{2*(p,q)}) \circ L = 1_{M_p \times N_q}$. The claim follows since both spaces have the same dimension.

EXERCISE 12. Let $U$ be an open set in $\mathbb{R}^n$, $f \in \mathcal{F}U$. Show that $F :$ $U \to \mathbb{R}^{n+1}$, where $F(a) = (a, f(a))$, is a differentiable imbedding. It follows that $F(U)$ is a differentiable $n$-submanifold of $\mathbb{R}^{n+1}$, called the *graph* of $f$. For example, if $U = \mathbb{R}^n$ and $f(a) = |a|^2$, the corresponding graph is called a paraboloid.

EXERCISE 13. Suppose $f : M \to N$ is differentiable, and let $Q$ denote a submanifold of $N$. $f$ is said to be *transverse regular* at $p \in f^{-1}(Q)$ if $f_{*p} M_p + Q_{f(p)} = N_{f(p)}$. Show that if $f$ is transverse regular at every point of $f^{-1}(Q) \neq \emptyset$, then $f^{-1}(Q)$ is a submanifold of $M$ of codimension equal to the codimension of $Q$ in $N$. Theorem 6.1 is the special case when $Q$ consists of a single point.

EXERCISE 14. For $p \in \mathbb{R}^{n+1}$, let $\mathcal{J}_p : \mathbb{R}^{n+1} \to (\mathbb{R}^{n+1})_p$ denote the canonical isomorphism. Use Proposition 6.2 to show that if $p \in S_r^n$, then

$$\imath_*(S_r^n)_p = \mathcal{J}_p(p^{\perp}),$$

where $p^{\perp} = \{a \in \mathbb{R}^{n+1} \mid \langle a, p \rangle = 0\}$ is the orthogonal complement of $p$.

EXERCISE 15. Prove that if $M$ is compact, then $f : M^n \to \mathbb{R}^n$ cannot have maximal rank everywhere. Show by means of an example that such an $f$ can nevertheless have maximal rank on a dense subset of $M$.

## 7. Vector Fields

In calculus, one defines a vector field on an open set $U \subset \mathbb{R}^n$ as a differentiable map $F = (f_1, \ldots, f_n) : U \to \mathbb{R}^n$. When graphing a vector field on, say, $\mathbb{R}^2$, one draws the vector $F(p)$ with its origin at $p$, in order to distinguish it from the values of $F$ at other points; in terms of tangent spaces, this means that $F(p)$ is considered to be a vector in the tangent space of $\mathbb{R}^n$ at $p$. It is now natural to generalize this concept to manifolds as follows:

DEFINITION 7.1. Let $U$ be an open set of the differentiable manifold $M^n$. A (differentiable) *vector field* on $U$ is a (differentiable) map $X : U \to TM$ such that $\pi \circ X = 1_U$. Here $\pi : TM \to M$ denotes the tangent bundle projection.

Thus, the value of $X$ at $p$, which we often denote by $X_p$, is a vector in $M_p$. Any $f \in \mathcal{F}U$ determines a new function $Xf$ on $U$ by setting $Xf(p) := X_p(f)$. If $(U, x)$ is a chart, the *coordinate vector fields* are the vector fields $\partial/\partial x^i$ whose value at $p \in U$ is $\partial/\partial x^i(p)$, cf. (3.1). Any vector field $X$ on $U$ can then be written as $X = \sum_i X(x^i)\partial/\partial x^i = \sum_i dx^i(X)\partial/\partial x^i$.

PROPOSITION 7.1. *Let $X : U \to TM$ be a map such that $\pi \circ X = 1_U$. The following statements are equivalent:*

(1) $X$ is a vector field on $U$ (i.e., $X$, as a map, is differentiable).

(2) If $(V, x)$ is a chart with $V \subset U$, then $X x^i \in \mathcal{F}V$.

(3) If $f \in \mathcal{F}V$, then $Xf \in \mathcal{F}V$.

PROOF. (1)$\Rightarrow$(2): Recall that $(V, x)$ induces a coordinate map $\bar{x}$ on $\pi^{-1}(V)$, where $\bar{x}(v) = (x \circ \pi(v), v(x^1), \ldots, v(x^n))$. Since $X$ is smooth, $\bar{x} \circ X|_V = (x \circ 1|_V, X|_V(x^1), \ldots, X|_V(x^n))$ also has that property. Thus, each component function $X x^i$ is differentiable on $V$.

(2)$\Rightarrow$(3): If each $X|_V(x^i) \in \mathcal{F}V$, then $X|_V(f) = \sum_i (X|_V x^i) \partial f / \partial x^i \in \mathcal{F}V$.

(3)$\Rightarrow$(1): $\bar{x} \circ X|_V = (x, X|_V(x^1), \ldots, X|_V(x^n))$ is smooth, and therefore so is $X|_V$. Since this is true for any chart $(V, x)$ with $V \subset U$, $X$ is differentiable. $\square$

EXAMPLE 7.1. A vector field $X$ on $\mathbb{R}^n$ induces a differentiable map $F = (f^1, \ldots, f^n) : \mathbb{R}^n \to \mathbb{R}^n$, where $f^i = du^i(X)$; conversely, any smooth map $F : U \to \mathbb{R}^n$ on an open subset $U$ of $\mathbb{R}^n$ determines a vector field $X$ on $U$, with $X(p) = \mathcal{J}_p F(p)$.

Let $\mathfrak{X}U$ denote the set of vector fields on $U$. $\mathfrak{X}U$ is a real vector space and a module over $\mathcal{F}U$ with the operations $(X + Y)_p = X_p + Y_p$, $(\phi X)_p = \phi(p)X_p$. If $f, g \in \mathcal{F}U$ and $\alpha, \beta \in \mathbb{R}$, then $X(\alpha f + \beta g) = \alpha(Xf) + \beta(Xg)$, and $X(fg) = (Xf)g + (Xg)f$.

We recall two theorems from the theory of ordinary differential equations:

THEOREM 7.1 (Existence of Solutions). Let $F : U \to \mathbb{R}^n$ be a differentiable map, where $U$ is open in $\mathbb{R}^n$. For any $a \in U$, there exists a neighborhood $W$ of $a$, an interval $I$ around $0$, and a differentiable map $\psi : I \times W \to U$ such that

(1) $\psi(0, u) = u$, and

(2) $D\psi(t, u)\mathbf{e}_1 = F \circ \psi(t, u)$

for $t \in I$ and $u \in W$.

Theorem 7.1 may be interpreted as follows: A curve $c : I \to U$ is called an integral curve of (the system of ordinary differential equations defined by) $F$ if $c^{i\prime} = F^i \circ c$, $1 \le i \le n$; in this case, $Dc = F \circ c$, and the restriction of $F$ to $c$ is the "velocity field" of $c$. Thus, 7.1 asserts that integral curves $t \mapsto c(t) := \psi(t, u)$ exist for arbitrary initial conditions $c(0) = u$, that they depend smoothly on the initial conditions, and that at least locally, they can be defined on a fixed common interval. Also notice that in manifold notation, $c$ is an integral curve of $F : \mathbb{R}^n \to \mathbb{R}^n$ iff $\dot{c} = X \circ c$, where $X = \mathcal{J}F$, cf. the example above.

THEOREM 7.2 (Uniqueness of Solutions). If $c, \tilde{c} : I \to U$ are two integral curves of $F : U \to \mathbb{R}^n$ with $c(t_0) = \tilde{c}(t_0)$ for some $t_0 \in I$, then $c = \tilde{c}$.

DEFINITION 7.2. Let $M$ be a manifold, $X \in \mathfrak{X}M$, and $I$ an interval. A curve $c : I \to M$ is called an integral curve of $X$ if $\dot{c} = X \circ c$.

THEOREM 7.3. Let $M$ be a manifold, $X \in \mathfrak{X}M$. For any $q \in M$, there exists a neighborhood $V$ of $q$, an interval $I$ around $0$, and a differentiable map $\Phi : I \times V \to M$ such that

(1) $\Phi(0, p) = p$, and

(2) $\Phi_* \frac{\partial}{\partial t}(t, p) = X \circ \Phi(t, p)$

*for all $t \in I$, $p \in V$. Here, $\partial/\partial t(t,p) := \imath_{p*} D(t)$ for the injection $\imath_p : I \to I \times V$ which maps $t$ to $(t,p)$.*

Notice that

$$\Phi_* \frac{\partial}{\partial t}(t,p) = \widetilde{\Phi \circ \imath_p}(t) = \dot{\Phi}_p(t),$$

where $\Phi_p(t) = \Phi(t,p)$. Theorem 7.3 asserts that for any $p \in V$, $\Phi_p : I \to M$ is an integral curve of $X$ passing through $p$ at $t = 0$. $\Phi$ is called a *local flow* of $X$.

PROOF. Let $(U,x)$ be a chart around $q$, and set $G := x(U)$, $a := x(q)$, and

$$F := (dx^1(X), \ldots, dx^n(X)) \circ x^{-1} : G \to \mathbb{R}^n.$$

By Theorem 7.1, there exists a neighborhood $W$ of $a$, an interval $I$ around 0, and a map $\psi : I \times W \to G$ such that (1) and (2) of 7.1 hold. Let $V := x^{-1}(W)$, and $\Phi : I \times V \to M$ be given by $\Phi(t,p) = x^{-1} \circ \psi(t, x(p))$. $\quad\square$

An argument similar to the one above generalizes the uniqueness theorem 7.2 to manifolds:

THEOREM 7.4. *If $c, \tilde{c} : I \to M$ are two integral curves of $X \in \mathfrak{X}M$ with $c(t_0) = \tilde{c}(t_0)$ for some $t_0 \in I$, then $c = \tilde{c}$.*

For each $p \in M$, let $I_p$ denote the maximal open interval around 0 on which the (unique by 7.4) integral curve $\Phi_p : I_p \to M$ of $X$ with $\Phi_p(0) = p$ is defined.

THEOREM 7.5. *Given any $X \in \mathfrak{X}M$, there exists a unique open set $W \subset \mathbb{R} \times M$ and a unique differentiable map $\Phi : W \to M$ such that*

(1) $I_p \times \{p\} = W \cap (\mathbb{R} \times \{p\})$ *for all $p \in M$, and*
(2) $\Phi(t,p) = \Phi_p(t)$ *if $(t,p) \in W$.*

$\Phi$ is called the *maximal flow* of $X$. By (2), $\{0\} \times M \subset W$, and (1), (2) of Theorem 7.3 are satisfied.

PROOF. (1) determines $W$ uniquely, while (2) does the same for $\Phi$. It thus remains to show that $W$ is open, and that $\Phi$ is differentiable.

Fix $p \in M$, and let $I$ denote the set of all $t \in I_p$ for which there exists a neighborhood of $(t,p)$ contained in $W$ on which $\Phi$ is differentiable. We will establish that $I$ is nonempty, open and closed in $I_p$, so that $I = I_p$: $I$ is nonempty because $0 \in I$ by Theorem 7.3, and is open by definition. To see that it is closed, consider $t_0 \in \bar{I}$; by 7.3, there exists a local flow $\Phi' : I' \times V' \to M$ with $0 \in I'$ and $\Phi_p(t_0) \in V'$. Let $t_1 \in I$ be small enough that $t_0 - t_1 \in I'$ (recall that $t_0$ belongs to the closure of $I$) and $\Phi_p(t_1) \in V'$ (by continuity of $\Phi_p$). Choose an interval $I_0$ around $t_0$ such that $t - t_1 \in I'$ for $t \in I_0$. Finally, by continuity of $\Phi$ at $(t_1, p)$, there exists a neighborhood $V$ of $p$ such that $\Phi(t_1 \times V) \subset V'$.

We claim that $\Phi$ is defined and differentiable on $I_0 \times V$, so that $t_0 \in I$: Indeed, if $t \in I_0$ and $q \in V$, then by definition of $I_0$ and $V$, $t - t_1 \in I'$ and $\Phi(t_1, q) \in V'$, so that $\Phi'(t - t_1, \Phi(t_1, q))$ is defined. The curve $s \mapsto \Phi'(s - t_1, \Phi(t_1, q))$ is an integral curve of $X$ which equals $\Phi(t_1, q)$ at $t_1$. By uniqueness, $\Phi(t, q) = \Phi'(t - t_1, \Phi(t_1, q))$ is defined, and $\Phi$ is therefore differentiable at $(t, q)$. $\quad\square$

DEFINITION 7.3. Let $\Phi : \mathbb{R} \times M \to M$ be differentiable, and define $\Phi_t : M \to M$ by $\Phi_t(p) := \Phi(t, p)$. $\{\Phi_t\}_{t \in \mathbb{R}}$ is called a *one-parameter group of diffeomorphisms* of $M$ if

    (1) $\Phi_0 = 1_M$, and

    (2) $\Phi_{t_1+t_2} = \Phi_{t_1} \circ \Phi_{t_2}, \quad t_1, t_2 \in \mathbb{R}$.

Observe that each $\Phi_t$ is indeed a diffeomorphism of $M$ with inverse $\Phi_{-t}$. If $\Phi$ is a one-parameter group of diffeomorphisms, then the vector field $X$ defined by $X_p := \Phi_* \frac{\partial}{\partial t}\big|_{(0,p)}$ has $\Phi$ as maximal flow (since integral curves are defined for all time). Conversely, if $X \in \mathfrak{X}M$, then the maximal flow of $X$ induces a one-parameter group of diffeomorphisms provided $X$ is *complete*; i.e., provided integral curves are defined for all time. The exercises at the end of the section establish that vector fields on compact manifolds are always complete.

EXAMPLE 7.2. Consider the vector field $X \in \mathfrak{X}\mathbb{R}^2$ whose value at $a = (a_1, a_2)$ is given by $-a_2 D_1|_a + a_1 D_2|_a$. Fix $p = (p_1, p_2) \in \mathbb{R}^2$, and let $c : \mathbb{R} \to \mathbb{R}^2$ denote the curve

$$c(t) = ((\cos t)p_1 - (\sin t)p_2, (\sin t)p_1 + (\cos t)p_2).$$

Then

$$\dot{c}(t) = (-(\sin t)p_1 - (\cos t)p_2)D_1|_{c(t)} + ((\cos t)p_1 - (\sin t)p_2)D_2|_{c(t)} = X \circ c(t).$$

Thus, $c$ is the integral curve of $X$ with $c(0) = p$, and $X$ is complete. The one-parameter group of $X$ is the rotation group

$$\Phi_t(p_1, p_2) = \begin{pmatrix} \cos t & -\sin t \\ \sin t & \cos t \end{pmatrix} \begin{pmatrix} p_1 \\ p_2 \end{pmatrix}.$$

EXERCISE 16. Show explicitly that $\Phi$ in Theorem 7.3 satisfies (1) and (2).

EXERCISE 17. With notation as in Theorem 7.5,

    (a) Show by means of an example that there need not exist an open interval $I$ around 0 such that $I \times M \subset W$. *Hint:* Let $M = \mathbb{R}$, $X_t = -t^2 D_t$.

    (b) Show that if such an interval exists, then it equals all of $\mathbb{R}$; i.e., $W = \mathbb{R} \times M$, and integral curves are defined for all time.

    (c) Prove that if $M$ is compact, then any vector field on $M$ is complete.

EXERCISE 18. Let $\phi : [\alpha, \beta) \to M$ be an integral curve of $X \in \mathfrak{X}M$, and suppose that for some sequence $t_n \to \beta$, $\phi(t_n) \to p$ for some $p \in M$.

    (a) Show that $\bar{\phi} : [\alpha, \beta] \to M$, where $\bar{\phi}|_{[\alpha,\beta)} = \phi$ and $\bar{\phi}(\beta) = p$, is continuous.

    (b) Prove that if $c : I \to M$ is the maximal integral curve of $X$ with $c(\beta) = p$, then $[\alpha, \beta] \subset I$, and $c_{|[\alpha,\beta]} = \bar{\phi}$.

    (c) Use parts (a) and (b) to recover the result from Exercise 17 (c): Namely, if $M$ is compact, then every integral curve of $X \in \mathfrak{X}M$ is defined on all of $\mathbb{R}$.

## 8. The Lie Bracket

Consider two vector fields $X$ and $Y$ on an open subset $U$ of $M$, with flows $\Phi_s$ and $\Psi_t$, respectively. It may well happen that these flows commute; i.e., that $\Phi_s \circ \Psi_t = \Psi_t \circ \Phi_s$ for small $s$ and $t$. This is the case for example when $X$ and $Y$ are coordinate vector fields, since the standard fields $D_i$ and $D_j$ in

Euclidean space have commuting flows. In general, the Lie bracket $[X, Y]$ of $X$ and $Y$ is a new vector field that detects noncommuting flows. This concept actually makes sense in the more general setting of an arbitrary vector space $E$:

DEFINITION 8.1. A *Lie bracket* on a real vector space $E$ is a map $[,]$ : $E \times E \to E$ satisfying:

(1) $[\alpha X + \beta Y, Z] = \alpha[X, Z] + \beta[Y, Z]$,
(2) $[X, Y] = -[Y, X]$, and
(3) $[X, [Y, Z]] + [Y, [Z, X]] + [Z, [X, Y]] = 0$

for all $X, Y, Z \in E$, $\alpha, \beta \in \mathbb{R}$. By (1) and (2), the Lie bracket is linear in the second component also. (3) is called the *Jacobi identity*. A vector space together with a Lie bracket is called a *Lie algebra*.

A trivial example of a Lie algebra is $\mathbb{R}^n$ with $[,] \equiv 0$. This is the so-called abelian $n$-dimensional Lie algebra. $\mathbb{R}^3$ is also a Lie algebra, if one takes the Lie bracket to be the classical cross-product of two vectors.

Let $M$ be a differentiable manifold, $p$ a point in an open set $U$ of $M$, and $X, Y \in \mathfrak{X}U$. Define $X_pY : \mathcal{F}_pU \to \mathbb{R}$ by setting $(X_pY)f := X_p(Yf)$. $X_pY$ is not a tangent vector at $p$, because although it is linear on functions, it is not a derivation. However, $X_pY - Y_pX$ is one:

$$
\begin{aligned}
(X_pY - Y_pX)(fg) &= X_p(Y(fg)) - Y_p(X(fg)) \\
&= X_p(f(Yg) + g(Yf)) - Y_p(f(Xg) + g(Xf)) \\
&= (X_pf)(Y_pg) + f(p)X_p(Yg) + (X_pg)(Y_pf) + g(p)X_p(Yf) \\
&\quad - (Y_pf)(X_pg) - f(p)Y_p(Xg) - (Y_pg)(X_pf) \\
&\quad - g(p)Y_p(Xf) \\
&= f(p)(X_pY - Y_pX)(g) + g(p)(X_pY - Y_pX)(f).
\end{aligned}
$$

Thus, $p \mapsto X_pY - Y_pX$ is a vector field on $U$.

DEFINITION 8.2. Let $X, Y \in \mathfrak{X}U$, where $U$ is open in $M$. The *Lie bracket* of $X$ with $Y$ is the vector field $[X, Y]$ on $U$ defined by $[X, Y]_p := X_pY - Y_pX$.

It is straightforward to check that $\mathfrak{X}U$ with the above bracket is a Lie algebra. One often denotes $X(Yf)$ by $XYf$, so that one may write

$$[X, Y] = XY - YX.$$

Observe also that for $f \in \mathcal{F}U$, $[fX, Y] = f[X, Y] - (Yf)X$.

PROPOSITION 8.1. *Let $(U, x)$ denote a chart of $M^n$. Then $[\partial/\partial x^i, \partial/\partial x^j] \equiv 0$ for $1 \leq i, j \leq n$.*

PROOF. For $\phi \in \mathcal{F}U$,

$$
\begin{aligned}
[\frac{\partial}{\partial x^i}, \frac{\partial}{\partial x^j}]\phi &= \frac{\partial}{\partial x^i}\frac{\partial}{\partial x^j}\phi - \frac{\partial}{\partial x^j}\frac{\partial}{\partial x^i}\phi \\
&= D_i\left(\frac{\partial}{\partial x^j}\phi \circ x^{-1}\right) \circ x - D_j\left(\frac{\partial}{\partial x^i}\phi \circ x^{-1}\right) \circ x \\
&= D_i(D_j(\phi \circ x^{-1})) \circ x - D_j(D_i(\phi \circ x^{-1})) \circ x = 0.
\end{aligned}
$$

$\square$

If $f : M \to N$ is differentiable and $X \in \mathfrak{X}M$, then the formula $Y_{f(p)} := f_* X_p$ does not, in general, define a vector field on $N$. We say $X \in \mathfrak{X}M$ and $Y \in \mathfrak{X}N$ are $f$-related if $Y_{f(p)} := f_* X_p$ for all $p \in M$; i.e., if $f_* X = Y \circ f$. When $f$ is a diffeomorphism, any $X \in \mathfrak{X}M$ is $f$-related to the vector field $f_* \circ X \circ f^{-1}$ on $N$.

PROPOSITION 8.2. *Let* $f : M \to N$ *be differentiable,* $X_i \in \mathfrak{X}M$, $Y_i \in \mathfrak{X}N$, $i = 1, 2$. *If* $X_i$ *and* $Y_i$ *are* $f$-*related, then* $[X_1, X_2]$ *and* $[Y_1, Y_2]$ *are* $f$-*related.*

PROOF. If $\phi \in \mathcal{F}N$, then for $p \in M$,

$$[Y_1, Y_2]_{f(p)} \phi = Y_{1|f(p)}(Y_2 \phi) - Y_{2|f(p)}(Y_1 \phi) = f_* X_{1|p}(Y_2 \phi) - f_* X_{2|p}(Y_1 \phi)$$
$$= X_{1|p}((Y_2 \phi) \circ f) - X_{2|p}((Y_1 \phi) \circ f).$$

Next, observe that $(Y_i \phi) \circ f = X_i(\phi \circ f)$, since

$$((Y_i \phi) \circ f)(q) = (Y_i \phi)(f(q)) = Y_{i|f(q)} \phi = (f_* X_{i|q}) \phi = X_{i|q}(\phi \circ f).$$

Thus,

$$[Y_1, Y_2]_{f(p)} \phi = X_{1|p}(X_2(\phi \circ f)) - X_{2|p}(X_1(\phi \circ f)) = [X_1, X_2]_p(\phi \circ f)$$
$$= (f_*[X_1, X_2]_p) \phi.$$

$\square$

DEFINITION 8.3. An $n$-dimensional manifold and group $G$ is called a *Lie group* if the group multiplication $G \times G \to G$ and the operation of taking the inverse $G \to G$ are differentiable.

It follows that for $h \in G$, left-translation $L_h : G \to G$ by $h$, $L_h g := hg$, is differentiable. A vector field $X \in \mathfrak{X}G$ is said to be *left-invariant* if it is $L_g$-related to itself for any $g \in G$. Such a vector field will be abbreviated l.i.v.f. The collection $\mathfrak{g}$ of all l.i.v.f. is a real vector space, and by Proposition 8.2, is also a Lie algebra. It is called the *Lie algebra* of $G$.

Any $X \in \mathfrak{g}$ is uniquely determined by its value at the identity $e$: indeed, $X_g = X \circ L_g(e) = L_{g*} X_e$. Thus, the linear map $\mathfrak{g} \to G_e$ which sends a l.i.v.f. to its value at the identity is one-to-one. It is actually an isomorphism: given $v \in G_e$, the vector field $X$ defined by $X_g := L_{g*} v$ is left-invariant, since

$$L_{h*} X_g = (L_h \circ L_g)_* v = L_{hg*} v = X \circ L_h(g).$$

We may therefore consider $G_e$ to be a Lie algebra by setting $[X_e, Y_e] := [X, Y]_e$ for l.i.v.f.'s $X$ and $Y$.

EXAMPLE 8.1. (i) $\mathbb{R}^n$ is a Lie group with the usual vector addition. Left translation by $v \in \mathbb{R}^n$ is just $L_v w = v + w$. Since the Jacobian matrix of $L_v$ is the identity, we have that $L_{v*} D_{i|a} = D_{i|L_v(a)}$; equivalently, the standard coordinate vector fields form a basis for the Lie algebra of $\mathbb{R}^n$; this Lie algebra is abelian by Proposition 8.1.

(ii) Let $G = GL(n)$ denote the collection of invertible $n$ by $n$ real matrices. It becomes a Lie group under matrix multiplication. As an open subset of the $n^2$-dimensional vector space $M_n$ of all $n$ by $n$ matrices, its Lie algebra $\mathfrak{gl}(n)$ may be identified with $M_n$ via

$$M_n \xrightarrow{\mathcal{J}_e} G_e \xrightarrow{\cong} \mathfrak{gl}(n),$$

where $e$ is the $n$ by $n$ identity matrix. We claim that under this identification, the Lie bracket is given by

$$(8.1) \qquad [\mathcal{J}_e M, \mathcal{J}_e N] = \mathcal{J}_e(MN - NM), \qquad M, N \in M_n.$$

To see this, let $X, Y$ be the left-invariant vector fields with $X(e) = \mathcal{J}_e M$, $Y(e) = \mathcal{J}_e N$. Since left translation by $A$ is a linear transformation, $X(A) = \mathcal{J}_A(AM)$. If $u^{ij} : G \to \mathbb{R}$ denotes the function that assigns to a matrix $A$ its $(i,j)$-th entry $A_{ij}$, then

$$(Yu^{ij})(A) = Y(A)(u^{ij}) = \mathcal{J}_A(AN)(u^{ij}) = (AN)_{ij},$$

so that $Yu^{ij} = u^{ij} \circ R_N$, where $R_N$ is right translation by $N$, $R_N(A) = AN$. Consider the curve $t \mapsto c(t) = e + tM$. Then

$$X(e)(Yu^{ij}) = \dot{c}(0)(u^{ij} \circ R_N) = D_0(t \mapsto u^{ij}(N + tMN)) = (MN)_{ij},$$

and similarly, $Y(e)(Xu^{ij}) = (NM)_{ij}$. Thus,

$$[\mathcal{J}_e M, \mathcal{J}_e N](u^{ij}) = [X, Y](e)(u^{ij}) = (MN - NM)_{ij} = \mathcal{J}_e(MN - NM)(u^{ij}).$$

Since $\mathcal{J}_e Q = \sum_{i,j}(\mathcal{J}_e Q)(u^{ij})(\partial/\partial u^{ij})_{|e}$ for any $Q \in M_n$, this establishes the claim.

(iii) Given a Lie group $G$, and $g \in G$, *conjugation* by $g$ is the map $\tau_g := L_g \circ R_{g^{-1}} : G \to G$. Under the identification $\mathfrak{g} = G_e$, the derivative $\tau_{g*e}$ belongs to $GL(\mathfrak{g})$, and is denoted $\mathrm{Ad}_g$. The map $\mathrm{Ad} : G \to GL(\mathfrak{g})$ which sends $g$ to $\mathrm{Ad}_g$ is then a Lie group homomorphism, and is called the *adjoint representation* of $G$. Notice that if $G$ is abelian, then this representation is trivial; in general, the kernel of $\mathrm{Ad}$ is the center $Z(G) = \{g \in G \mid gh = hg, h \in G\}$ of $G$.

As an example, consider the Lie group $G = GL(n)$. We claim that $\mathrm{Ad}_g$ is just $\tau_g$; more precisely, viewing $G$ as an open subset of the space $M_n$ of all $n$ by $n$ matrices, we have the identification $\mathcal{J}_e : M_n \to \mathfrak{gl}(n)$ as in (ii). Linearity of $\tau_g$ then implies that the diagram

$$
\begin{array}{ccc}
\mathfrak{gl}(n) & \xrightarrow{\;\mathrm{Ad}_g\;} & \mathfrak{gl}(n) \\[4pt]
\mathcal{J}_e \uparrow & & \uparrow \mathcal{J}_e \\[4pt]
M_n & \xrightarrow[\;\tau_g\;]{} & M_n
\end{array}
$$

commutes.

(iv) The set $\mathbb{H}$ of *quaternions* is just $\mathbb{R}^4 = \{\sum_{i=1}^4 \alpha_i \mathbf{e}_i \mid \alpha_i \in \mathbb{R}\}$; in addition to the vector space structure, there is an associative and distributive multiplication which generalizes that of complex numbers: write $\sum_{i=1}^4 \alpha_i \mathbf{e}_i$ as $\alpha_1 + \alpha_2 i + \alpha_3 j + \alpha_4 k$, and define $i^2 = j^2 = k^2 = -1$, $ij = -ji = k$, $jk = -kj = i$, $ki = -ik = j$, and $1u = u$ for any quaternion $u$. The set $\mathbb{H}^*$ of nonzero quaternions is then a Lie group with the above multiplication. Furthermore, it is straightforward to check that multiplication is norm-preserving in the sense that $|uv| = |u||v|$ for quaternions $u, v$ (with the usual Euclidean norm), so that $\mathbb{H}^*$ contains $S^3$ as a subgroup.

Recall the canonical isomorphism $\mathcal{J}_u : \mathbb{H} \to \mathbb{H}_u$ with $\mathcal{J}_u \mathbf{e}_i = D_{i|u}$, $u \in \mathbb{H}$. Since left translation by $u$ is a linear transformation of $\mathbb{R}^4$, we have that $L_{u*}\mathcal{J}_{\mathbf{e}_1} a = \mathcal{J}_u(ua)$. Thus, the l.i.v.f. $X$ with $X_{\mathbf{e}_1} = \mathcal{J}_{\mathbf{e}_1} a$ is given by $X_u =$

$\mathcal{J}_u(ua)$. Applying this to $a = \mathbf{e}_i$, we obtain a basis $X_i$ of l.i.v.f. with $X_{i|\mathbf{e}_1} = D_{i|\mathbf{e}_1}$, where

$$X_1 = u^1 D_1 + u^2 D_2 + u^3 D_3 + u^4 D_4,$$
$$X_2 = -u^2 D_1 + u^1 D_2 + u^4 D_3 - u^3 D_4,$$
$$X_3 = -u^3 D_1 - u^4 D_2 + u^1 D_3 + u^2 D_4,$$
$$X_4 = -u^4 D_1 + u^3 D_2 - u^2 D_3 + u^1 D_4.$$

Observe that $X_1$ is the "position" vector field, $X_{1|p} = \mathcal{J}_p p$, and that for $i > 1$, the $X_i$'s form a basis of the orthogonal complement $\mathcal{J}_p(p^\perp)$ at $p$. Thus, there are vector fields $I$, $J$, and $K$ on $S^3$ which are $\imath$-related to $X_2$, $X_3$, and $X_4$ for the inclusion $\imath : S^3 \hookrightarrow \mathbb{H}^*$. They are left-invariant and form a basis of the Lie algebra of $S^3$. This Lie algebra is actually isomorphic to the Lie algebra of $\mathbb{R}^3 = \{\alpha i + \beta j + \gamma k \mid \alpha, \beta, \gamma \in \mathbb{R}\}$ with the cross product, via $I \mapsto 2i$, $J \mapsto 2j$, $K \mapsto 2k$. It is well known that $S^1$ and $S^3$ are the only spheres that admit a Lie group structure.

DEFINITION 8.4. Let $X \in \mathfrak{X}M$ have flow $\Phi_t$. The *Lie derivative* of a vector field $Y$ with respect to $X$ at $p$ is the tangent vector at $p$ given by

$$(L_X Y)_p = \lim_{t \to 0} \frac{\Phi_{-t*} Y_{\Phi_t(p)} - Y_p}{t}.$$

Notice that $(L_X Y)_p = c'(0)$, where $c$ is the curve in $M_p$ given by $c(t) = \Phi_{-t*} Y_{\Phi_t(p)}$.

Recall that as a special case of Lemma 3.1, any smooth function $f : I \to \mathbb{R}$ with $f(0) = 0$ may be written as $f(t) = t\psi(t)$, where $\psi(0) = f'(0)$. In fact, $\psi(t_0) = \int_0^1 f'(st_0) \, ds$.

LEMMA 8.1. *Let $I$ denote an interval around $0$, $U$ an open set of $M$, and $f : I \times U \to \mathbb{R}$ a differentiable function such that $f(0,p) = 0$ for all $p \in U$. Then there exists a differentiable function $g : I \times U \to \mathbb{R}$ satisfying*

$$f(t,p) = tg(t,p), \qquad \frac{\partial}{\partial t}(0,p)(f) = g(0,p), \qquad t \in I, \quad p \in U,$$

*where $\partial/\partial t$ is the vector field on $I \times U$ that is $\imath_p$-related to $D$; i.e., $\imath_{p*} D = \partial/\partial t \circ \imath_p$ for the imbedding $\imath_p : I \to I \times U$, $\imath_p(t) = (t,p)$.*

PROOF. Set $g(t_0, p) := \int_0^1 (\partial/\partial t(st_0, p)(f)) \, ds$. $\square$

THEOREM 8.1. *For vector fields $X$ and $Y$ on $M$, $L_X Y = [X, Y]$.*

PROOF. Let $p \in M$, $f : M \to \mathbb{R}$, and $\Phi : I \times U \to M$ be a local flow of $X$ with $p \in U$. Apply Lemma 8.1 to the function $I \times U \to \mathbb{R}$ which maps $(t, q)$ to $(f \circ \Phi)(t, q) - f(q)$, and deduce that there exists a one-parameter family $g_t$ of functions on $U$ such that $f \circ \Phi_t = f + tg_t$, and $g_0 = Xf$. Now,

$$\Phi_{-t*} Y_{\Phi_t(p)}(f) = Y_{\Phi_t(p)}(f \circ \Phi_{-t}) = Y_{\Phi_t(p)}(f - tg_{-t})$$
$$= (Yf) \circ \Phi_t(p) - t(Yg_{-t}) \circ \Phi_t(p).$$

Observe that for a function $\phi$ on $U$,

$$(8.2) \qquad X_p(\phi) = (\phi \circ c)'(0) = \lim_{t \to 0} \frac{\phi \circ \Phi_t(p) - \phi(p)}{t},$$

where $c(t) = \Phi_t(p)$. Therefore,

$$(L_X Y)_p f = \lim_{t \to 0} \frac{(Yf) \circ \Phi_t(p) - (Yf)(p)}{t} - \lim_{t \to 0} (Yg_{-t}) \circ \Phi_t(p)$$
$$= X_p(Yf) - (Yg_0)(p) = [X, Y]_p f,$$

as claimed.                                                                                    □

The Lie bracket of two vector fields measures the extent to which their flows fail to commute:

PROPOSITION 8.3. *Let $\Phi_t$ and $\Psi_s$ denote local flows of $X$ and $Y \in \mathfrak{X}M$. Then $[X, Y] \equiv 0$ iff $\Phi_t \circ \Psi_s = \Psi_s \circ \Phi_t$ for all $s$, $t$.*

PROOF. Suppose that $\Phi_t \circ \Psi_s = \Psi_s \circ \Phi_t$. By Exercise 21 below, $Y$ is $\Phi_t$-related to itself; i.e., $\Phi_{t*} Y = Y \circ \Phi_t$, so that $\Phi_{-t*} Y \circ \Phi_t = Y$. $L_X Y$ then vanishes by definition.

Conversely, suppose that the Lie bracket of $X$ and $Y$ vanishes. For any fixed $p$ in $M$, the curve $c$ in $M_p$ given by $c(t) = \Phi_{-t*} Y \circ \Phi_t(p)$ then satisfies $c'(0) = 0$. We will show that $c$ is the constant curve $c(t) = Y_p$ for all $t$, or equivalently, that $c' \equiv 0$. Fix any $t$, and set $q = \Phi_t(p)$. Then

$$c'(t) = \lim_{h \to 0} \frac{c(t+h) - c(t)}{h}$$
$$= \lim_{h \to 0} \frac{1}{h} [\Phi_{-(t+h)*} \circ Y \circ \Phi_{t+h}(p) - \Phi_{-t*} \circ Y \circ \Phi_t(p)]$$
$$= \lim_{h \to 0} \frac{1}{h} \Phi_{-t*} [(\Phi_{-h*} \circ Y \circ \Phi_h)(\Phi_t p) - Y(\Phi_t p)]$$
$$= \Phi_{-t*} \lim_{h \to 0} \frac{1}{h} [\Phi_{-h*} \circ Y \circ \Phi_h(q) - Y(q)] = 0,$$

as claimed.                                                                                    □

THEOREM 8.2. *Let $V$ be open in $M^n$, and consider $k$ vector fields $X_1, \ldots, X_k$ on $V$ that are linearly independent at some $p \in V$. If $[X_i, X_j] \equiv 0$ for all $i$ and $j$, then there exists a coordinate chart $(U, x)$ around $p$ such that $\partial/\partial x^i = X_{i|U}$, for $i = 1, \ldots, k$. As a special case, if $X$ is a vector field that is nonzero at some point, then there is a coordinate chart $(U, x)$ around that point such that $\partial/\partial x^1 = X_{|U}$.*

PROOF. Recall that if $(U, x)$ is a chart, then $\partial/\partial x^i$ is the unique vector field on $U$ that is $x$-related to $D_i$. The theorem states that under the given hypotheses, there exists a chart $(U, x)$ such that $x_* \circ X_i \circ x^{-1} = D_i$ on $x(U)$.

It actually suffices to consider the case when $M = \mathbb{R}^n$, $p = 0$, and $X_{i|0} = D_{i|0}$: For the first two assertions, notice that if $z$ is a coordinate map taking $p$ to $0 \in \mathbb{R}^n$, then the vector fields $Y_i := z_* \circ X_i \circ z^{-1}$ have vanishing bracket, being $z$-related to $X_i$. Furthermore, if $y$ is a local diffeomorphism of $\mathbb{R}^n$ such that $y_* \circ Y_i \circ y^{-1} = D_i$, then $x := y \circ z$ is a chart satisfying the claim of the theorem. The last assertion follows from the fact that if $x : \mathbb{R}^n \to \mathbb{R}^n$ is an isomorphism that maps the basis $\{\mathcal{J}_0^{-1} X_{i|0}\}$ to the standard one, then $\tilde{X}_{i|0} = D_{i|0}$, where $\tilde{X}_i := x_* \circ X_i \circ x^{-1}$.

Let $\Phi_t^i$ denote the flow of $X_i$, and define in a small enough neighborhood $W$ of 0 a map $f : W \to \mathbb{R}^n$ by

$$f(a_1, \ldots, a_n) = (\Phi_{a_1}^1 \circ \cdots \circ \Phi_{a_k}^k)(0, \ldots, 0, a_{k+1}, \ldots, a_n).$$

Then for a smooth function $\phi$ on $\mathbb{R}^n$,

$$f_* D_{1|a}(\phi) = D_{1|a}(\phi \circ f) = \lim_{h \to 0} \frac{1}{h}[(\phi \circ f)(a_1 + h, a_2, \ldots, a_n) - (\phi \circ f)(a)]$$

$$= \lim_{h \to 0} \frac{1}{h}[(\phi \circ \Phi_{a_1+h}^1 \circ \Phi_{a_2}^2 \circ \cdots \circ \Phi_{a_k}^k)(0, \ldots, 0, a_{k+1}, \ldots, a_n) - (\phi \circ f)(a)]$$

$$= \lim_{h \to 0} \frac{1}{h}[(\phi \circ \Phi_h^1)(f(a)) - \phi(f(a))] = X_{1|f(a)}(\phi)$$

by (8.2), so that $f_* D_1 = X_1 \circ f$. Since $\Phi_{a_1}^1 \circ \cdots \circ \Phi_{a_i}^i \circ \cdots \circ \Phi_{a_k}^k = \Phi_{a_i}^i \circ \cdots \circ \Phi_{a_k}^k$, $D_i$ and $X_i$ are $f$-related for all $i \leq k$. Moreover,

$$f_* D_{k+i|0}(\phi) = D_{k+i|0}(\phi \circ f) = \lim_{h \to 0} \frac{1}{h}[(\phi \circ f)(0, \ldots, 0, h, 0, \ldots, 0) - \phi(0)]$$

$$= \lim_{h \to 0} \frac{1}{h}[\phi(0, \ldots, 0, h, 0 \ldots, 0) - \phi(0)] = D_{k+i|0}(\phi).$$

Thus, the derivative of $f$ at 0 is the identity, and by the inverse function theorem, there exists a chart $(U, x)$ around 0 with $x = f^{-1}$. The equation $f_* D_i = X_i \circ f$ is equivalent to $x_* X_i = D_i \circ x$.                                    □

The last theorem of this section provides a useful characterization of the Lie bracket that generalizes Proposition 8.3:

THEOREM 8.3. *Let $\Phi_t$ and $\Psi_s$ denote local flows of the vector fields $X$ and $Y$ respectively. Given $p \in M$, consider the curve $\bar{c} : [0, \epsilon) \to M$ given by*

$$\bar{c}(t) = (\Psi_{-\sqrt{t}} \circ \Phi_{-\sqrt{t}} \circ \Psi_{\sqrt{t}} \circ \Phi_{\sqrt{t}})(p),$$

*which is defined for small enough $\epsilon > 0$. If $f \in \mathcal{F}U$, where $U$ is a neighborhood of $p$, then*

$$[X, Y]_p(f) = \lim_{t \to 0^+} \frac{(f \circ \bar{c})(t) - (f \circ \bar{c})(0)}{t}.$$

The curve $\bar{c}$ is, in general, not (even right-) differentiable at 0. The theorem states that if we formally define a tangent vector $\bar{c}_* D_0$ by setting $\bar{c}_* D_0(f)$ equal to the right derivative of $f \circ \bar{c}$ at 0, then this vector equals $[X, Y]_p$.

PROOF. It is more convenient to work with the (smooth) curve $c(t) = \bar{c}(t^2)$. We will show that

    (1) $(f \circ c)'(0) = 0$, and
    (2) $\frac{1}{2}(f \circ c)''(0) = [X, Y]_p(f)$.

Once this is established, it follows from Taylor's theorem that

$$[X, Y]_p(f) = \frac{1}{2}(f \circ c)''(0) = \lim_{t \to 0} \frac{(f \circ c)(t) - (f \circ c)(0)}{t^2}$$

$$= \lim_{t \to 0^+} \frac{(f \circ c)(\sqrt{t}) - (f \circ c)(0)}{t} = \lim_{t \to 0^+} \frac{(f \circ \bar{c})(t) - (f \circ \bar{c})(0)}{t}.$$

In order to prove (1), we introduce "variational rectangles" $V_1$, $V_2$, and $V_3$ defined on a small enough rectangle $R$ around $0 \in \mathbb{R}^2$, given by

$$V_1(s,t) = (\Psi_s \circ \Phi_t)(p),$$
$$V_2(s,t) = (\Phi_{-s} \circ \Psi_t \circ \Phi_t)(p),$$
$$V_3(s,t) = (\Psi_{-s} \circ \Phi_{-t} \circ \Psi_t \circ \Phi_t)(p).$$

Observe that $c(t) = V_3(t,t)$, $V_3(0,t) = V_2(t,t)$, and $V_2(0,t) = V_1(t,t)$. By the chain rule,

$$\begin{aligned}
(f \circ c)'(0) &= D_1(f \circ V_3)(0,0) + D_2(f \circ V_3)(0,0) \\
&= D_1(f \circ V_3)(0,0) + D_1(f \circ V_2)(0,0) + D_2(f \circ V_2)(0,0) \\
&= D_1(f \circ V_3)(0,0) + D_1(f \circ V_2)(0,0) + D_1(f \circ V_1)(0,0) \\
&\quad + D_2(f \circ V_1)(0,0) \\
&= -Y_p f - X_p f + Y_p f + X_p f = 0,
\end{aligned}$$

which establishes (1). For (2), we have

$$(f \circ c)''(0) = D_{11}(f \circ V_3)(0,0) + 2D_{21}(f \circ V_3)(0,0) + D_{22}(f \circ V_3)(0,0).$$

Using the identity $D_1(f \circ V_3) = -(Yf) \circ V_3$, the first term on the right becomes

$$D_{11}(f \circ V_3)(0,0) = D_1(-Yf \circ V_3)(0,0) = Y_p Y f.$$

A lengthy but straightforward calculation using in addition the fact that $D_1(f \circ V_1) = (Yf) \circ V_1$, $D_1(f \circ V_2) = -(Xf) \circ V_2$, and $D_2(f \circ V_1)(0,h) = (Xf) \circ V_1(0,h)$ yields

$$2D_{21}(f \circ V_3)(0,0) = -2Y_p Y f, \quad D_{22}(f \circ V_3)(0,0) = Y_p Y f + 2[X,Y]_p f.$$

Substituting into the expression for $(f \circ c)''(0)$ now yields (2).  $\square$

EXERCISE 19. Let $(U,x)$ denote a chart of $M$, $X,Y \in \mathfrak{X}U$, so that

$$X = \sum \phi_i \frac{\partial}{\partial x^i}, \quad Y = \sum \psi_i \frac{\partial}{\partial x^i}, \quad \phi_i = Xx^i, \quad \psi_i = Yx^i.$$

Show that

$$[X,Y] = \sum_{i,j} \left( \phi_j \frac{\partial \psi_i}{\partial x^j} - \psi_j \frac{\partial \phi_i}{\partial x^j} \right) \frac{\partial}{\partial x^j}.$$

EXERCISE 20. Recall that the orthogonal group $O(n)$ consists of all matrices $A$ in $GL(n)$ such that $AA^t = I_n$. Apply Theorem 6.1 to the map $F : GL(n) \to GL(n)$ given by $F(A) = AA^t$ to deduce that $O(n)$ is a Lie subgroup of $GL(n)$ of dimension $\binom{n}{2}$. Show that its Lie algebra $\mathfrak{o}(n)$ is isomorphic to the algebra of skew-symmetric matrices $(A = -A^t)$ with the usual bracket.

EXERCISE 21. Let $f : M \to M$ be a diffeomorphism. Show that if $X \in \mathfrak{X}M$ has local flow $\Phi_t$, then the vector field $f_* \circ X \circ f^{-1}$ on $M$ has local flow $f \circ \Phi_t \circ f^{-1}$. Conclude that $X$ is $f$-related to itself iff $\Phi_t \circ f = f \circ \Phi_t$ for all $t$.

EXERCISE 22. Fill in the details of the proof of (2) in Theorem 8.3.

## 9. Distributions and Frobenius Theorem

Consider a nowhere-zero vector field $X$ on a manifold $M$. The map $p \mapsto \mathrm{span}\{X(p)\}$ assigns to each point $p$ in $M$ a one-dimensional subspace of $M_p$. The theory of ordinary differential equations guarantees that any point of $M$ belongs to an immersed submanifold—the flow line of $X$ through that point— that is everywhere tangent to these subspaces.

If we now replace the one-dimensional subspace by a $k$-dimensional one at each point (where $k > 1$), a little experimenting with the case $M = \mathbb{R}^3$ and $k = 2$ will convince the reader that is not always possible to find $k$-dimensional submanifolds that are everywhere tangent to these subspaces. In this section, we will describe conditions guaranteeing the existence of such manifolds. Although they are formulated in terms of Lie brackets, they actually reflect a classical theorem from the theory of partial differential equations.

DEFINITION 9.1. Given an $n$-dimensional manifold $M^n$ and $k \leq n$, a $k$-dimensional distribution $\Delta$ on $M$ is a map $p \mapsto \Delta_p$, which assigns to each point $p \in M$ a $k$-dimensional subspace $\Delta_p$ of $M_p$. This map is smooth in the sense that for any $q \in M$, there exists a neighborhood $U$ of $q$, and vector fields $X_1, \ldots, X_k$ on $U$, such that $X_{1|r}, \ldots, X_{k|r}$ span $\Delta_r$ for any $r \in U$.

We say a vector field $X$ on $M$ belongs to $\Delta$ ($X \in \Delta$) if $X_p \in \Delta_p$ for all $p \in M$. $\Delta$ is said to be *integrable* if $[X, Y] \in \Delta$ for all $X, Y \in \Delta$.

DEFINITION 9.2. A $k$-dimensional immersed submanifold $N$ of $M$ is said to be an *integral manifold* of $\Delta$ if $\imath_* N_p = \Delta_p$ for all $p \in N$, where $\imath : N \hookrightarrow M$ denotes inclusion.

PROPOSITION 9.1. *If for every $p \in M$ there exists an integral manifold $N(p)$ of $\Delta$ with $p \in N(p)$, then $\Delta$ is integrable.*

PROOF. Let $X, Y \in \Delta$, $p \in M$. We must show that $[X, Y]_p \in \Delta_p$. Since $\imath_{*q} : N(p)_q \to \Delta_q$ is an isomorphism for every $q \in N(p)$, there exist vector fields $\tilde{X}$ and $\tilde{Y}$ on $N(p)$ that are $\imath$-related to $X$ and $Y$. By Proposition 8.2, $[X, Y]_p = \imath_* [\tilde{X}, \tilde{Y}]_p \in \Delta_p$. □

An important special case is that of a one-dimensional distribution; any nowhere-zero vector field on $M$ defines one such, and conversely a one-dimensional distribution yields at least locally a vector field on $M$. Such a distribution $\Delta$ is always integrable (why?). Moreover, the converse of Proposition 9.1 holds: In fact, given $p \in M$, there exists a chart $(U, x)$ around $p$, an interval $I$ around 0, such that $x(p) = 0$, $x(U) = I^n$, and for any $a_2, \ldots, a_n \in I$, the *slice*

$$\{q \in U \mid x^2(q) = a_2, \ldots, x^n(q) = a_n\}$$

is an integral manifold of $\Delta$. Any connected integral manifold of $\Delta$ in $U$ is of this form. To see this, let $X$ be a vector field that spans $\Delta$ on some neighborhood $V$ of $p$. Since $X_p \neq 0$, there exists by Theorem 8.2 a chart $(U, x)$ around $p$ such that $X_{|U} = \partial/\partial x^1$.

What we have just described holds for *any* integrable distribution, and is the essence of the following theorem:

THEOREM 9.1. *Let $\Delta$ denote a $k$-dimensional integrable distribution on $M$. For every $p \in M$, there exists a chart $(U, x)$ with $x(p) = 0$, $x(U) = (-1, 1)^n$, and such that for any $a_{k+1}, \ldots, a_n \in I = (-1, 1)$, the slice $\{q \in U \mid x^{k+1}(q) = a_{k+1}, \ldots, x^n(q) = a_n\}$ is an integral manifold of $\Delta$. Furthermore, any connected integral manifold of $\Delta$ contained in $U$ is of this form.*

PROOF. The statement being a local one, we may assume that $M = \mathbb{R}^n$, $p = 0$, and $\Delta_0$ is spanned by $D_{i|0}$, $1 \le i \le k$. Let $\pi : \mathbb{R}^n \to \mathbb{R}^k$ denote the canonical projection. Then $\pi_{*|\Delta_0} : \Delta_0 \to \mathbb{R}_0^k$ is an isomorphism, and therefore so is $\pi_{*|\Delta_q} : \Delta_q \to \mathbb{R}_{\pi(q)}^k$ for all $q$ in some neighborhood $\tilde{U}$ of $0$. It follows that there are unique vector fields $X_i$ on $\tilde{U}$ that belong to $\Delta$, and are $\pi$-related to $D_i$, $1 \le i \le k$. Thus, $\pi_*[X_i, X_j] = 0$. But $[X_i, X_j] \in \Delta$ and $\pi_*$ is one-to-one on $\Delta$, so that $[X_i, X_j] \equiv 0$. By Theorem 8.2, there exists a chart $(U, x)$ around the origin, with $x(U) = I^n$ and $X_{i|U} = \partial/\partial x^i$.

Let $f = \pi_2 \circ x : U \to I^{n-k}$, where $\pi_2 : \mathbb{R}^n \to \mathbb{R}^{n-k}$ denotes projection. $f$ has maximal rank everywhere, and the above slices are the manifolds $f^{-1}(a)$, $a \in I^{n-k}$. If $N$ is the slice containing $q \in U$, then by Proposition 6.2,

$$\imath_* N_q = \{v \in M_q \mid f_* v = 0\} = \{v \in M_q \mid v(x^{k+j}) = 0, j = 1, \ldots, n-k\}$$

$$= \operatorname{span}\left\{\frac{\partial}{\partial x^i}\big|_q\right\}_{1 \le i \le k},$$

so that $N$ is an integral manifold of $\Delta$.

Conversely, suppose $N$ is an integral manifold of $\Delta$ contained in $U$. Given $v \in N_q$, $\imath_* v$ belongs to $\Delta_q = \operatorname{span}\{\partial/\partial x^i|_q\}_{1 \le i \le k}$, so that $\imath_* v(x^{k+j}) = 0$. Thus, $(x^{k+j} \circ \imath)_{*q} = 0$ for every $q \in N$. Since $N$ is connected, $x^{k+j} \circ \imath$ is constant by Exercise 10.                                                            □

DEFINITION 9.3. A $k$-dimensional *foliation* $\mathcal{F}$ of $M$ is a partition of $M$ into $k$-dimensional connected immersed submanifolds, called *leaves* of $\mathcal{F}$, such that

(1) the collection of tangent spaces to the leaves defines a distribution $\Delta$, and

(2) any connected integral manifold of $\Delta$ is contained in some leaf of $\mathcal{F}$.

A leaf of $\mathcal{F}$ is then also referred to as a *maximal integral manifold* of $\Delta$, and $\Delta$ is said to be *induced* by $\mathcal{F}$.

THEOREM 9.2 (Frobenius Theorem). *Every integrable distribution of $M$ is induced by a foliation of $M$.*

PROOF. By Theorem 9.1 and the fact that $M$ is second-countable, there exists a countable collection $\mathcal{C}$ of charts whose domains cover $M$, such that for any $(U, x) \in \mathcal{C}$, the slices

$$\{q \in U \mid x^{k+1}(q) = a_{k+1}, \ldots, x^n(q) = a_n\}$$

are integral manifolds of the distribution $\Delta$. Let $\mathcal{S}$ denote the collection of all such slices, and define an equivalence relation on $\mathcal{S}$ by $S \sim S'$ if there exists a finite sequence $S_0 = S, \ldots, S_l = S'$ of slices such that $S_i \cap S_{i+1} \ne \emptyset$ for $i = 0, \ldots, l - 1$. Each equivalence class contains only countably many slices because a slice $S$ of $U$ can intersect the domain $V$ of another chart in $\mathcal{C}$ in only countably many components of $V$, since $S$ is a manifold. The union of all slices in a given equivalence class is then an immersed connected integral manifold of $\Delta$, and two such are either equal or disjoint. By definition, any connected integral manifold of $\Delta$ is contained in such a union.                    $\square$

EXAMPLES AND REMARKS 9.1. (i) Leaves need not share the same topology: Let $\bar{S}^3 = \{(z_1, z_2) \in \mathbb{C}^2 \mid |z_1|^2 + |z_2|^2 = 1\}$, and for $\alpha \in \mathbb{R}$, consider the one-dimensional foliation $\mathcal{F}$ of $S^3$ defined as follows: the leaf through $(z_1, z_2)$ is the image of the curve $c : \mathbb{R} \to S^3$, $c(t) = (z_1 e^{it}, z_2 e^{i\alpha t})$. When $\alpha$ is irrational, some leaves will be immersed copies of $\mathbb{R}$, while others (the ones through $(1, 0)$ and $(0, 1)$) are imbedded circles. The foliation corresponding to $\alpha = 1$ is known as the *Hopf fibration*.

(ii) Let $M$ be the torus $S^1_{1/\sqrt{2}} \times S^1_{1/\sqrt{2}} = \{(z_1, z_2) \in \mathbb{C}^2 \mid |z_1|^2 = |z_2|^2 = 1/2\}$. $M$ is a submanifold of $S^3$, and the foliation from (i) above induces one on $M$. If $\alpha$ is irrational, it is easy to see that each leaf is dense in $M$.

EXERCISE 23. Define an inner product on the tangent space of $\mathbb{R}^n$ at any point $p$ so that $\mathcal{J}_p : \mathbb{R}^n \to \mathbb{R}^n_p$ becomes a linear isometry; i.e., $\langle u, v \rangle := \langle \mathcal{J}_p^{-1} u, \mathcal{J}_p^{-1} v \rangle$ for $u, v \in \mathbb{R}^n_p$, with the right side being the standard Euclidean inner product. Let $\Delta$ denote the two-dimensional distribution of $S^3$ which is orthogonal to the one-dimensional distribution induced by the Hopf fibration in Example 9.1(i). Show that $\Delta$ is not integrable.

EXERCISE 24. Let $\pi : M^n \to N^{n-k}$ be a surjective map of maximal rank everywhere. Show that the collection of pre-images $\pi^{-1}(q)$, as $q$ ranges over $N$, is a $k$-dimensional foliation of $M$.

## 10. Multilinear Algebra and Tensors

The material in this section is fairly algebraic in nature. The modern interpretation of many of the important results in differential geometry requires some knowledge of multilinear algebra; Stokes' theorem, Chern-Weil theory among others are formulated in terms of differential forms, which are tensor-valued functions on a manifold. Here, we have chosen Warner's approach [36], which is in a sense more thorough than Spivak's [34].

The *free vector space generated by a set* $A$ is the set $\mathcal{F}(A)$ of all functions $f : A \to \mathbb{R}$ which are 0 except at finitely many points of $A$, together with the usual addition and scalar multiplication of functions. The *characteristic function* $f_a$ of $a \in A$ is the function which assigns 1 to $a$ and 0 to every other element. If we identify elements of $A$ with their characteristic functions, then any $v \in \mathcal{F}(A)$ can be uniquely written as a finite sum $v = \sum \alpha_i a_i$, with $\alpha_i = v(a_i) \in \mathbb{R}$. In other words, $A$ is a basis of $\mathcal{F}(A)$.

Let $V$ and $W$ be finite-dimensional real vector spaces, and consider the subspace $\bar{\mathcal{F}}(V \times W)$ of $\mathcal{F}(V \times W)$ generated by all elements of the form

$$(v_1 + v_2, w) - (v_1, w) - (v_2, w), \quad (v, w_1 + w_2) - (v, w_1) - (v, w_2),$$

$$(\alpha v, w) - \alpha(v, w), \quad (v, \alpha w) - \alpha(v, w), \quad \alpha \in \mathbb{R}, v, v_i \in V, w, w_i \in W.$$

DEFINITION 10.1. The *tensor product* $V \otimes W$ is the quotient vector space $\mathcal{F}(V \times W)/\bar{\mathcal{F}}(V \times W)$.

The equivalence class $(v, w) + \mathcal{F}(V \times W)$ is denoted $v \otimes w$. The first relation above implies that $(v_1 + v_2) \otimes w = v_1 \otimes w + v_2 \otimes w$, and similar identities follow from the others.

When $W = \mathbb{R}$, $V \otimes \mathbb{R}$ is isomorphic to $V$, by mapping $v \otimes a$ to $av$ for $a \in \mathbb{R}$, $v \in V$, and extending linearly. Yet another simple example is the complexification of a vector space: Recall that the set $\mathbb{C}$ of complex numbers is a real 2-dimensional vector space. The *complexification* of a real vector space $V$ is $V \otimes \mathbb{C}$. Given $v \in V$, $z = a + bi \in \mathbb{C}$, the element $v \otimes z = v \otimes a + v \otimes bi$ is usually written as $av + ibv$.

Given vector spaces $V_1, \ldots, V_n$, and $Z$, a map $m : V_1 \times \cdots \times V_n \to Z$ is said to be *multilinear* if

$$m(v_1, \ldots, av_i + w, \ldots, v_n) = am(v_1, \ldots, v_i, \ldots, v_n) + m(v_1, \ldots, w, \ldots, v_n)$$

for all $v_i, w \in V_i$, $a \in \mathbb{R}$. When $n = 2$, a multilinear map is also called *bilinear*. The next lemma characterizes such maps as *linear* maps from the tensor product $V_1 \otimes \cdots \otimes V_n$ to $Z$:

LEMMA 10.1. *Let $\pi : V \times W \to V \otimes W$ denote the bilinear map $\pi(v, w) = v \otimes w$. For any vector space $Z$ and bilinear map $b : V \times W \to Z$, there exists a unique linear map $L : V \otimes W \to Z$ such that $L \circ \pi = b$. Conversely, if $X$ is a vector space that satisfies the above property (namely, there exists a bilinear map $p : V \times W \to X$ such that if $b : V \times W \to Z$ is any bilinear map into some space $Z$, then there exists a unique linear map $T : X \to Z$ with $T \circ p = b$), then $X$ is isomorphic to $V \otimes W$.*

PROOF. Since $V \times W$ is a basis of $V \otimes W$, $b$ induces a unique linear map $f : \mathcal{F}(V \times W) \to Z$ such that $f \circ \imath = b$, where $\imath : V \times W \to \mathcal{F}(V \times W)$ is inclusion. Since $b$ is bilinear, the kernel of $f$ contains $\bar{\mathcal{F}}(V \otimes W)$, and $f$ induces a unique linear map $L : V \otimes W \to Z$ such that $L \circ \bar{\pi} = f$, where $\bar{\pi} : \mathcal{F}(V \times W) \to V \otimes W$ denotes the projection. Thus, $L \circ \pi = L \circ \bar{\pi} \circ \imath = f \circ \imath = b$.

Conversely, if $X$ is a space as in the statement, then there exist linear maps $T : X \to V \otimes W$ and $L : V \otimes W \to X$ such that the diagrams

$$
\begin{array}{ccc}
V \times W & = & V \times W \\
\downarrow{\scriptstyle p} & & \downarrow{\scriptstyle \pi} \\
X & \xrightarrow{\;T\;} & V \otimes W
\end{array}
$$

and

$$
\begin{array}{ccc}
V \times W & = & V \times W \\
\downarrow{\scriptstyle p} & & \downarrow{\scriptstyle \pi} \\
X & \xleftarrow{\;L\;} & V \otimes W
\end{array}
$$

commute. Thus, $T$ and $L$ are mutual inverses. $\square$

For vector spaces $V$ and $W$, $\mathrm{Hom}(V, W)$ denotes the space of all linear transformations from $V$ to $W$ with the usual addition and scalar multiplication. Choosing bases for $V$ and $W$ (which amounts to choosing isomorphisms $V \to \mathbb{R}^n$, $W \to \mathbb{R}^m$) yields isomorphisms

$$\mathrm{Hom}(V, W) \cong \mathrm{Hom}(\mathbb{R}^n, \mathbb{R}^m) \cong M_{m,n}$$

with the space $M_{m,n}$ of $m \times n$ real matrices. In particular, $\dim \operatorname{Hom}(V, W) = \dim V \cdot \dim W$. The *dual* $V^*$ of $V$ is the space $\operatorname{Hom}(V, \mathbb{R}) \cong M_{1,n} \cong \mathbb{R}^n \cong V$. This noncanonical (because it depends on the choice of basis) isomorphism between $V^*$ and $V$ is equivalent to saying that if $\{e_i\}$ is a basis of $V$, then $\{\alpha^i\}$ is a basis of $V^*$ (called the dual basis to $\{e_i\}$), where $\alpha^i$ is the unique element of $V^*$ such that $\alpha^i(e_j) = \delta_{ij}$.

Notice, however, that there is a canonical isomorphism $L : V \to V^{**}$, given by

$$(Lv)(\alpha) = \alpha(v), \qquad v \in V, \quad \alpha \in V^*.$$

PROPOSITION 10.1. $V^* \otimes W$ *is canonically isomorphic to* $\operatorname{Hom}(V, W)$. *In particular,* $\dim(V \otimes W) = \dim V \cdot \dim W$. *In fact if* $\{e_i\}$ *and* $\{f_j\}$ *are bases of* $V$ *and* $W$ *respectively, then* $\{e_i \otimes f_j\}$ *is a basis of* $V \otimes W$.

PROOF. The map

$$V^* \times W \to \operatorname{Hom}(V, W),$$

$$(\alpha, w) \mapsto (v \mapsto (\alpha v) \cdot w)$$

is bilinear, and by Lemma 10.1 induces a unique linear map $L : V^* \otimes W \to \operatorname{Hom}(V, W)$. $L$ is easily seen to be an isomorphism with inverse $T \mapsto \sum \alpha^i \otimes T(e_i)$, where $\{e_i\}$ and $\{\alpha^i\}$ are any dual bases of $V$ and $V^*$ respectively. As to the second statement, if $v = \sum a_i e_i \in V$ and $w = \sum b_j f_j \in W$, then $v \otimes w = \sum_{i,j} a_i b_j e_i \otimes f_j$, so that $\{e_i \otimes f_j\}$ is a spanning set for $V \otimes W$. It must then be a basis by dimension considerations. $\square$

Thus, for example, $V \otimes \mathbb{R} \cong V^{**} \otimes \mathbb{R} \cong \operatorname{Hom}(V^*, \mathbb{R}) \cong V^{**} \cong V$.

DEFINITION 10.2. A *tensor* of type $(r, s)$ is an element of the space

$$T_{r,s}(V) := \underbrace{V \otimes \cdots \otimes V}_{r} \otimes \overbrace{V^* \otimes \cdots \otimes V^*}^{s}.$$

Our next aim is to show that $T_{r,s}(V)$ may be naturally identified with the space $M_{s,r}(V)$ of multilinear maps $V \times \cdots \times V \times V^* \times \cdots \times V^* \to \mathbb{R}$ ($s$ copies of $V$, $r$ copies of $V^*$). For example, a bilinear form on $V$ (e.g., an inner product) is a tensor of type $(0, 2)$.

Recall that a *nonsingular pairing* of $V$ with $W$ is a bilinear map $b : V \times W \to \mathbb{R}$ such that if the restriction of $b$ to $\{v\} \times W$, respectively $V \times \{w\}$, is identically $0$, then $v$, respectively $w$, is $0$ for any $v \in V$ and $w \in W$. When $V$ and $W$ are finite-dimensional, such a pairing induces isomorphisms $V \to W^*$ and $W \to V^*$: Define $L : V \to W^*$ by $(Lv)w = b(v, w)$. $L$ is one-to-one, and since $b$ induces a similar map from $W$ to $V^*$, $V$ and $W$ must have the same dimension, and $L$ is an isomorphism. The isomorphism $V \cong V^{**}$ above comes from the pairing $b : V \times V^* \to \mathbb{R}$, $b(\alpha, v) = \alpha(v)$.

PROPOSITION 10.2. $T_{r,s}(V)$ *is canonically isomorphic to* $M_{s,r}(V)$.

PROOF. Define a nonsingular pairing $b$ of $T_{r,s}(V)$ with $T_{r,s}(V^*)$ as follows: for $u = u_1 \otimes \cdots \otimes u_r \otimes v^*_{r+1} \otimes \cdots \otimes v^*_{r+s} \in T_{r,s}(V)$ and $v^* = v^*_1 \otimes \cdots \otimes v^*_r \otimes u_{r+1} \otimes \cdots \otimes u_{r+s} \in T_{r,s}(V^*)$ (such elements are called *decomposable*), set $b(u, v^*) := \prod v_i^*(u_i)$, and extend bilinearly to all elements. Together with Lemma 10.1, this yields the sequence of isomorphisms

$$T_{r,s}(V) \cong (T_{r,s}(V^*))^* \cong M_{r,s}(V^*) \cong M_{s,r}(V).$$

It follows that under the above identification, an element $u_1 \otimes \cdots \otimes u_r \otimes u_1^* \otimes \cdots \otimes u_s^* \in T_{r,s}(V)$ is a multilinear map $V^s \times V^{*r} \to \mathbb{R}$ satisfying

$$u_1 \otimes \cdots \otimes u_r \otimes u_1^* \otimes \cdots \otimes u_s^*(v_1, \ldots, v_s, v_1^*, \ldots, v_r^*) = \prod_{i=1}^r v_i^*(u_i) \prod_{j=1}^s u_j^*(v_j).$$

$\square$

As an elementary example, consider a manifold $M^n$, $p \in M$, and a chart $(U, x)$ around $p$. Set $V = M_p$. Then $\{dx_{|p}^i \otimes dx_{|p}^j\}_{1 \le i,j \le n}$ is a basis of $V_{0,2}$. By Proposition 10.2, $dx^i \otimes dx^j(v, w) = dx^i(v) dx^j(w)$.

It is convenient to group all the tensor spaces into one: The *tensor algebra* of $V$ is the graded algebra $T(V) = \oplus_{r,s \ge 0} T_{r,s}(V)$ with the multiplication $\otimes$: $T_{r_1,s_1}(V) \times T_{r_2,s_2}(V) \to T_{r_1+s_1, r_2+s_2}(V)$. The subalgebra $T_0(V) = \oplus_r T_{r,0}(V)$ is called the algebra of *contravariant tensors*. Let $I(V)$ denote the ideal of $T_0(V)$ generated by the set of elements of the form $v \otimes v$, $v \in V$. We may write $I(V) = \oplus_r I_r(V)$, where $I_r(V) = I(V) \cap T_{r,0}(V)$.

DEFINITION 10.3. The *exterior algebra* of $V$ is the graded algebra $\Lambda(V) = T_0(V)/I(V)$.

Observe that $\Lambda(V) = \oplus_{k \ge 0} \Lambda_k(V)$, where $\Lambda_0(V) = \mathbb{R}$, $\Lambda_1(V) = V$, and $\Lambda_k(V) = T_{k,0}(V)/I_k(V)$ for $k > 1$. The multiplication in $T_0(V)$ induces one in $\Lambda(V)$, called the *wedge product*, and denoted $\wedge$: If $\pi : T_0(V) \to \Lambda(V)$ denotes the projection, then

$$\pi(v_1 \otimes \cdots \otimes v_k) = v_1 \wedge \cdots \wedge v_k.$$

PROPOSITION 10.3. *If $\{e_1, \ldots, e_n\}$ is a basis of $V$, then $\{e_{i_1} \wedge \cdots \wedge e_{i_k} \mid 1 \le i_1 < \cdots < i_k \le n\}$ is a basis of $\Lambda_k(V)$. Thus, $\dim \Lambda_k(V) = \binom{n}{k}$ for $k \le n$, $\Lambda_n(V) \cong \mathbb{R}$, and $\Lambda_{n+k}(V) = 0$.*

PROOF. Since $(e_i + e_j) \wedge (e_i + e_j) = 0$, $e_i \wedge e_j = -e_j \wedge e_i$, so the set in the statement spans $\Lambda_k(V)$. To see that it is linearly independent, suppose that $\sum_j \alpha_j e_{i_1 j} \wedge \cdots \wedge e_{i_k j} = 0$. Fix $j$, and let $e_{j_1}, \ldots, e_{j_{n-k}}$ be those basis vectors that do not appear in the expression $e_{i_1 j} \wedge \cdots \wedge e_{i_k j}$. Observe that for $l \ne j$, the set $\{e_{i_1 l}, \ldots, e_{i_k l}\}$ intersects $\{e_{j_1}, \ldots, e_{j_{n-k}}\}$. Thus,

$$0 = \left( \sum_l \alpha_l e_{i_1 l} \wedge \cdots \wedge e_{i_k l} \right) \wedge (e_{j_1} \wedge \cdots \wedge e_{j_{n-k}}) = \pm \alpha_j e_1 \wedge \cdots \wedge e_n,$$

and it only remains to show that $e_1 \wedge \cdots \wedge e_n \ne 0$, or equivalently, that $e_1 \otimes \cdots \otimes e_n \notin I_n(V)$. But any $w \in I_n(V)$ can be uniquely written as a linear combination of terms $e_{j_1} \otimes \cdots \otimes e_{j_n}$, each of which either satisfies $j_i = j_{i+1}$ for some $i$, or else comes paired with another term of the form $e_{j_1} \otimes \cdots \otimes e_{j_{i+1}} \otimes e_{j_i} \otimes \cdots \otimes e_{j_n}$ bearing the same scalar coefficient. This is clearly not the case for $e_1 \otimes \cdots \otimes e_n$. $\square$

Given a space $W$, a multilinear map $m : V^k \to W$ is said to be *alternating* if $m(\ldots, v, \ldots, v, \ldots) = 0$. When $W = \mathbb{R}$, the space of such maps will be denoted $A_k(V)$. Just as multilinear maps can be viewed as linear maps from a tensor algebra, Exercise 26 establishes that alternating multilinear maps from $V^k$ can be considered as linear maps from $\Lambda_k(V)$.

PROPOSITION 10.4. *There are canonical isomorphisms* $\Lambda_k(V^*) \cong \Lambda_k(V)^* \cong A_k(V)$.

PROOF. The second isomorphism is the one induced from Exercise 26. For the first one, there is a unique bilinear map $b : \Lambda_k(V^*) \times \Lambda_k(V) \to \mathbb{R}$ which is given on decomposable elements by

$$b(v_1^* \wedge \cdots \wedge v_k^*, v_1 \wedge \cdots \wedge v_k) = \det(v_i^* v_j).$$

It determines a nonsingular pairing, and therefore an isomorphism $\Lambda_k(V^*) \cong \Lambda_k(V)^*$. $\qquad \square$

Observe that under the identification $\Lambda_k(V^*) \cong A_k(V)$,

$$(v_1^* \wedge \cdots \wedge v_k^*)(v_1, \ldots, v_k) = \det(v_i^* v_j).$$

Moreover, $A(V) := \oplus_k A_k(V) \cong \oplus \Lambda_k(V^*) = \Lambda(V^*)$, so that $A(V)$ is a graded algebra. Now, if $u \in \Lambda_k(V)$, $v \in \Lambda_l(V)$, then $u \wedge v = (-1)^{kl} v \wedge u$, as follows by writing $u$ and $v$ in terms of decomposable elements and considering the case $k = l = 1$. It follows that

$$\alpha \wedge \beta = (-1)^{kl} \beta \wedge \alpha, \qquad \alpha \in A_k(V), \quad \beta \in A_l(V).$$

PROPOSITION 10.5. *For* $\alpha \in A_k(V)$ *and* $\beta \in A_l(V)$,

$$\alpha \wedge \beta(v_1, \ldots, v_{k+l}) = \sum_{\sigma \in \bar{P}_{k+l}} (\operatorname{sgn} \sigma) \alpha(v_{\sigma(1)}, \ldots, v_{\sigma(k)}) \beta(v_{\sigma(k+1)}, \ldots, v_{\sigma(k+l)})$$

$$= \sum_{\sigma \in P_{k+l}} \frac{1}{k! l!} (\operatorname{sgn} \sigma) \alpha(v_{\sigma(1)}, \ldots, v_{\sigma(k)}) \beta(v_{\sigma(k+1)}, \ldots, v_{\sigma(k+l)}).$$

*(Here, $P_{k+l}$ is the group of all permutations of $\{1, \ldots, k+l\}$, and $\bar{P}_{k+l}$ is the subset of all $(k,l)$-shuffles; i.e., those permutations $\sigma$ with $\sigma(1) < \cdots < \sigma(k)$ and $\sigma(k+1) < \cdots < \sigma(k+l)$).*

PROOF. It suffices to establish the result for decomposable elements $\alpha = u_1^* \wedge \cdots \wedge u_k^*$ and $\beta = w_1^* \wedge \cdots \wedge w_l^*$. Notice that

$$\alpha(v_1, \ldots, v_k) = \det(u_i^* v_j) = \sum_{\tau \in P_k} (\operatorname{sgn} \tau) u_1^* v_{\tau(1)} \ldots u_k^* v_{\tau(k)}$$

by definition of the determinant. Given $\sigma \in P_{k+l}$, let $w_i = v_{\sigma(i)}$, so that

$$\alpha(v_{\sigma(1)}, \ldots, v_{\sigma(k)}) = \alpha(w_1, \ldots, w_k) = \sum_{\tau \in P_k} (\operatorname{sgn} \tau) u_1^* w_{\tau(1)} \ldots u_k^* w_{\tau(k)}$$

$$= \sum_{\tau \in P_k} (\operatorname{sgn} \tau) u_1^* v_{\sigma \tau(1)} \ldots u_k^* v_{\sigma \tau(k)}.$$

Now, for any $\sigma \in P_{k+l}$, there exist unique $\tau_k \in P_k, \tau_l \in P_l$, and $(k,l)$-shuffle $\tilde{\sigma}$ such that $\sigma(i) = \tilde{\sigma}\tau_k(i)$ for $i \leq k$ and $\sigma(k+i) = \tilde{\sigma}(k + \tau_l(i))$ for $i \leq l$. Thus,

$$\sum_{\sigma \in \bar{P}_{k+l}} (\operatorname{sgn} \sigma) \alpha(v_{\sigma(1)}, \ldots, v_{\sigma(k)}) \beta(v_{\sigma(k+1)}, \ldots, v_{\sigma(k+l)})$$

$$= \sum_{\sigma \in \bar{P}_{k+l}} (\operatorname{sgn} \sigma) \sum_{\tau \in P_k} (\operatorname{sgn} \tau) u_1^* v_{\sigma\tau(1)} \cdots u_k^* v_{\sigma\tau(k)}$$

$$\cdot \sum_{\tau \in P_l} (\operatorname{sgn} \tau) w_1^* v_{\sigma(k+\tau(1))} \cdots w_l^* v_{\sigma(k+\tau(l))}$$

$$= \sum_{\sigma \in P_{k+l}} (\operatorname{sgn} \sigma) u_1^* v_{\sigma(1)} \cdots u_k^* v_{\sigma(k)} w_1^* v_{\sigma(k+1)} \cdots w_l^* v_{\sigma(k+l)}$$

$$= (u_1^* \wedge \cdots \wedge u_k^*) \wedge (w_1^* \wedge \cdots \wedge w_l^*)(v_1, \ldots, v_{k+l}) = (\alpha \wedge \beta)(v_1, \ldots, v_{k+l}).$$

The second identity is left as an exercise.                    $\square$

EXAMPLE 10.1. Let $M^n$ be a manifold, and $(U, x)$ a chart around some $p \in M$. Then $\{dx^i_{|p} \wedge dx^j_{|p}\}_{1 \leq i < j \leq n}$ is a basis of $\Lambda_2(M_p^*) \cong A_2(M_p)$, and

$$dx^i \wedge dx^j(u, v) = dx^i(u) \, dx^j(v) - dx^i(v) \, dx^j(u), \qquad u, v \in M_p.$$

Any 2-form $\omega \in A_2(M_p)$ can be written as $\omega = \sum_{1 \leq i < j \leq n} \omega_{ij} dx^i_{|p} \wedge dx^j_{|p}$, with $\omega(\partial/\partial x^k_{|p}, \partial/\partial x^l_{|p}) = \sum \omega_{ij} dx^i \wedge dx^j(\partial/\partial x^k_{|p}, \partial/\partial x^l_{|p}) = \omega_{kl}$ for $k < l$.

EXERCISE 25. Show that $V \otimes W$ is canonically isomorphic to $W \otimes V$.

EXERCISE 26. (a) Prove that a multilinear map $m : V^k \to W$ is alternating iff

$$m(v_{\sigma(1)}, \ldots, v_{\sigma(k)}) = (\operatorname{sgn} \sigma) m(v_1, \ldots, v_k), \qquad v_i \in V, \quad \sigma \in P_k.$$

(b) Let $\pi : V^k \to \Lambda_k(V)$ denote the alternating multilinear map sending $(v_1, \ldots, v_k)$ to $v_1 \wedge \cdots \wedge v_k$. Show that if $m : V^k \to W$ is alternating multilinear, then there exists a unique linear $L : \Lambda_k(V) \to W$ such that $L \circ \pi = m$.

EXERCISE 27. Prove the second identity in Proposition 10.5.

EXERCISE 28. Show that vectors $v_1, \ldots, v_k \in V$ are linearly independent iff $v_1 \wedge \cdots \wedge v_k \neq 0$.

EXERCISE 29. Let $(V, \langle,\rangle)$ denote an $n$-dimensional inner product space, and $\mathfrak{o}(V)$ the Lie algebra of all skew-symmetric linear endomorphisms of $V$ from Exercise 20. Consider the linear map $L : \Lambda_2(V) \to \mathfrak{o}(V)$ defined on decomposable elements by

$$(L(u \wedge v))(w) = \langle v, w \rangle u - \langle u, w \rangle v, \qquad u, v, w \in V.$$

Prove that $L$ is an isomorphism.

EXERCISE 30. (a) Show that if $\dim V = 3$, then any element of $\Lambda_2(V)$ is decomposable.

(b) Show that (a) is false if $\dim V > 3$. (*Hint:* consider $u_1 \wedge u_2 + u_3 \wedge u_4$, where the $u_i$'s are linearly independent.)

(c) Prove that nevertheless, any element in $\Lambda_2(V)$ can be written as $u_1 \wedge u_2 + \cdots + u_{k-1} \wedge u_k$, where the $u_i$'s are linearly independent.

## 11. Tensor Fields and Differential Forms

We can now do with tensors what we did with vectors when defining vector fields; i.e., assign to each point $p$ of a manifold $M$ a tensor of a given type in the tangent space $M_p$ of $M$ at $p$ in a differentiable way. This is conveniently done by introducing the following concepts:

DEFINITION 11.1. Let $M^n$ denote a manifold. The three sets $T_{r,s}(M) := \cup_{p \in M} T_{r,s}(M_p)$, $\Lambda_k^*(M) := \cup_{p \in M} \Lambda_k(M_p^*)$, and $\Lambda^*(M) := \cup_{p \in M} \Lambda(M_p^*)$ are called the *bundle of tensors of type (r,s) over M*, the *exterior k-bundle over M*, and the *exterior algebra bundle over M*, respectively.

The term "bundle" will be explored further in the next chapter. For now, observe that each of these sets admits a natural map $\pi$ onto $M$, called the *projection*, which sends a tensor on $M_p$ to the point $p$ itself. The following proposition should be compared to our construction of the tangent bundle of $M$.

PROPOSITION 11.1. *The differentiable structure on $M$ induces differentiable structures on $T_{r,s}(M)$, $\Lambda_k^*(M)$, and $\Lambda^*(M)$ for which the projection onto $M$ are submersions.*

PROOF. Given a chart $(U, x)$ of $M$, consider bases $\{\partial/\partial x^i_{|p}\}$ and $\{dx^i_{|p}\}$ of $M_p$ and $M_p^*$ respectively. These in turn induce bases for the vector spaces $T_{r,s}(M_p)$, $\Lambda_k(M_p^*)$, and $\Lambda^*(M_p)$; i.e., isomorphisms $L_p$ between these spaces and the Euclidean space $\mathbb{R}^l$ for appropriate choices of $l$. This yields bijective maps $(x \circ \pi, L) : \pi^{-1}(U) \to x(U) \times \mathbb{R}^l$, where $L(u) := L_{\pi(u)}(u)$. Endow each space with the topology for which these maps become local homeomorphisms. It follows that the collections of such maps are differentiable atlases, and the projections onto $M$ are differentiable submersions. $\square$

We will denote by the same letter $\pi$ the projections of all three bundles onto $M$.

DEFINITION 11.2. A *tensor field of type (r,s) on M* is a (smooth) map $T : M \to T_{r,s}(M)$ such that $\pi \circ T = 1_M$. A *differential k-form on M* is a map $\alpha : M \to \Lambda_k^*(M)$ such that $\pi \circ \alpha = 1_M$.

Notice that a vector field on $M$ is just a tensor field of type $(1,0)$. The following proposition is proved in the same way as we did for vector fields:

PROPOSITION 11.2. *$T$ is a tensor field of type (r,s) on $M$ iff for any chart $(U, x)$ of $M$,*

$$T_{|U} = \sum T_{i_1,\ldots,i_r;j_1,\ldots,j_s} \frac{\partial}{\partial x^{i_1}} \otimes \cdots \otimes \frac{\partial}{\partial x^{i_r}} \otimes dx^{j_1} \otimes \cdots \otimes dx^{j_s}$$

*for smooth functions $T_{i_1,\ldots,i_r;j_1,\ldots,j_s}$ on $U$. Similarly, $\alpha$ is a differential k-form on $M$ iff*

$$\alpha_{|U} = \sum_{1 \le i_1 < \cdots < i_k \le n} \alpha_{i_1 \ldots i_k} dx^{i_1} \wedge \cdots \wedge dx^{i_k}$$

*for smooth functions $\alpha_{i_1 \ldots i_k}$ on $U$.*

Observe that the functions in Proposition 11.2 satisfy

$$T_{i_1,\ldots,i_r;j_1,\ldots,j_s} = T\left(dx^{i_1},\ldots,dx^{i_r},\frac{\partial}{\partial x^{j_1}},\ldots,\frac{\partial}{\partial x^{j_s}}\right),$$

$$\alpha_{i_1,\ldots,i_k} = \alpha\left(\frac{\partial}{\partial x^{i_1}},\ldots,\frac{\partial}{\partial x^{i_k}}\right).$$

EXAMPLES AND REMARKS 11.1. (i) Let $(U,x)$ and $(V,y)$ be charts of $M^n$. Then on $U \cap V$,

$$dy^1 \wedge \cdots \wedge dy^n = \det D(y \circ x^{-1})dx^1 \wedge \cdots \wedge dx^n$$

because $dy^1 \wedge \cdots \wedge dy^n(\partial/\partial x^1,\ldots,\partial/\partial x^n) = \det(dy^i(\partial/\partial x^j)) = \det D_j(y^i \circ x^{-1})$.

(ii) A *Riemannian metric* on a manifold $M$ is a tensor field $g$ of type $(0,2)$ on $M$ such that for every $p \in M$, $g(p)$ is an inner product on $M_p$.

We denote by $A_k(M)$ the set of all differential $k$-forms on $M$, and by $A(M)$ the collection of all forms. Given $\alpha, \beta \in A(M)$, $a \in \mathbb{R}$, define $a\alpha, \alpha + \beta, \alpha \wedge \beta \in A(M)$ by setting

$$(a\alpha)(p) = a\alpha(p), (\alpha+\beta)(p) = \alpha(p) + \beta(p), (\alpha \wedge \beta)(p) = \alpha(p) \wedge \beta(p), \qquad p \in M.$$

$A(M)$ then becomes a graded algebra with these operations.

Since $\Lambda_0^*(M) = \cup_{p \in M}\Lambda_0(M_p^*) = \cup_{p \in M}\mathbb{R} = M \times \mathbb{R}$, $A_0(M)$ is naturally isomorphic to $\mathcal{F}(M)$ if we identify $\alpha \in A_0(M)$ with $f = \pi_2 \circ \alpha$, where $\pi_2 : M \times \mathbb{R} \to \mathbb{R}$ is projection. For $f \in A_0(M)$, we write $f\alpha$ instead of $f \wedge \alpha$. Thus, $A(M)$ is a module over $\mathcal{F}(M)$.

Any $\alpha \in A_k(M)$ is an alternating multilinear map

$$\alpha : \mathfrak{X}(M) \times \cdots \times \mathfrak{X}(M) \to \mathcal{F}(M), \qquad \alpha(X_1,\ldots,X_k)(p) = \alpha(p)(X_{1|p},\ldots,X_{k|p}).$$

Moreover, $\alpha$ is linear over $\mathcal{F}(M)$; i.e.,

$$\alpha(X_1,\ldots,fX_i,\ldots,X_k) = f\alpha(X_1,\ldots,X_i,\ldots,X_k).$$

The converse is also true:

PROPOSITION 11.3. *A multilinear map* $T : \mathfrak{X}(M) \times \cdots \times \mathfrak{X}(M) \to \mathbb{R}$ *is a tensor field iff it is linear over* $\mathcal{F}(M)$.

PROOF. The condition is necessary by the above remark. Conversely, suppose $T$ is multilinear and linear over $\mathcal{F}(M)$. We claim that $T$ "lives pointwise": If $X_{i|p} = Y_{i|p}$ for all $i$, then $T(X_1,\ldots,X_k)(p) = T(Y_1,\ldots,Y_k)(p)$. To see this, assume for simplicity that $k = 1$, the general case being analogous. It suffices to establish that if $X_p = 0$, then $T(X)(p) = 0$. Consider a chart $(U,x)$ around $p$, and write $X_{|U} = \sum f_i \partial/\partial x^i$ with $f_i(p) = 0$. Let $V$ be a neighborhood of $p$ whose closure is contained in $U$, and $\phi$ a nonnegative function with support in $U$ which equals 1 on the closure of $V$. Define vector fields $X_i$ on $M$ by setting them equal to $\phi\partial/\partial x^i$ on $U$ and to 0 outside $U$. Similarly, let $g_i$ be the functions that equal $\phi f_i$ on $U$ and 0 outside $U$. Then $X = \phi^2 X + (1 - \phi^2)X = \sum g_i X_i + (1 - \phi^2)X$, and

$$(TX)(p) = \sum g_i(p)(TX_i)(p) + (1 - \phi^2(p))(TX)(p) = 0,$$

establishing the claim.

We may therefore define for each $p \in M$ an element $T_p \in T_{0,k}(M_p)$ by

$$T_p(u_1, \ldots, u_k) := T(X_1, \ldots, X_k)(p)$$

for any vector fields $X_i$ with $X_{i|p} = u_i$. The map $p \mapsto T_p$ is clearly smooth.  $\square$

EXAMPLES AND REMARKS 11.2. (i) Given $\alpha, \beta \in A_1(M)$ and $X, Y \in \mathfrak{X}(M)$,

$$(\alpha \wedge \beta)(X, Y) = \alpha(X)\beta(Y) - \alpha(Y)\beta(X).$$

(ii) Recall that for $f \in \mathcal{F}(M)$ and $p \in M$, $df(p)$ is the element of $M_p^*$ given by $df(p)u = u(f)$ for $u \in M_p$. The assignment $p \mapsto df(p)$ defines a differential 1-form $df$, since in a chart $(U, x)$, $df_{|U} = \sum_i (\partial f/\partial x^i)dx^i$ with $\partial f/\partial x^i \in \mathcal{F}(U)$. $df$ is called the *differential* of $f$. Observe that $d : A_0(M) \to A_1(M)$ and that $d(fg) = f\,dg + g\,df$.

THEOREM 11.1. *There is a unique linear map* $d : A(M) \to A(M)$, *called the* exterior derivative operator *such that*

(i) $d : A_k(M) \to A_{k+1}(M), \qquad k \in \mathbb{N}$;
(ii) $d(\alpha \wedge \beta) = d\alpha \wedge \beta + (-1)^k \alpha \wedge d\beta, \qquad \alpha \in A_k(M), \quad \beta \in A(M)$;
(iii) $d^2 = 0$; and
(iv) *for* $f \in A_0(M)$, $df$ *is the differential of* $f$.

PROOF. We will first define $d$ locally in terms of charts, and then show that the definition is independent of the chosen chart. An invariant formula for $d$ will be given later on.

Given $p \in M$, and a chart $(U, x)$ around $p$, any form $\alpha$ defined on a neighborhood of $p$ may be locally written as $\alpha = \sum \alpha_I dx^I$, where $I$ ranges over subsets of $\{1, \ldots, n\}$, $dx^I = dx^{i_1} \wedge \cdots \wedge dx^{i_k}$ if $I = \{i_1, \ldots, i_k\}$ with $i_1 < \cdots < i_k$ (or $dx^I = 1$ if $I = \emptyset$), and the $\alpha_I$ are smooth functions on a neighborhood of $p$. Define

$$d\alpha(p) = \sum d\alpha_I(p) \wedge dx^I(p).$$

We first check that $d$ satisfies the following properties at $p$: Given $\alpha \in A_k(M)$,

(1) $d\alpha(p) \in \Lambda_{k+1}(M_p^*)$;
(2) if $\alpha = \beta$ on a neighborhood of $p$, then $d\alpha(p) = d\beta(p)$;
(3) $d(a\alpha + b\beta)(p) = a\,d\alpha(p) + b\,d\beta(p), \qquad a, b \in \mathbb{R}, \quad \beta \in A(M)$;
(4) $d(\alpha \wedge \beta)(p) = d\alpha \wedge \beta(p) + (-1)^k \alpha \wedge d\beta(p)$;
(5) $d(df)(p) = 0, \qquad f \in A_0(M)$.

Properties (1)–(3) are immediate. To establish (4), we may, by (3), assume that $\alpha = f dx^I$ and $\beta = g dx^J$. In case $I$ and/or $J$ are empty, the statement is clear from the definition and the fact that (4) is true for functions. Otherwise,

$$d(\alpha \wedge \beta)(p) = d(fg\,dx^I \wedge dx^J)(p) = (df(p)g(p) + f(p)dg(p)) \wedge (dx^I \wedge dx^J)(p)$$

$$= (df(p) \wedge dx^I(p)) \wedge (g(p)dx^J(p))$$

$$+ (-1)^k (f(p)dx^I(p)) \wedge (dg(p) \wedge dx^J(p))$$

$$= d\alpha(p) \wedge \beta(p) + (-1)^k \alpha(p) \wedge d\beta(p).$$

For (5), write $df$ locally as $\sum (\partial f / \partial x^i) dx^i$. Then

$$d(df)(p) = \sum_i d\left( \frac{\partial}{\partial x^i}(f) \right)(p) \wedge dx^i(p) = \sum_{i,j} \frac{\partial^2 f}{\partial x^j \partial x^i}(p) dx^j(p) \wedge dx^i(p)$$

$$= \sum_{i<j} \left( \frac{\partial^2 f}{\partial x^i \partial x^j}(p) - \frac{\partial^2 f}{\partial x^j \partial x^i}(p) \right) dx^i(p) \wedge dx^j(p) = 0.$$

Next, we verify that $d$ is well defined; i.e., independently of the chosen chart. Suppose $d'$ is defined in the same way relative to some other chart around $p$. Then $df(p) = d'f(p)$ for functions $f$. Furthermore, $d'$ satisfies properties (1) through (5), so

$$d'\alpha(p) = \sum d'(\alpha_I dx^I) = \sum d'\alpha_I(p) \wedge dx^I(p) + \alpha_I(p) d'(dx^I)(p)$$

$$= \sum d\alpha_I(p) \wedge dx^I(p) + \alpha_I(p) d'(dx^I)(p),$$

and it only remains to show that $d'(dx^I)(p) = 0$. But this follows immediately from applying (4) to $d'(dx^{i_1} \wedge \cdots \wedge dx^{i_k})(p)$, together with the fact that $d'(dx^i)(p) = d'(d'x^i)(p) = 0$. Thus, $d = d'$, and $d$ is well defined. Moreover, $d$ clearly satisfies the statements (i)–(iv) of the theorem.

It finally remains to establish uniqueness. Let $d'$ be any operator satisfying the properties of the theorem. By the previous argument, it suffices to show that $d'$ satisfies properties (1)–(5). All but (2) are immediate. For (2), it is enough to show that if $\alpha \in A(M)$ is 0 on a neighborhood $U$ of $p$, then $d'\alpha(p) = 0$. To see this, let $\phi$ be a nonnegative function which is 0 on a neighborhood of $p$ whose closure is contained in $U$ and 1 on $M \setminus U$. Then $\phi\alpha = \alpha$, and $d'\alpha(p) = d'(\phi\alpha)(p) = d'\phi(p) \wedge \alpha(p) + \phi(p)d'\alpha(p) = 0$. This establishes uniqueness.                    $\square$

DEFINITION 11.3. $\alpha \in A(M)$ is said to be *closed* if $d\alpha = 0$, and is said to be *exact* if $\alpha = d\beta$ for some $\beta \in A(M)$.

It follows from Theorem 11.1 that every exact form is closed. We will later see that the converse is true locally.

EXAMPLE 11.1. Let $M = \mathbb{R}^2 \setminus \{0\}$. We claim there exists a closed 1-form $\alpha$ on $M$ which is not exact, even though any point of $M$ has a neighborhood such that the restriction of $\alpha$ to this neighborhood is exact: Roughly speaking, if $\theta$ is the classical polar coordinate angle on $M$, then $\alpha = d\theta$ locally. To make this more precise, let $L$ denote the half-line $[0, \infty) \times 0$ in $\mathbb{R}^2$, and consider the map $F : (0, \infty) \times (0, 2\pi) \to \mathbb{R}^2 \setminus L$ given by $F(a_1, a_2) = (a_1 \cos a_2, a_1 \sin a_2)$. Since $F$ is a diffeomorphism, we may define polar coordinate functions $r$ and $\theta$ on $\mathbb{R}^2 \setminus L$ by $r = u^1 \circ F^{-1}$, $\theta = u^2 \circ F^{-1}$.

In order to extend $d\theta$ to all of $M$, fix any $a \in (0, 2\pi)$, and define $\tilde{F} : (0, \infty) \times (a, 2\pi + a) \to \mathbb{R}^2 \setminus \tilde{L}$ by the same formula as for $F$; here $\tilde{L}$ denotes the half-line from the origin passing through $(\cos a, \sin a)$. $\tilde{F}$ is also a diffeomorphism, and has an inverse $\tilde{F}^{-1} = (r, \tilde{\theta})$.

On their common domain, $\theta = \tilde{\theta}$ or $\theta = \tilde{\theta} + 2\pi$. We may therefore define a global 1-form $\alpha$ on $M$ by setting it equal to $d\theta$ on $\mathbb{R}^2 \setminus L$, and to $d\tilde{\theta}$ on $\mathbb{R}^2 \setminus \tilde{L}$. $\alpha$ is closed since it is locally exact, but there is no function $f$ on $M$ such that

$\alpha = df$: For otherwise, $d(f - \theta) = 0$ on $\mathbb{R}^2 \setminus L$ implies that $f = \theta + c$ on $\mathbb{R}^2 \setminus L$ for some constant $c$. Similarly, $f = \tilde{\theta} + \tilde{c}$ on $\mathbb{R}^2 \setminus \tilde{L}$ for some constant $\tilde{c}$. This would imply that $\theta + c = \tilde{\theta} + \tilde{c}$ on all of $\mathbb{R}^2 \setminus (L \cup \tilde{L})$, which is false.

In general, the collection $Z_k(M)$ of closed $k$-forms on $M$ is a vector space, and the collection $B_k(M)$ of exact $k$-forms is a subspace.

DEFINITION 11.4. The $k$-th (de Rham) cohomology vector space of $M$ is the space $H^k(M) = Z_k(M)/B_k(M)$.

Thus, $H^k(M) = 0$ iff every closed $k$-form on $M$ is exact. When $k = 0$, we define $H^0(M) := Z_0(M)$. If $M$ is connected, it follows from Exercise 10 that $H^0(M) \cong \mathbb{R}$.

DEFINITION 11.5. Let $f : M \to N$ be differentiable. For $\alpha \in A_k(N)$, define the *pullback of $\alpha$ via $f$* to be the $k$-form $f^*\alpha$ on $M$ given by

$$(f^*\alpha)(p)(v_1, \ldots, v_k) = \alpha(f(p))(f_*v_1, \ldots, f_*v_k), \qquad p \in M, \quad v_i \in M_p.$$

In the special case that $k = 0$, i.e., when $\alpha$ is a function $\phi$ on $M$, define $f^*\phi = \phi \circ f$.

Clearly, $f^* : A(N) \to A(M)$ is linear.

THEOREM 11.2. *If $f : M \to N$ is differentiable, then*

(1) $f^* : A(N) \to A(M)$ *is an algebra homomorphism,*
(2) $df^* = f^*d$, *and*
(3) $f^*$ *induces a linear transformation $f^* : H^k(N) \to H^k(M)$.*

PROOF. In order to establish (1), notice that because $f^*$ is linear, it suffices to check that $f^*(\alpha_1 \wedge \cdots \wedge \alpha_k) = f^*\alpha_1 \wedge \cdots \wedge f^*\alpha_k$ for 1-forms $\alpha_i$ on $N$. But if $X_1, \ldots, X_k \in \mathfrak{X}(M)$, then

$$f^*(\alpha_1 \wedge \cdots \wedge \alpha_k)(X_1, \ldots, X_k) = (\alpha_1 \wedge \cdots \wedge \alpha_k) \circ f(f_*X_1, \ldots, f_*X_k)$$
$$= \det((\alpha_i \circ f)(f_*X_j)) = \det(f^*\alpha_i(X_j))$$
$$= f^*\alpha_1 \wedge \cdots \wedge f^*\alpha_k(X_1, \ldots, X_k).$$

(2) We first prove the statement for functions. If $\phi \in \mathcal{F}(N)$ and $X \in \mathfrak{X}(M)$, then

$$f^*d\phi(X) = (d\phi) \circ f(f_*X) = (f_*X)\phi = X(\phi \circ f) = d(\phi \circ f)(X) = d(f^*\phi)(X),$$

so (2) holds on $A_0(N)$. In the general case, $\alpha \in A(N)$ may locally be written as $\alpha = \sum \alpha_I dx^{i_1} \wedge \cdots \wedge dx^{i_k}$. By the above and (1),

$$f^*\alpha = \sum (\alpha_I \circ f) f^*dx^{i_1} \wedge \cdots \wedge f^*dx^{i_k} = \sum (\alpha_I \circ f) df^*x^{i_1} \wedge \cdots \wedge df^*x^{i_k},$$

so that

$$df^*\alpha = \sum d(\alpha_I \circ f) \wedge f^*dx^{i_1} \wedge \cdots \wedge f^*dx^{i_k}$$
$$= \sum f^*d\alpha_I \wedge f^*dx^{i_1} \wedge \cdots \wedge f^*dx^{i_k}$$
$$= f^*(\sum d\alpha_I \wedge dx^{i_1} \wedge \cdots \wedge dx^{i_k}) = f^*d\alpha.$$

The last statement in the theorem is a direct consequence of the first two. $\square$

We end this section with a coordinate-free characterization of the exterior derivative operator $d$. The proof uses concepts and results from Exercises 33 and 34 below.

THEOREM 11.3. *If* $\omega \in A_k(M)$ *and* $X_0, X_1, \ldots, X_k \in \mathfrak{X}(M)$, *then*

$$d\omega(X_0, \ldots, X_k) = \sum_{i=0}^{k} (-1)^i X_i(\omega(X_0, \ldots, \hat{X}_i, \ldots, X_k))$$

$$+ \sum_{i<j} (-1)^{i+j} \omega([X_i, X_j], X_0, \ldots, \hat{X}_i, \ldots, \hat{X}_j, \ldots, X_k).$$

*(The "hat" over a vector field means the latter is deleted.)*

PROOF. We shall consider the case $k = 1$, the general case being a straightforward induction. It follows from Exercise 33 that, in general,

$$(L_{X_0}\omega)(X_1, \ldots, X_k) = L_{X_0}(\omega(X_1, \ldots, X_k)) - \sum_{i=1}^{k} \omega(X_1, \ldots, L_{X_0}X_i, \ldots, X_k).$$

Together with Exercise 34, this implies that

$$\begin{aligned} d\omega(X_0, X_1) &= (i(X_0)d\omega)(X_1) = (L_{X_0}\omega)(X_1) - d(i(X_0)\omega)(X_1) \\ &= L_{X_0}(\omega(X_1)) - \omega[X_0, X_1] - d(\omega(X_0))(X_1) \\ &= X_0(\omega X_1) - X_1(\omega X_0) - \omega[X_0, X_1]. \end{aligned}$$

$\square$

EXERCISE 31. Show that the form $\alpha$ in Example 11.1 is equal to

$$\frac{-u^2}{(u^1)^2 + (u^2)^2} du^1 + \frac{u^1}{(u^1)^2 + (u^2)^2} du^2.$$

EXERCISE 32. Let $\alpha$ be a 1-form on $\mathbb{R}^2$, so that we may write $\alpha = f_1 du^1 + f_2 du^2$ for smooth functions $f_i$ on $\mathbb{R}^2$.
(a) Show that $d\alpha = 0$ iff $D_2 f_1 = D_1 f_2$.
(b) Show that if $\alpha$ is closed, then it is exact. Thus, $H^1(\mathbb{R}^2) = 0$.
*Hint:* Fix any $(a, b) \in \mathbb{R}^2$, and show that $\alpha = df$, where $f$ is defined by $f(x, y) = \int_a^x f_1(t, b)dt + \int_b^y f_2(x, t)dt$.

EXERCISE 33. If $\omega$ is a $k$-form on $M$, and $X$ a vector field with flow $\Phi_t$, one defines the *Lie derivative of* $\omega$ *with respect to* $X$ to be the $k$-form given by

$$(L_X\omega)(p) = \lim_{t \to 0} \frac{1}{t}[(\Phi_t^*\omega)(p) - \omega(p)], \qquad p \in M.$$

(a) Show that $L_X f = X f$ for $f \in A_0(M)$.
(b) Show that $L_X(\omega_1 \wedge \cdots \wedge \omega_k) = \sum_i \omega_1 \wedge \cdots \wedge L_X\omega_i \wedge \cdots \wedge \omega_k$ for 1-forms $\omega_i$.
(c) Show that $L_X \circ d = d \circ L_X$ on $A_0(M)$.
(d) Use (a) through (c) to show that $L_X \circ d = d \circ L_X$ on $A(M)$.

EXERCISE 34. Given a vector field $X$ on $M$, *interior multiplication* $i(X) : A_k(M) \to A_{k-1}(M)$ by $X$ is defined by

$$(i(X)\omega)(X_1, \ldots, X_{k-1}) := \omega(X, X_1, \ldots, X_{k-1}), \qquad w \in A_k(M), \quad X_i \in \mathfrak{X}(M).$$

Prove that $L_X = i(X) \circ d + d \circ i(X)$ (see Exercise 33).

## 12. Integration on Chains

We are now ready to generalize integration on Euclidean space to manifolds. Thanks to the work done in the preceding sections, we will be able to integrate differential forms rather than functions; one advantage lies in that the change of variables formula for integrals is particularly simple for differential forms.

DEFINITION 12.1. A *singular $k$-cube* in a manifold $M^n$ is a differentiable map $c : [0,1]^k \to M$. (For $k = 0$, define $[0,1]^0 = \{0\}$, so that a singular 0-cube is determined by one point $c(0) \in M$). The *standard $k$-cube* is the inclusion map $I^k : [0,1]^k \hookrightarrow \mathbb{R}^k$.

DEFINITION 12.2. Let $\omega$ be a $k$-form on $[0,1]^k$, and write $\omega = f du^1 \wedge \cdots \wedge du^k$, where $f = \omega(D_1, \ldots, D_k)$. The *integral of $\omega$ over $[0,1]^k$* is defined to be

$$(12.1) \qquad \int_{[0,1]^k} \omega = \int_{[0,1]^k} f.$$

If $\omega$ is a $k$-form on a manifold $M$, $k > 0$, and $c$ is a singular $k$-cube in $M$, the *integral of $\omega$ over $c$* is

$$(12.2) \qquad \int_c \omega = \int_{[0,1]^k} c^* \omega,$$

where the right side is defined in (12.1). For $k = 0$, $\int_c \omega := \omega(c(0))$.

EXAMPLES AND REMARKS 12.1. (i) In classical calculus, a vector field on the plane is a differentiable map $F = (f, g) : \mathbb{R}^2 \to \mathbb{R}^2$. If $c : [0,1] \to \mathbb{R}^2$ is a curve (or a singular 1-cube) in the plane, the integral of the vector field along $c$ is defined to be $\int_0^1 \langle F \circ c, c' \rangle$. In the current context, it equals $\int_c \omega$, where $\omega$ is the 1-form dual to $F$: $\omega = f du^1 + g du^2$. This is because

$$c^* \omega(D) = (\omega \circ c)(\dot{c}) = \langle F \circ c, c' \rangle.$$

(ii) (Change of Variables) Consider a singular $n$-cube $c$ in $\mathbb{R}^n$ with $\det D(c) \neq 0$, and an $n$-form $\omega = f du^1 \wedge \cdots \wedge du^n$. The change of variables formula for multiple integrals translates into

$$\int_c \omega = \pm \int_{c[0,1]^n} f,$$

with the sign depending on whether the Jacobian matrix $D(c)$ of $c$ has nonnegative determinant. Indeed,

$$\begin{aligned}
c^* \omega(D_1, \ldots, D_n) &= \omega \circ c(c_* D_1, \ldots, c_* D_n) \\
&= (f \circ c) du^1 \wedge \cdots \wedge du^n (c_* D_1, \ldots, c_* D_n) \\
&= (f \circ c) \det(du^i(c_* D_j)) = (f \circ c) \det(D_j(u^i \circ c)) \\
&= (f \circ c) \det D(c).
\end{aligned}$$

Thus,

$$\int_c \omega = \int_{[0,1]^n} (f \circ c) \det D(c) = \pm \int_{[0,1]^n} (f \circ c) |\det D(c)| = \pm \int_{c[0,1]^n} f.$$

(iii) (Independence of Parametrization) Let $c$ be a singular $k$-cube in $M$. If $F$ is a diffeomorphism of $[0,1]^k$ with positive determinant, then the singular $k$-cube $\tilde{c} := c \circ F$ is called an *orientation preserving reparametrization* of $c$. In this case, $\int_c \omega = \int_{\tilde{c}} \omega$ for any $k$-form $\omega$ on $M$:

$$\int_{c \circ F} \omega = \int_{[0,1]^k} (c \circ F)^* \omega = \int_{[0,1]^k} F^* c^* \omega = \int_F c^* \omega^* = \int_{F[0,1]^k} c^* \omega(D_1, \ldots, D_k)$$

$$= \int_{[0,1]^k} c^* \omega(D_1, \ldots, D_k) = \int_c \omega,$$

where we used (ii) in the fourth equality.

DEFINITION 12.3. A $k$-*chain* in $M$ is an element of the free vector space generated by the collection of all singular $k$-cubes in $M$. Thus, a $k$-chain has the form $c = \sum_{i=1}^n a_i c_i$, where $a_i \in \mathbb{R}$ and each $c_i$ is a singular k-cube. The *integral* of a $k$-form $\omega$ over the $k$-chain $c = \sum a_i c_i$ is defined to be

$$\int_c \omega = \sum a_i \int_{c_i} \omega.$$

We will need the concept of boundary of a chain. For $1 \le i \le n$, the $(i,0)$ and $(i,1)$ *faces* of the standard $n$-cube $I^n$ are the singular $(n-1)$-cubes $I_{i,0}^n$ and $I_{i,1}^n$ defined by

$$I_{i,0}^n(a) = (a_1, \ldots, a_{i-1}, 0, a_i, \ldots, a_{n-1}),$$
$$I_{i,1}^n(a) = (a_1, \ldots, a_{i-1}, 1, a_i, \ldots, a_{n-1})$$

for $a = (a_1, \ldots, a_{n-1}) \in [0,1]^{n-1}$. Similarly, the $(i,j)$ *face* of the singular $n$-cube $c$ is defined to be $c_{i,j} = c \circ I_{i,j}^n$, $1 \le i \le n$, $j = 0, 1$.

The *boundary* of a singular $n$-cube $c$ is the singular $(n-1)$-chain

$$\partial c = \sum_{i=1}^n \sum_{j=0,1} (-1)^{i+j} c_{i,j}.$$

For example, the boundary of a 1-cube $c$ is the chain $\partial c = c(1) - c(0)$ if we identify 0-cubes with their values. The boundary of a 0-cube $c : \{0\} \to M$ is defined to be $\partial c = 1 \in \mathbb{R}$. Notice that for 1-cubes, $\partial \partial c = 0$. We extend $\partial$ linearly to the space of $n$-chains.

EXAMPLE 12.1 (The Fundamental Theorem of Calculus, or Stokes' Theorem). If $f : \mathbb{R} \to \mathbb{R}$ is differentiable, and $c$ is a singular 1-chain in $\mathbb{R}$, then

$$\int_c df = \int_{\partial c} f.$$

Indeed, by linearity of the integral, we may, without loss of generality, assume that $c$ is a singular 1-cube. By the fundamental theorem of calculus,

$$\int_c df = \int_0^1 c^* df(D) = \int_0^1 (f \circ c)' = (f \circ c)(1) - (f \circ c)(0) = \int_{c(1)} f - \int_{c(0)} f$$

$$= \int_{\partial c} f.$$

PROPOSITION 12.1. $\partial^2 = 0$.

PROOF. Since $\partial$ is linear, it suffices to show that $\partial\partial c = 0$ for a singular $n$-cube $c$. Now,

$$\partial(\partial c) = \partial\left(\sum_{i=1}^{n}\sum_{k=0,1}(-1)^{i+k}c_{i,k}\right) = \sum_{i=1}^{n}\sum_{k=0,1}\sum_{j=1}^{n-1}\sum_{l=0,1}(-1)^{i+k+j+l}(c_{i,k})_{j,l}.$$

The terms in this sum cancel pairwise because of the identity in Exercise 35 below.                                                                                    □

EXERCISE 35. Show that if $i \leq j \leq n-1$, and $k, l = 0, 1$, then $I_{i,k}^{n} \circ I_{j,l}^{n-1} = I_{j+1,l}^{n} \circ I_{i,k}^{n-1}$. Deduce that $(c_{i,k})_{j,l} = (c_{j+1,l})_{i,k}$ for a singular $n$-cube $c$.

## 13. The Local Version of Stokes' Theorem

We saw in the previous section that the fundamental theorem of calculus, translated in the notation of chains and forms, says that $\int_c df = \int_{\partial c} f$ for a smooth function $f$ and a singular 1-cube in $\mathbb{R}$. This generalizes to the following:

THEOREM 13.1 (Stokes' Theorem, Local Version). *If $c$ is a $k$-chain in $M$, and $\omega$ is a $(k-1)$-form on $M$, then*

$$\int_c d\omega = \int_{\partial c} \omega.$$

PROOF. We first consider the case $M = \mathbb{R}^k$ and $c = I^k$. Then

$$\int_{\partial c} \omega = \sum_{j,\alpha}(-1)^{j+\alpha}\int_{[0,1]^{k-1}} I_{j,\alpha}^{k*}\omega(D_1,\ldots,D_{k-1}).$$

By linearity, we may assume that $\omega = f\,du^1 \wedge \cdots \wedge \widehat{du^i} \wedge \cdots \wedge du^k$ for some $i$, so that

$$I_{j,\alpha}^{k*}\omega(D_1,\ldots,D_{k-1})$$
$$= (f \circ I_{j,\alpha}^k)du^1 \wedge \cdots \wedge \widehat{du^i} \wedge \cdots \wedge du^k(I_{j,\alpha*}^k D_1,\ldots,I_{j,\alpha*}^k D_{k-1})$$
$$= (f \circ I_{j,\alpha}^k)\det(D_l(u^p \circ I_{j,\alpha}^k)),$$

$1 \leq l \leq k-1$, $p = 1,\ldots,i-1,i+1,\ldots,k$. This determinant is 1 if $i = j$, and 0 otherwise, because

$$u^p \circ I_{j,\alpha}^k(a_1,\ldots,a_{k-1}) = u^p(a_1,\ldots,a_{j-1},\alpha,a_j,\ldots,a_{k-1}),$$

so that $u^j \circ I_{j,\alpha}^k \equiv \alpha$. Thus,

$$\int_{\partial c} \omega = (-1)^{i+1}\int_{[0,1]^{k-1}} f(u^1,\ldots,u^{i-1},1,u^i,\ldots,u^{k-1})du^1\ldots du^{k-1}$$
$$+ (-1)^i\int_{[0,1]^{k-1}} f(u^1,\ldots,u^{i-1},0,u^i,\ldots,u^{k-1})du^1\ldots du^{k-1}.$$

On the other hand,

$$\int_c d\omega = \int_{I^k} df \wedge du^1 \wedge \cdots \wedge \widehat{du^i} \wedge \ldots du^{k-1}(D_1, \ldots, D_k)$$

$$= \int_{I^k} \sum_j (D_j f) du^j \wedge du^1 \wedge \cdots \wedge \widehat{du^i} \wedge \ldots du^{k-1}(D_1, \ldots, D_k)$$

$$= (-1)^{i-1} \int_{I^k} D_i f$$

$$= (-1)^{i-1} \int_0^1 \cdots \int_0^1 [f(u^1, \ldots, 1, \ldots, u^k)$$
$$- f(u^1, \ldots, 0, \ldots, u^k)] \, du^1 \ldots \widehat{du^i} \ldots du^k$$

$$= \int_{[0,1]^k} (-1)^{i-1}[f(u^1, \ldots, 1, \ldots, u^k)$$
$$+ (-1)^i f(u^1, \ldots, 0, \ldots, u^k)] \, du^1 \ldots du^k,$$

which establishes the result when $M = \mathbb{R}^k$ and $c = I^k$. In the general case, if $\omega \in A_{k-1}(M)$ and $c$ is a singular $k$-cube in $M$, we have

$$\int_{\partial c} \omega = \sum_{i=1}^k \sum_{j=0,1} (-1)^{i+j} \int_{c \circ I_{i,j}^{k-1}} \omega = \sum_{i,j} (-1)^{i+j} \int_{I^{k-1}} (c \circ I_{i,j}^{k-1})^* \omega$$

$$= \sum_{i,j} (-1)^{i+j} \int_{I^{k-1}} I_{i,j}^{k-1*} c^* \omega = \sum_{i,j} (-1)^{i+j} \int_{I_{i,j}^{k-1}} c^* \omega = \int_{\partial I^k} c^* \omega.$$

Thus,

$$\int_{\partial c} \omega = \int_{\partial I^k} c^* \omega = \int_{I^k} dc^* \omega = \int_{I^k} c^* d\omega = \int_c d\omega.$$

By linearity, the formula then also holds for singular $k$-chains.    □

EXAMPLES AND REMARKS 13.1. (i) Consider the closed 1-form

$$\alpha = \frac{-u^2}{(u^1)^2 + (u^2)^2} du^1 + \frac{u^1}{(u^1)^2 + (u^2)^2} du^2$$

on $M = \mathbb{R}^2 \setminus \{0\}$ from Exercise 31. If $c$ is the 1-cube on $M$ given by $c(t) = (\cos(2\pi t), \sin(2\pi t))$, then $\int_c \alpha = 2\pi$. Indeed,

$$\int_c \alpha = \int_0^1 c^* \alpha(D) = \int_0^1 \alpha \circ c(t)(c_* D(t)) \, dt$$

$$= \int_0^1 \alpha \circ c(t)((u^1 \circ c)'(t) D_1(c(t)) + (u^2 \circ c)'(t) D_2(c(t))) \, dt$$

$$= \int_0^1 -u^2 \circ c(t)(u^1 \circ c)'(t) + u^1 \circ c(t)(u^2 \circ c)'(t) \, dt$$

$$= \int_0^1 2\pi(\sin^2(2\pi t) + \cos^2(2\pi t)) \, dt = 2\pi.$$

This shows once again that $\alpha$ is not exact: For if $\alpha = df$, then

$$\int_c \alpha = \int_c df = \int_{\partial c} f = f(c(1)) - f(c(0)) = 0.$$

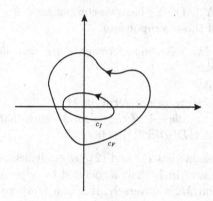

FIGURE 5

(ii) Let $M, \alpha$ be as in (i), and consider positive functions $f, F : \mathbb{R} \to \mathbb{R}$ of period $2\pi$, with $f < F$. Define singular 1-cubes $c_f$ and $c_F$ by

$$c_f(t) = (f(2\pi t)\cos(2\pi t), f(2\pi t)\sin(2\pi t)),$$
$$c_F(t) = (F(2\pi t)\cos(2\pi t), F(2\pi t)\sin(2\pi t));$$

see Figure 5. We claim $\int_{c_f} \alpha = \int_{c_F} \alpha = 2\pi$. To see this, consider the singular 2-cube $c$ in $M$ defined by

$$c(a_1, a_2) = (1 - a_2)c_f(a_1) + a_2 c_F(a_1).$$

Since $\alpha$ is closed, Stokes' theorem implies

$$0 = \int_c d\alpha = \int_{\partial c} \alpha = \int_{c \circ I_{1,1}^2} \alpha - \int_{c \circ I_{1,0}^2} \alpha - \int_{c \circ I_{2,1}^2} \alpha + \int_{c \circ I_{2,0}^2} \alpha.$$

Now, $c \circ I_{1,1}^2(t) = c \circ I_{1,0}^2(t) = (1-t)(f(0), 0) + t(F(0), 0)$, whereas $c \circ I_{2,1}^2 = c_F$, and $c \circ I_{2,0}^2 = c_f$. Thus $0 = \int_{\partial c} \alpha = \int_{c_f} \alpha - \int_{c_F} \alpha$. By (i), $\int_{c_f} \alpha = 2\pi$ when $f \equiv 1$. This establishes the claim.

## 14. Orientation and the Global Version of Stokes' Theorem

As the title of this section suggests, the global version of Stokes' theorem uses the concept of orientation. Recall that if $V$ is an $n$-dimensional vector space, then $\Lambda_n^*(V) \cong \mathbb{R}$. An *orientation* of $V$ is a choice of one of the two components of $\Lambda_n^*(V) \setminus \{0\}$. An ordered basis $v_1, \ldots, v_n$ of $V$ is said to be *positively oriented* (resp., *negatively oriented*) if $\omega(v_1, \ldots, v_n) > 0$ (resp., $< 0$) for some $\omega$ in the chosen component. An orientation is therefore also specified by an ordered basis $v_1, \ldots, v_n$, since the dual basis $\alpha_1, \ldots, \alpha_n$ determines a nonzero $n$-form $\omega = \alpha_1 \wedge \cdots \wedge \alpha_n$ satisfying $\omega(v_1, \ldots, v_n) > 0$.

If $M$ is an $n$-dimensional manifold, then each $\Lambda_n^*(M_p) \cong \mathbb{R}$, but the bundle $\Lambda_n^*(M)$ itself is not always identifiable with $M \times \mathbb{R}$. As usual, $\pi : \Lambda_n^*(M) \to M$ will denote the bundle projection.

DEFINITION 14.1. $M^n$ is said to be *orientable* if there is a map $L : \Lambda_n^*(M) \to \mathbb{R}$, which is an isomorphism on $\Lambda_n^*(M_p)$ for each $p \in M$, such that $(\pi, L) : \Lambda_n^*(M) \to M \times \mathbb{R}$ is a diffeomorphism.

In this case, the set $\Sigma_0 = \{0 \in \Lambda_n^*(M_p) \mid p \in M\}$ corresponds to $M \times 0$ under $(\pi, L)$, so that $\Lambda_n^*(M) \setminus \Sigma_0$ has two components. An *orientation* of $M$ is then a choice of one of these components.

THEOREM 14.1. *The following statements are equivalent for a connected $n$-dimensional manifold:*

(1) *$M$ is orientable.*
(2) *There is a nowhere-zero $n$-form on $M$.*
(3) *There exists an atlas $\mathcal{A}$ of charts on $M$ such that for any $(U, x)$ and $(V, y)$ in $\mathcal{A}$, $\det(D(y \circ x^{-1})) > 0$.*

PROOF. We first show that (1) and (2) are equivalent: If $(\pi, L)$ is a diffeomorphism as in Definition 14.1, then $\omega$, defined by $\omega(p) = (\pi, L)^{-1}(p, 1)$, is a nowhere-zero $n$-form on $M$. Conversely, if $\omega$ is a nowhere-zero $n$-form on $M$, let $F : M \times \mathbb{R} \to \Lambda_n^*(M)$ be given by $F(p, t) = t\omega(p)$. Then $F^{-1}$ is the desired map $\Lambda_n^*(M) \to M \times \mathbb{R}$.

To complete the proof, we now show that (2) and (3) are equivalent: If $\omega$ is a nowhere-zero $n$-form on $M$, consider the collection $\mathcal{A}$ of all charts $(U, x)$ on $M$ such that $\omega(\partial/\partial x^1, \ldots, \partial/\partial x^n) > 0$. $\mathcal{A}$ is an atlas, since the components of any coordinate map may be reordered. Furthermore, if $(U, x)$ and $(V, y)$ belong to $\mathcal{A}$, then $\omega_{|U} = f dx^1 \wedge \cdots \wedge dx^n$ and $\omega_{|V} = g dy^1 \wedge \cdots \wedge dy^n$ for positive functions $g$ and $h$. Thus on $U \cap V$, $dy^1 \wedge \ldots dy^n = h dx^1 \wedge \ldots dx^n$, where $h = f/g > 0$. But $h = \det(D(y \circ x^{-1}))$. Conversely, if $\mathcal{A}$ is an atlas as in (3), choose a partition $\phi_i$ of unity subordinate to the domains $U_i$ of the charts $(U_i, x_i)$ in $\mathcal{A}$, and define $n$-forms $\omega_i$ on $M$ by setting $\omega_i(p) = \phi_i(p) dx_i^1 \wedge \cdots \wedge dx_i^n(p)$ if $p \in U_i$, and $\omega_i(p) = 0$ otherwise. Then $\omega = \sum_i \omega_i$ is a nowhere-zero $n$-form on $M$. $\square$

If $M^n$ is an oriented manifold, a nowhere-zero $n$-form on $M$ is said to be a *volume form* if it belongs to the component of $\Lambda_n^*(M) \setminus \Sigma_0$ determined by the orientation. An imbedding $\iota : M \to N$ between two oriented manifolds of the same dimension is said to be *orientation-preserving* if $\iota^*\omega$ is a volume form on $M$ for any volume form $\omega$ on $N$.

As an application of the above theorem, we show that the $n$-sphere $S^n$ is orientable: Let $\iota : S^n \to \mathbb{R}^{n+1}$ denote inclusion, and consider the "position" vector field $P$ on $\mathbb{R}^{n+1}$ given by $P(p) = \mathcal{J}_p p$. By Exercise 14, if $\omega = du^1 \wedge \cdots \wedge du^{n+1}$, then $\iota^*(i(P)\omega)$ is a nowhere-zero $n$-form on the sphere (here $i(P)$ is interior multiplication by $P$ as defined in Exercise 34).

In order to exhibit a nonorientable example, we must disgress and briefly consider group actions on manifolds. If $G$ is a Lie group, a (smooth, left) *action of $G$ on a manifold $M$* is a differentiable map $\mu : G \times M \to M$ such that $\mu(g_1, \mu(g_2, p)) = \mu(g_1 g_2, p)$ and $\mu(e, p) = p$ for all $g_i \in G$, $p \in M$. We denote $\mu(g, p)$ simply by $g(p)$. Observe that $g : M \to M$ is then a diffeomorphism of $M$ with inverse $g^{-1}$. The *orbit $G(p)$* of $p \in M$ is the set $\{g(p) \mid g \in G\}$, and the collection of orbits is denoted $M/G$.

Two orbits $G(p)$ and $G(q)$ are said to have the *same type* if there exists an equivariant bijection between them; i.e., a bijection $f$ such that $f(gm) = gf(m)$

for all $g \in G$ and $m$ in an orbit. When all orbits have the same type, we say the action is by *principal orbits*. This is the case for example when the action is *free*; i.e., when the condition $g(p) = p$ for some $p \in M$ and $g \in G$ implies that $g = e$. For in this case, the map $h : G \to G(p)$ given by $h(g) = g(p)$ is an equivariant bijection for any $p \in M$. Finally, $G$ is said to act *properly discontinuously* on $M$ if for any two points $p$ and $q$ in $M$, there exist neighborhoods $U$ of $p$ and $V$ of $q$ such that $g(U) \cap V = \emptyset$ for all $g \in G$.

THEOREM 14.2. (1) *If $G$ acts freely and properly discontinuously on $M$, then there exists a unique differentiable structure on the space $M/G$ of orbits for which the projection $\pi : M \to M/G$ becomes a local diffeomorphism.*

(2) *If a compact Lie group $G$ acts on $M$ with principal orbits, then there exists a unique differentiable structure on $M/G$ for which $\pi : M \to M/G$ becomes a submersion; i.e., $\pi$ is onto $M/G$, and its derivative has maximal rank everywhere.*

PROOF. Part (2) will be proved in Chapter 5, once we have discussed Riemannian metrics on manifolds. For (1), notice that since the action is free and properly discontinuous, the orbit of any point, and hence $G$ itself, is discrete. For the same reason, $M/G$ is a Hausdorff space in the quotient topology. Consider an orbit $q$ in $M/G$ and a point $p \in M$ in this orbit. There exists a neighborhood $V$ of $p$ on which $\pi$ is a homeomorphism. If $(U, x)$ is a chart of $M$ around $p$, then $(\pi(U \cap V), x \circ (\pi_{|U \cap V})^{-1})$ can be taken as a chart of $M/G$ around $q$. It is straightforward to check that this induces a differentiable structure on the quotient $M/G$. Uniqueness follows from the fact that $\pi$ is a local diffeomorphism. $\square$

Under the hypotheses of Theorem 14.2 (1), $\pi : M \to M/G$ is called a *covering map*. As an example, let $M = S^n$, $G = \{\pm 1_M\}$. The quotient $M/G$ is called *real projective $n$-space* $\mathbb{R}P^n$. Observe that if $n$ is even, then the antipodal map $-1_M : p \mapsto -p$, when extended to $\mathbb{R}^{n+1}$, is orientation-reversing: Since $-1_M$ is linear, $(-1_M)_* \mathcal{J}_p u = \mathcal{J}_{-1_M(p)}(-1_M(u)) = -\mathcal{J}_{-p} u$ for $p \in \mathbb{R}^{n+1}$, $u \in \mathbb{R}^{n+1}$. But if $\mathcal{J}_p u_1, \ldots, \mathcal{J}_p u_{n+1}$ is a positively oriented basis of $\mathbb{R}_p^{n+1}$, then $-\mathcal{J}_{-p} u_1, \ldots, -\mathcal{J}_{-p} u_{n+1}$ is a negatively oriented basis at the antipodal point. This implies that the antipodal map is orientation-reversing on the sphere, since the position vector field is $-1_M$-related to itself. Thus, $(-1_M)^* \omega \neq \omega$ for any $n$-form on $M$. Now, if $\mathbb{R}P^n$ were orientable, it would admit a nonvanishing $n$-form $\eta$, and $\pi^* \eta$ would be a nonvanishing $n$-form on $M$. This is impossible, because $\pi \circ (-1_M) = \pi$, so that $(-1_M)^* \pi^* \eta = \pi^* \eta$.

We now define integration on $M$ in terms of integration on orientation-preserving cubes. The following lemma ensures that it will be independent of the particular cube chosen:

LEMMA 14.1. *Let $M^n$ be an oriented manifold, $c_1, c_2 : [0,1]^n \to M$ two orientation-preserving imbeddings. If $\omega \in A_n(M)$ has support in $c_1[0,1]^n \cap c_2[0,1]^n$, then $\int_{c_1} \omega = \int_{c_2} \omega$.*

PROOF. This follows from Examples and Remarks 12.1(iii), since $c_2^{-1} \circ c_1$ is orientation-preserving, so that $\int_{c_2} \omega = \int_{c_2 \circ c_2^{-1} \circ c_1} \omega = \int_{c_1} \omega$. Notice that $c_2^{-1} \circ c_1$

is not necessarily defined on all of $[0,1]^n$, but the support of $\omega$ is contained in $c_1[0,1]^n \cap c_2[0,1]^n$.                    □

DEFINITION 14.2. Let $\omega \in A_n(M)$ have support in the image of an orientation-preserving imbedding $c : [0,1]^n \to M$. Define $\int_M \omega = \int_c \omega$.

In general, there exists an open cover of $M$ such that for any open set $U$ in the cover, there exists an orientation-preserving imbedding $c : [0,1]^n \to M$ the image of which contains $U$. Let $\Phi$ be a partition of unity subordinate to this cover.

DEFINITION 14.3. If $\omega \in A_n(M)$ has compact support, define the *integral of $\omega$ over $M$* to be

$$\int_M \omega = \sum_{\phi \in \Phi} \int_M \phi\omega.$$

Notice that the above sum is finite, because $\omega$ has compact support. Furthermore, the definition is independent of the chosen partition of unity: If $\Psi$ is a partition of unity subordinate to some other open cover, then

$$\sum_{\phi \in \Phi} \int_M \phi\omega = \sum_{\phi \in \Phi} \int_M \sum_{\psi \in \Psi} \psi\phi\omega = \sum_{\psi \in \Psi} \int_M \sum_{\phi \in \Phi} \phi\psi\omega = \sum_{\psi \in \Psi} \int_M \psi\omega.$$

One can also define integration on manifolds with boundary, which are locally modeled on the closed half-space $H^n = \{p \in \mathbb{R}^n \mid u^n(p) \geq 0\}$ instead of $\mathbb{R}^n$. A map from $H^n$ to $\mathbb{R}^n$ is defined to be differentiable if it can be extended to a differentiable map on a neighborhood of $H^n$.

DEFINITION 14.4. A *topological n-manifold with boundary* is a second-countable Hausdorff space $M$, with the property that for any $p \in M$, there exists a neighborhood $U$ of $p$ and a homeomorphism $x : U \to x(U)$, where $x(U)$ is an open set of either $\mathbb{R}^n$ or $H^n$, and $x(p) = 0$. The *boundary* $\partial M$ of $M$ consists of those points that get mapped to the boundary $\partial H^n$ of $H^n$ under some (and hence any) coordinate map, and is an $(n-1)$-manifold. One defines a differentiable structure as usual by requiring the transition maps to be differentiable.

For example, $H^n$ itself is an $n$-manifold with boundary $\mathbb{R}^{n-1}$; the same is true of the closed $n$-disk $D^n = \{p \in \mathbb{R}^n \mid |p| \leq 1\}$, which has boundary $S^{n-1}$.

It follows from the above definition that if $p \in \partial M$, and $(U,x)$ is a chart of $M$ with $x(p) = 0$, then $U \cap \partial M = \{q \in M \mid x^n(q) = 0\}$, and $(U \cap \partial M, \pi \circ x)$ is a chart of $\partial M$, where $\pi : \mathbb{R}^n \to \mathbb{R}^{n-1}$ denotes the projection. Furthermore, a function $f$ defined on a neighborhood of $p$ is smooth iff $f \circ x^{-1} : H^n \to \mathbb{R}$ is smooth; i.e., extendable to a smooth function on $\mathbb{R}^n$. Thus, one has a well-defined $n$-dimensional tangent space

$$M_p = (x^{-1})_{*0}(\mathbb{R}_0^n)$$

at $p$, and if $\imath : \partial M \hookrightarrow M$, $j : \mathbb{R}^{n-1} \times 0 \hookrightarrow \mathbb{R}^n$ denote inclusions, then

$$\imath_*(\partial M)_p = x_*^{-1}(j_*(\mathbb{R}^{n-1} \times 0)_0).$$

If $M$ is oriented, it induces an orientation of the boundary $\partial M$: For $p \in \partial M$, $v \in M_p$ is said to be *outward-pointing* (resp., *inward-pointing*) if $v(x^n) < 0$

(resp., $> 0$). A basis $v_1, \ldots, v_{n-1}$ of $\partial M_p$ is said to be *positively oriented* if $v, v_1, \ldots, v_{n-1}$ is a positively oriented basis of $M_p$ for any outward-pointing $v \in M_p$. It is easily checked that this does indeed define an orientation on $\partial M$: If $(U, x)$ and $(V, y)$ are charts of $M$ sending $p$ to $0$, and such that $\det D(y \circ x^{-1}) > 0$, then for the induced coordinate maps $\tilde{x} = \pi \circ x$ and $\tilde{y} = \pi \circ y$ on $\partial M$, one has

$$\det D(y \circ x^{-1})(0) = \det \begin{pmatrix} D(\tilde{y} \circ \tilde{x}^{-1})(0) & 0 \\ 0 & D_n(y^n \circ x^{-1})(0) \end{pmatrix},$$

where $D_n(y^n \circ x^{-1})(0) > 0$ by Exercise 41.

Now, let $\omega$ be an $n$-form on an oriented manifold $M^n$. There is an open cover $\mathcal{U}$ of $M$ such that each $U \in \mathcal{U}$ is contained inside some $c[0,1]^n$, where $c$ is an orientation-preserving imbedding with either $c[0,1]^n \subset M \setminus \partial M$ or else $\partial M \cap c[0,1]^n = c_{n,0}[0,1]^{n-1}$. If $\Phi$ is a partition of unity subordinate to $\mathcal{U}$ and $\omega$ has compact support, we define as before

$$\int_M \omega = \sum_{\phi \in \Phi} \int_M \phi \omega.$$

THEOREM 14.3 (Stokes' Theorem, Global Version). *If $\omega$ is an $(n-1)$-form with compact support on an oriented $n$-manifold $M$ with boundary $\partial M$, then*

$$\int_{\partial M} \omega = \int_M d\omega.$$

PROOF. Suppose first that the support of $\omega$ is contained in some $U \in \mathcal{U}$, where $U$ itself lies inside the image of $c$, for one of the above cubes $c$. If the image of $c$ does not intersect the boundary of $M$, then by the local version of Stokes' theorem,

$$\int_M d\omega = \int_c d\omega = \int_{\partial c} \omega = \int_{\partial M} \omega.$$

Indeed, $\omega$ has support in $U$, and the latter does not intersect the image of $\partial c$, so that the integral of $\omega$ on the boundary of $c$ is zero. Similarly, the integral of $\omega$ on the boundary of $M$ vanishes. If on the other hand $c$ is of the second type, then $U$, and hence the support of $\omega$, may intersect the boundary of $M$. Now, $c_* D_1, \ldots, c_* D_n$ is positively oriented, so $-c_* D_n, c_* D_1, \ldots, c_* D_{n-1}$ is positively oriented iff $n$ is even. Observe that if $(a, 0) \in [0,1]^{n-1} \times 0$, then $c_{*(a,0)} D_i = (c_{n,0})_{*a} D_i$ for $1 \leq i \leq n-1$. Thus, $c_{n,0} : [0,1]^{n-1} \to \partial M$ is orientation-preserving iff $n$ is even. In any case, by Exercise 38,

$$\int_M d\omega = \int_c d\omega = \int_{\partial c} \omega = \int_{(-1)^n c_{n,0}} \omega = (-1)^n \int_{c_{n,0}} \omega$$
$$= (-1)^n (-1)^n \int_{\partial M} \omega = \int_{\partial M} \omega.$$

In general,

$$\int_{\partial M} \omega = \sum_{\phi \in \Phi} \int_{\partial M} \phi \omega = \sum_{\phi \in \Phi} \int_M d(\phi \omega) = \sum_{\phi \in \Phi} \int_M d\phi \wedge \omega + \phi d\omega$$
$$= \int_M d\omega + \sum_{\phi \in \Phi} \int_M d\phi \wedge \omega.$$

The last term is actually a finite sum $\int_M \sum_{i=1}^{k} d\phi_i \wedge \omega$, with $\sum \phi_i \equiv 1$ on a neighborhood of the support of $\omega$. Thus, $\sum d\phi_i = d(\sum \phi_i) = 0$, which concludes the argument.  $\square$

We end this section by discussing some additional properties of integration when $M$ is a compact, connected Lie group $G$. By Exercise 36 below, $G$ is orientable. Choosing an orientation on $G$, there exists a unique $\omega \in A_k(M)$ which is left-invariant in the sense that $L_g^* \omega = \omega$ for $g \in G$, and satisfies $\int_G \omega = 1$. We then define the integral of a function $f : G \to \mathbb{R}$ by

$$\int_G f := \int_G f\omega.$$

PROPOSITION 14.1. For $f : G \to \mathbb{R}$, $g \in G$, $\int_G f = \int_G f \circ L_g = \int_G f \circ R_g$.

PROOF. For the first identity, notice that $\int_G L_g^*(f\omega) = \int_G f\omega$ by Exercise 39 below, since $L_g$ is an orientation-preserving diffeomorphism of $G$. But $\omega$ is left-invariant, so that

$$\int_G f = \int_G f\omega = \int_G L_g^*(f\omega) = \int_G (f \circ L_g) L_g^* \omega = \int_G (f \circ L_g)\omega = \int_G f \circ L_g.$$

The second identity follows by a similar argument, once we establish that (i) $R_g$ is orientation-preserving, and (ii) $\omega$ is right-invariant. Since (i) follows from (ii), we only need to prove the latter. Now, for any $g \in G$, the form $R_g^* \omega$ is left-invariant, since $L_a^* R_g^* \omega = R_g^* L_a^* \omega = R_g^* \omega$. In particular, this form is nowhere zero, and there exists a function $h : G \to \mathbb{R} \setminus \{0\}$ such that $R_g^* \omega = h(g)\omega$, $g \in G$. Furthermore, $h : G \to (\mathbb{R} \setminus \{0\}, \cdot)$ is a Lie group homomorphism, because

$$h(ab)\omega = R_{ab}^* \omega = R_b^* R_a^* \omega = R_b^*(h(a)\omega) = h(a) R_b^* \omega = h(a)h(b)\omega.$$

Since $G$ is both compact and connected, $h \equiv h(e) = 1$.  $\square$

EXERCISE 36. Show that any connected Lie group is orientable.

EXERCISE 37. Prove the statement preceding Definition 14.3; namely, that there exists an open cover $\mathcal{U}$ of $M$, such that any $U \in \mathcal{U}$ is contained inside the image of some orientation-preserving imbedding $c : [0,1]^n \to M^n$.

EXERCISE 38. Let $M$ be an oriented $n$-manifold, and denote by $-M$ the manifold with the opposite orientation. Show that $\int_M \omega = -\int_{-M} \omega$ for any $n$-form $\omega$ with compact support on $M$.

EXERCISE 39. Let $f : M^n \to N^n$ be a diffeomorphism of oriented manifolds, and $\omega$ an $n$-form on $N$ with compact support. Prove that $\int_M f^* \omega = \pm \int_M \omega$, with the plus sign occuring iff $f$ is orientation-preserving.

EXERCISE 40. Show explicitly that the closed $n$-disk $D^n = \{p \in \mathbb{R}^n \mid |p| \leq 1\}$ is a $n$-manifold with boundary.

EXERCISE 41. Let $p \in \partial M$, and $(U, x)$, $(V, y)$ be two charts around $p$ with $x(p) = y(p) = 0$. Prove that $D_n(y^n \circ x^{-1})(0) > 0$. Deduce that the definition we gave of an outward-pointing vector $v \in M_p$ is independent of the chosen chart.

## 15. Some Applications of Stokes' Theorem

In this section, we derive the Poincaré Lemma, which implies that any closed form is locally exact, and also yields important topological applications. But we can already establish, as an immediate consequence of Stokes' theorem, that the top cohomology class of a (compact, oriented) manifold is nontrivial:

PROPOSITION 15.1. *Let $M^n$ be an oriented, compact manifold without boundary, $n > 0$. Then $H^n(M) \neq 0$.*

PROOF. Let $\omega$ be a *volume form* on $M$; i.e., $\omega$ is a nowhere-zero $n$-form with $\omega(v_1, \ldots, v_n) > 0$ for any positively oriented basis $v_1, \ldots, v_n$ of $M_p$, $p \in M$. If $c : [0,1]^n \to M$ is an orientation-preserving imbedding, then $c^*\omega = f \, du^1 \wedge \cdots \wedge du^n$ for some positive function $f$. Thus, $\int_c \omega$, and therefore also $\int_M \omega$ is positive. This means that the closed $n$-form $\omega$ is not exact, for if $\omega = d\eta$, then $\int_M \omega = \int_M d\eta = \int_{\partial M} \eta = 0$, since $\partial M$ is empty. $\qquad\square$

Proposition 15.1 will later be refined to show that $H^n(M)$ is isomorphic to $\mathbb{R}$ if $M$ is connected. In order to derive further consequences from Stokes' theorem, we will need the Poincaré Lemma, which deals with homotopic maps.

Recall that a manifold $M$ is said to be (smoothly) contractible if there exists some point $p_0$ in $M$ and a differentiable map $H : M \times [0,1] \to M$ such that $H \circ \iota_0 \equiv p_0$ and $H \circ \iota_1 \equiv 1_M$, where $\iota_t : M \to M \times [0,1]$ is given by $\iota_t(p) = (p,t)$ for $t \in [0,1]$ and $p \in M$. More generally, two maps $f_0, f_1 : M \to N$ are said to be (smoothly) *homotopic* if there exists a differentiable map $H : M \times [0,1] \to N$ such that $H \circ \iota_0 = f_0$ and $H \circ \iota_1 = f_1$. $H$ is then called a *homotopy* between $f_0$ and $f_1$. Thus, $M$ is contractible if the identity map $1_M$ is homotopic to a constant map. Our next aim is to show that homotopic maps induce the same cohomology homomorphism. Let $\pi_M$ and $t$ denote the projections of the product $M \times [0,1]$ onto the factors, so that $(\pi_{M*}, t_*) : (M \times [0,1])_{(p,t_0)} \to M_p \times [0,1]_{t_0}$ is an isomorphism. We will denote by $\tilde{D}$ the vector field on $M \times [0,1]$ given by $\tilde{D}(p,a) = (\pi_{M*}, t_*)_{(p,a)}^{-1} D_a$.

LEMMA 15.1. *Any $\omega \in A_k(M \times [0,1])$ can be uniquely written as $\omega = \omega_1 + dt \wedge \eta$, where $\omega_1$ and $\eta$ are $k$ and $(k-1)$-forms on $M \times [0,1]$ respectively such that $i(\tilde{D})\omega_1 = i(\tilde{D})\eta = 0$.*

PROOF. Because of the isomorphism $(\pi_{M*}, t_*)$, it suffices to check that if a vector space $V$ decomposes as $W \times \mathbb{R}$, then any $\omega \in \Lambda_k(V^*)$ can be uniquely written as $\omega_1 + \alpha \wedge \eta$, where $i(t)\omega_1, i(t)\eta = 0$ for $t \in \mathbb{R}$ and $\alpha \in \mathbb{R}^* \setminus \{0\}$. But this is clear, since if $\alpha_1, \ldots, \alpha_n$ is a basis of $W^*$, then $\alpha_1, \ldots, \alpha_n, \alpha$ is a basis of $V^*$. $\qquad\square$

Writing $\omega = \omega_1 + dt \wedge \eta \in A_k(M \times [0,1])$ as in Lemma 15.1, define a linear operator $I : A_k(M \times [0,1]) \to A_{k-1}(M)$ by

$$(15.1) \qquad I\omega(p)(v_1, \ldots, v_{k-1}) = \int_0^1 \eta(p,t)(\iota_{t*}v_1, \ldots, \iota_{t*}v_{k-1}) dt.$$

PROPOSITION 15.2 (Poincaré Lemma). *If $\omega$ is a $k$-form on $M \times [0,1]$, then $\iota_1^*\omega - \iota_0^*\omega = d(I\omega) + I(d\omega)$.*

PROOF. Let $x$ be a coordinate map on $M$ so that $(\bar{x}, t)$ is a coordinate map on $M \times [0, 1]$, where $\bar{x} = x \circ \pi_M$. By Lemma 15.1, the restriction of $\omega$ can be written as a sum of terms of two types: $f d\bar{x}^I = f d\bar{x}^{i_1} \wedge \cdots \wedge d\bar{x}^{i_k}$ (type (1)), and $f dt \wedge d\bar{x}^I = f dt \wedge d\bar{x}^{i_1} \wedge \cdots \wedge d\bar{x}^{i_{k-1}}$ (type (2)). Since $I$ is linear, it suffices to consider the case when $\omega$ is one of the above types.

If $\omega = f d\bar{x}^I$, then $d\omega = df \wedge d\bar{x}^I = $ (terms not involving $dt$) $+ (\partial f / \partial t) dt \wedge d\bar{x}^I$. Now, for any $t \in [0, 1]$, $\pi_M \circ \iota_t = 1_M$, so that $\iota_t^* d\bar{x}^i = d(\bar{x}^i \circ \iota_t) = dx^i$, and

$$I(d\omega)(p) \left( \frac{\partial}{\partial x^{j_1}}, \ldots, \frac{\partial}{\partial x^{j_k}} \right) = \int_0^1 \frac{\partial f}{\partial t}(p, t) d\bar{x}^I_{|(p,t)} \left( \iota_{t*} \frac{\partial}{\partial x^{j_1}}, \ldots, \iota_{t*} \frac{\partial}{\partial x^{j_k}} \right) dt$$

$$= \int_0^1 \frac{\partial f}{\partial t}(p, t) dx^I_{|p} \left( \frac{\partial}{\partial x^{j_1}}, \ldots, \frac{\partial}{\partial x^{j_k}} \right) dt.$$

Thus,

$$I(d\omega)(p) = \left[ \int_0^1 \frac{\partial f}{\partial t}(p, t) \, dt \right] dx^I(p) = [f(p, 1) - f(p, 0)] dx^I(p)$$

$$= \iota_1^* \omega(p) - \iota_0^* \omega(p),$$

which proves the result in this case, since $I\omega = 0$. Similarly, if $\omega = f dt \wedge d\bar{x}^I$, then $\iota_1^* \omega = \iota_0^* \omega = 0$, since $\iota_{t_0}^* dt = 0$ for any $t_0$. On the other hand,

$$I(d\omega)(p) = I\left( -\sum_{\alpha=1}^n \frac{\partial f}{\partial \bar{x}^\alpha} dt \wedge d\bar{x}^\alpha \wedge d\bar{x}^I \right)(p)$$

$$= -\sum_\alpha \left[ \int_0^1 \frac{\partial f}{\partial \bar{x}^\alpha}(p, t) \, dt \right] dx^\alpha \wedge dx^I,$$

while

$$d(I\omega)(p) = d\left( \left[ \int_0^1 f(p, t) \, dt \right] dx^I \right) = \sum_\alpha \frac{\partial}{\partial x^\alpha} \left( \int_0^1 f(p, t) \, dt \right) dx^\alpha \wedge dx^I,$$

so that $I(d\omega) + d(I\omega) = 0$. $\square$

The Poincaré Lemma is particularly useful when dealing with homotopic maps:

THEOREM 15.1. *If $f_0, f_1 : M \to N$ are homotopic, then the induced maps $f_0^*, f_1^* : H^k(N) \to H^k(M)$ are equal for all $k$.*

PROOF. Let $\omega \in A_k(N)$ be closed, and denote by $[\omega]$ its equivalence class in $H^k(N)$. If $H$ is a homotopy between $f_0$ and $f_1$, then

$$f_1^* \omega - f_0^* \omega = \iota_1^* H^* \omega - \iota_0^* H^* \omega = dIH^* \omega + IdH^* \omega = dIH^* \omega + IH^* d\omega = dIH^* \omega$$

is exact, so that $[f_1^* \omega] - [f_0^* \omega] = [f_1^* \omega - f_0^* \omega] = 0$. $\square$

If $M$ is contractible, then the identity map $1_M$ is homotopic to a constant map (sending all of $M$ to some point $p \in M$). Since the latter induces the trivial map at the cohomology level, we have as an immediate consequence of Theorem 15.1:

COROLLARY 15.1. *If $M^n$ is contractible, then $H^k(M) = 0$ for $1 \leq k \leq n$.*

In particular, any closed $k$-form $\alpha$ on a manifold is locally exact, as we stated earlier: Given $p \in M$, $H^k(U) = 0$ for any contractible neighborhood $U$ of $p$, so that the restriction of $\alpha$ to $U$ is exact. Several texts refer to this property as the Poincaré Lemma.

Corollary 15.1 and Proposition 15.1 now yield:

COROLLARY 15.2. *A compact oriented $n$-manifold with boundary is not contractible if $n > 0$.*

Corollary 15.1 says that cohomology cannot help us distinguish between contractible spaces of different dimensions. For this, we need the following concept:

DEFINITION 15.1. Let $M^n$ be a manifold. For $k \leq n$, the *de Rham cohomology vector spaces with compact support* $H_c^k(M)$ are the spaces $Z_c^k(M)/B_c^k(M)$, where $Z_c^k(M)$ denotes the space of all closed $k$-forms with compact support, and $B_c^k(M)$ the space of all $k$-forms $d\alpha$, where $\alpha \in A_{k-1}(M)$ has compact support.

$H^k(M)$ and $H_c^k(M)$ coincide of course when $M$ is compact. In general, though, not every exact $k$-form with compact support belongs to $B_c^k(M)$: If $f$ is a nonnegative function on $\mathbb{R}^n$ which is positive at some point and has ·compact support, then the $n$-form $\omega = f\,du^1 \wedge \cdots \wedge du^n$ is exact (because it is closed and $H^n(\mathbb{R}^n) = 0$). Since $\int_{\mathbb{R}^n} \omega > 0$, the following proposition shows it does not equal $d\alpha$ for any $\alpha$ with compact support:

PROPOSITION 15.3. *Let $M^n$ be connected and orientable (without boundary). Given $\omega \in Z_c^n(M)$, $\omega$ belongs to $B_c^n(M)$ iff $\int_M \omega = 0$.*

PROOF. If $\omega$ belongs to $B_c^n(M)$, then $\omega = d\alpha$, where $\alpha$ has compact support. By Stokes' theorem,

$$\int_M \omega = \int_M d\alpha = \int_{\partial M} \alpha = 0.$$

We merely illustrate the proof of the converse in the case $M = \mathbb{R}$: Suppose $\omega$ is a 1-form on $\mathbb{R}$ with compact support such that $\int_{\mathbb{R}} \omega = 0$. Since $H^1(\mathbb{R}) = 0$, $\omega = df$ for some function $f$ (which need not, a priori, have compact support). However, $df$ must vanish outside some interval $[-N, N]$, so that $f(t) = c_1$ when $t < -N$ and $f(t) = c_2$ when $t > N$ for some constants $c_1$ and $c_2$. Then

$$0 = \int_{\mathbb{R}} \omega = \int_{[-N-1,N+1]} df = \int_{\partial[-N-1,N+1]} f = c_2 - c_1,$$

so that $c_1 = c_2 = c$. Then $f - c$ has compact support, and $\omega = d(f - c)$. $\square$

THEOREM 15.2. *If $M^n$ is connected and orientable, then $H_c^n(M) \cong \mathbb{R}$.*

PROOF. Consider the linear transformation from $Z_c^n(M)$ to $\mathbb{R}$ which maps a closed $n$-form $\omega$ with compact support to $\int_M \omega \in \mathbb{R}$. This map is nontrivial (and hence onto) since for example if $(U, x)$ is a positively oriented chart around some $p \in M$, and $f$ a nonnegative function which is positive at $p$ and has support in $U$, then $\int_M \omega > 0$, where $\omega = f\,dx^1 \wedge \cdots \wedge dx^n$. By Proposition 15.3, its kernel is $B_c^n(M)$, and the statement follows. $\square$

Thus, for example, if $M^n$ is compact, connected, and orientable, then $H^n(M) \cong \mathbb{R}$. It can be shown that for $n \geq 1$, $H_c^n(M) = 0$ if $M$ is not orientable, and in general, $H^n(M) = H_c^n(M)$ if $M$ is compact and equals 0 otherwise.

Let $M_1^n$, $M_2^n$ be connected orientable, $f : M_1 \to M_2$ differentiable. We then have a linear transformation $\mathbb{R} \to \mathbb{R}$ such that the diagram

$$
\begin{array}{ccc}
H_c^n(M_2) & \xrightarrow{\;f^*\;} & H_c^n(M_1) \\
\cong \downarrow & & \downarrow \cong \\
\mathbb{R} & \longrightarrow & \mathbb{R}
\end{array}
$$

commutes. For $[\omega] \in H^n(M_2)$, the above diagram reads

$$
\begin{array}{ccc}
[\omega] & \longrightarrow & [f^*\omega] \\
\downarrow & & \downarrow \\
\int_{M_2} \omega & \longrightarrow & \int_{M_1} f^*\omega
\end{array}
$$

The bottom map must then be multiplication by some number $\deg f$, called the *degree of $f$*; i.e.,

$$
\int_{M_1} f^*\omega = (\deg f) \int_{M_2} \omega.
$$

This number can in many cases be computed as follows:

THEOREM 15.3. *Let $M_i^n$ be connected and orientable, $i = 1, 2$, and consider a proper map $f : M_1 \to M_2$. Suppose $q \in M_2$ is a regular value of $f$. For each $p \in f^{-1}(q)$, define the sign of $f$ at $p$ to be the number $\operatorname{sgn}_p f = +1$ if $f_{*p}$ is orientation-preserving, and $-1$ if it is orientation-reversing. Then*

$$
\deg f = \sum_{p \in f^{-1}(q)} \operatorname{sgn}_p f.
$$

EXAMPLES AND REMARKS 15.1. (i) Recall that $f$ is proper if the preimage of a compact set is compact. In particular, $f$ is proper whenever $M_1$ is compact.

(ii) Regular values always exist by Sard's theorem; in fact, their complement has measure 0. Notice that $\deg f = 0$ if $f$ is not onto.

PROOF OF THEOREM 15.3. By the inverse function theorem, $f^{-1}(q)$ consists of isolated points; being compact, it is a finite collection $\{p_1, \ldots, p_k\}$. Choose charts $(U_i, x_i)$ around $p_i$ such that each restriction $f : U_i \to V_i := f(U_i)$ is a diffeomorphism, and $U_i \cap U_j = \emptyset$. Then $V := \cap_i V_i$ is the domain of a chart $(V, y)$ around $q$, and redefining $U_i$ to be $U_i \cap f^{-1}(V)$, we still have diffeomorphisms $f : U_i \to V$. Let $g$ be a nonnegative function with compact support in $V$, and set $\omega = g\, dy^1 \wedge \cdots \wedge dy^n$. $f^*\omega$ then has support in $U_1 \cup \cdots \cup U_k$, so that by Exercise 39,

$$
\int_{M_1} f^*\omega = \sum_{i=1}^k \int_{U_i} f^*\omega = \sum_{i=1}^k (\operatorname{sgn}_{p_i} f) \int_V \omega = \sum_{i=1}^k (\operatorname{sgn}_{p_i} f) \int_{M_2} \omega.
$$

$\square$

We end this chapter with a couple of topological applications of Theorem 15.3:

COROLLARY 15.3. *If $n$ is even, then any vector field on $S^n$ vanishes somewhere.*

PROOF. We have seen in Section 14 that the antipodal map $f : S^n \to S^n$ is orientation-reversing when $n$ is even, so that $f$ has degree $-1$. By Theorems 15.1 and 15.3, $f$ is not homotopic to the identity $I$. But if $X$ were a nowhere-zero vector field on the sphere, it would induce a homotopy between $f$ and $I$: Recall that there is a canonical inner product on each tangent space (the one for which $\mathcal{J}_p : \mathbb{R}^n \to \mathbb{R}^n_p$ becomes a linear isometry for each $p$), so that we may assume $|X| \equiv 1$. Given $p \in S^n$, let $c_p$ denote the great circle $c_p(t) = (\cos \pi t)p + (\sin \pi t)\mathcal{J}_p^{-1}X(p)$. The desired homotopy $H$ is then given by $H(p,t) = c_p(t)$. $\square$

COROLLARY 15.4. $H^k(\mathbb{R}^n \setminus \{0\}) \cong H^k(S^{n-1})$ *for all $k$.*

PROOF. Let $r : \mathbb{R}^n \setminus \{0\} \to S^{n-1}$ denote the retraction $r(p) = p/|p|$, and $\imath : S^{n-1} \hookrightarrow \mathbb{R}^n \setminus \{0\}$ the inclusion. Then $r \circ \imath$ is the identity map on the sphere, and $\imath \circ r$ is homotopic to the identity map on $\mathbb{R}^n \setminus \{0\}$ via $H(p,t) = tp + (1-t)(\imath \circ r)(p)$. By Theorem 15.1, $(r \circ \imath)^* = \imath^* \circ r^*$ and $(\imath \circ r)^* = r^* \circ \imath^*$ are the identity on the respective cohomology spaces, so that $\imath^* : H^k(\mathbb{R}^n \setminus \{0\}) \to H^k(S^{n-1})$ is an isomorphism. $\square$

EXERCISE 42. Let $\omega$ be a 1-form on $M$ such that $\int_c \omega = 0$ for any closed curve $c$ in $M$. Show that $\omega$ is exact.

EXERCISE 43. $M$ is said to be *simply connected* if any closed curve $c : S^1 \to M$ is homotopic to a constant map. Use Exercise 42 to prove that if $M$ is simply connected, then $H^1(M) = 0$.

EXERCISE 44. Let $U = \mathbb{R}^3 \setminus \{(0,0,z) \mid z \geq 0\}$, $V = \mathbb{R}^3 \setminus \{(0,0,z) \mid z \leq 0\}$.
(a) Show that $U$ and $V$ are contractible.
(b) Suppose $\omega$ is a closed 1-form on $\mathbb{R}^3 \setminus \{0\}$. Show that there are functions $f : U \to \mathbb{R}$ and $g : V \to \mathbb{R}$, such that $\omega_{|U} = df$, $\omega_{|V} = dg$. Conclude that $\omega = dh$ for some function $h$, so that $H^1(\mathbb{R}^3 \setminus \{0\}) = 0$.
(c) Prove that $H^1(S^2) = 0$.

EXERCISE 45. This exercise generalizes Example 11.1, exhibiting a closed $(n-1)$-form on $\mathbb{R}^n \setminus \{0\}$ which is not exact. Let $P$ be the position vector field on $\mathbb{R}^n$, $P(p) = \mathcal{J}_p p$, and $\omega$ the $(n-1)$-form on $\mathbb{R}^n \setminus \{0\}$ given by $\omega = i(P)du^1 \wedge \cdots \wedge du^n$.
(a) Show that if $\imath : S^{n-1} \to \mathbb{R}^n \setminus \{0\}$ is inclusion, then $\tilde{\omega} := \imath^*\omega$ is the form which gives the standard orientation on the sphere.
(b) Let $r : \mathbb{R}^n \setminus \{0\} \to S^{n-1}$ denote the retraction $r(p) = p/|p|$, and $N : \mathbb{R}^n \setminus \{0\} \to \mathbb{R}$ the norm function $N(p) = |p|$. Prove that $r^*\tilde{\omega} = (1/N^n)\omega$.
(c) Show that $r^*\tilde{\omega}$ is a closed form on $\mathbb{R}^n \setminus \{0\}$ which is not exact.

EXERCISE 46. Show that the wedge product of forms induces a ring structure on $H^*(M) := \oplus_{i=0}^\infty H^i(M)$. The corresponding product is called the *cup product*. Prove that if $f : M \to N$ is differentiable, then $f^* : H^*(N) \to H^*(M)$ is a ring homomorphism.

# CHAPTER 2

# Fiber Bundles

## 1. Basic Definitions and Examples

We have already encountered examples of manifolds that possess some additional structure, such as the tangent bundle $TM$ of an $n$-dimensional manifold $M$. In this case, each point of $TM$ has a neighborhood diffeomorphic to a product $U \times \mathbb{R}^n$, where $U$ is an open set in $M$. Of course, $TM$ itself need not be diffeomorphic to $M \times \mathbb{R}^n$. In most of the sequel, we will be concerned with manifolds that, roughly speaking, look locally like products.

As usual, all maps are assumed to be differentiable.

DEFINITION 1.1. Let $F$, $M$, $B$ denote manifolds, $G$ a Lie group acting effectively on $F$ (i.e., if $g(p) = p$ for all $p \in F$, then $g = e$). A *coordinate bundle over the base space $B$ with total space $M$, fiber $F$, and structure group $G$* is a surjective map $\pi : M \to B$, called the *bundle projection*, together with a *bundle atlas* $\mathcal{A} = \{(\pi^{-1}(U_\alpha), (\pi, \phi_\alpha)\}_{\alpha \in A}$ on $M$; i.e.,

(1) $\{U_\alpha\}_{\alpha \in A}$ is an open cover of $B$.
(2) $(\pi, \phi_\alpha) : \pi^{-1}(U_\alpha) \to U_\alpha \times F$ is a diffeomorphism, called a *bundle chart*. Notice that for $p \in U_\alpha$, $\phi_{\alpha|\pi^{-1}(p)} : \pi^{-1}(p) \to F$ is a diffeomorphism. If $p$ also belongs to $U_\beta$, then $\phi_{\beta|\pi^{-1}(p)} : \pi^{-1}(p) \to F$ need not coincide with $\phi_\alpha$; however, they must differ by the operation of some element in $G$. To be specific:
(3) For $\alpha, \beta \in A$, there is a smooth map $f_{\alpha,\beta} : U_\alpha \cap U_\beta \to G$, called the *transition function from $\phi_\alpha$ to $\phi_\beta$* given by $f_{\alpha,\beta}(p) = \phi_\beta \circ (\phi_{\alpha|\pi^{-1}(p)})^{-1} : F \to F$; equivalently, $\phi_{\beta|\pi^{-1}(U_\alpha \cap U_\beta)} = (f_{\alpha,\beta} \circ \pi) \cdot \phi_{\alpha|\pi^{-1}(U_\alpha \cap U_\beta)}$.

Statement (3) says that the diagram

$$
\begin{array}{ccc}
\pi^{-1}(U_\alpha \cap U_\beta) & = \!\!=\!\!= & \pi^{-1}(U_\alpha \cap U_\beta) \\
(\pi, \phi_\alpha) \downarrow & & \downarrow (\pi, \phi_\beta) \\
(U_\alpha \cap U_\beta) \times F & \longrightarrow & (U_\alpha \cap U_\beta) \times F \\
(p, m) & \longrightarrow & (p, f_{\alpha,\beta}(p)\, m)
\end{array}
$$

commutes. Roughly speaking, the total space $M$ consists of a collection $\cup U_\alpha \times F$, where the $U_\alpha$'s cover $B$, and copies of $F$ belonging to intersecting $U_\alpha$'s are identified by means of elements of $G$. The projection $\pi$ is a submersion by (2). The set $\pi^{-1}(p)$ is called the *fiber over $p$*. Notice that (3) implies that $f_{\alpha,\alpha} = e$, $(f_{\alpha,\beta}(p))^{-1} = f_{\beta,\alpha}(p)$, and $f_{\alpha,\gamma}(p) = f_{\beta,\gamma}(p) \cdot f_{\alpha,\beta}(p)$.

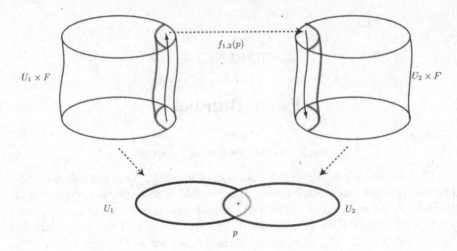

FIGURE 1

DEFINITION 1.2. A (real) *coordinate vector bundle of rank n* is a coordinate bundle with fiber $\mathbb{R}^n$ and structure group $GL(n)$ (or a subgroup of $GL(n)$).

We will often use Greek letters such as $\xi$ to denote a bundle $\pi : M \to B$. If $\xi$ is a rank $n$ vector bundle, then each fiber $\pi^{-1}(b)$, $b \in B$, is a vector space: Given a bundle chart $(\pi^{-1}(U), (\pi, \phi))$ with $b \in U$, define the vector space operations on the fiber over $b$ so that $\phi_{|\pi^{-1}(b)} : \pi^{-1}(b) \to \mathbb{R}^n$ becomes an isomorphism. The vector space structure is independent of the chosen chart because any transition function $f_{\phi,\psi}(p)$ at $p$ is an isomorphism of $\mathbb{R}^n$.

EXAMPLES AND REMARKS 1.1. (i) The *trivial bundle* with base space $B$ and fiber $F$ is the projection $\pi : B \times F \to B$ onto the first factor. The structure group is $\{1_F\}$. In general, the size of the structure group measures how twisted the bundle is.

(ii) The tangent bundle $TM$ of an $n$-dimensional manifold $M$ is the total space of a rank $n$ vector bundle over $M$ with the bundle projection $\pi : TM \to M$ from Chapter 1: If $\{(U_\alpha, x_\alpha)\}$ is an atlas on $M$, then $\{(U_\alpha, (\pi, \phi_\alpha)\}$ is a bundle atlas on $TM$, where

$$\phi_\alpha = (dx_\alpha^1, \ldots, dx_\alpha^n) : \pi^{-1}(U_\alpha) \to \mathbb{R}^n.$$

The transition function from $\phi_\alpha$ to $\phi_\beta$ is $f_{\alpha,\beta} = D(x_\beta \circ x_\alpha^{-1}) \circ x_\alpha : U_\alpha \cap U_\beta \to GL(n)$. A similar argument applies to the tensor bundles $T_{r,s}(M)$ and the exterior bundles $\Lambda_k^*(M)$. The tangent bundle of $M$ will be denoted $\tau M$ to distinguish it from its total space $TM$ (which, to confuse things further, is traditionally also referred to by the same name).

(iii) The *Hopf fibration* (see also Chapter 1, Examples and Remarks 9.1(i)): View $S^{2n+1} \subset \mathbb{R}^{2n+2} = \mathbb{C}^{n+1}$ as the set of all $(z_1, \ldots, z_{n+1}) \in \mathbb{C}^{n+1}$ such that $\sum |z_i|^2 = 1$, and consider the free action of $S^1 = \{z \in \mathbb{C} \mid |z| = 1\}$ on $S^{2n+1}$ given by $z(z_1, \ldots, z_{n+1}) = (z_1 z, \ldots, z_{n+1} z)$. Since $S^1$ is compact, this action is proper, and by Chapter 1, Theorem 14.2, there exists a unique differentiable structure on the quotient $S^{2n+1}/S^1$ for which the projection becomes a submersion. The quotient is called *complex projective n-space* $\mathbb{C}P^n$ of dimension $2n$.

We claim that $\pi : S^{2n+1} \to \mathbb{C}P^n$ is the projection of a bundle with fiber and group $S^1$, called the *Hopf fibration*. In order to establish this, we exhibit an atlas of bundle charts satisfying Definition 1.1: for $i = 1, \ldots, n+1$, define $\hat{U}_i = \{(z_1, \ldots, z_{n+1}) \in S^{2n+1} \mid z_i \neq 0\}$, and $U_i = \pi(\hat{U}_i) \subset \mathbb{C}P^n$. It is easily checked that $\hat{U}_i = \pi^{-1}(U_i)$, so that $\{U_i\}$ is an open cover of $\mathbb{C}P^n$. Define $\phi_i : \hat{U}_i \to S^1$ by $\phi_i(z_1, \ldots, z_{n+1}) = z_i/|z_i|$. Then $(\pi, \phi_i) : \hat{U}_i \to U_i \times S^1$ is a diffeomorphism with inverse $(\pi(w_1, \ldots, w_{n+1}), z) \mapsto (z|w_i|/w_i)(w_1, \ldots, w_{n+1})$, and the transition function $f_{i,j} : U_i \cap U_j \to S^1$ is given by $f_{i,j}(\pi(z_1, \ldots, z_{n+1})) = z_j z_i^{-1} |z_i| |z_j|^{-1}$.

DEFINITION 1.3. Let $\pi_i : M_i \to B_i$ be two coordinate bundles with fiber $F$ and group $G$. A differentiable map $h : M_1 \to M_2$ is said to be a *bundle map* if

(1) $h$ maps each fiber $\pi_1^{-1}(p_1)$ diffeomorphically onto a fiber $\pi_2^{-1}(p_2)$, thereby inducing a differentiable map $\bar{h} : B_1 \to B_2$ such that $\pi_2 \circ h = \bar{h} \circ \pi_1$; and

(2) for any bundle charts $(\pi_1^{-1}(U_\alpha), (\pi_1, \phi_\alpha))$ and $(\pi_2^{-1}(V_\beta), (\pi_2, \psi_\beta))$ of $\pi_1$ and $\pi_2$, respectively, $p \in U_\alpha \cap \bar{h}^{-1}(V_\beta)$, the map $\psi_\beta \circ h \circ (\phi_{\alpha|\pi_1^{-1}(p)})^{-1}$ from $F$ to $F$ coincides with the operation of an element of $G$, and the resulting map

$$f_{\alpha,\beta} : U_\alpha \cap \bar{h}^{-1}(V_\beta) \to G,$$

$$p \mapsto \psi_\beta \circ h \circ (\phi_{\alpha|\pi_1^{-1}(p)})^{-1}$$

is differentiable.

The two coordinate bundles are said to be *equivalent* if $B_1 = B_2$ and the induced map is the identity on the base. A *fiber bundle* is then defined to be an equivalence class of coordinate bundles. Alternatively, one could define it to be a coordinate bundle with a maximal atlas.

Notice that if $h$ is a bundle map, then by the second condition above, the coordinate bundle over $B_1$ with bundle charts of the form $(\pi_1^{-1}(\bar{h}^{-1}(U)), (\pi_1, \phi \circ h))$, where $(\pi_2^{-1}(U), (\pi_2, \phi))$ is a bundle chart of $\pi_2$, is equivalent to $\pi_1$. Its transition functions $f_{\phi \circ h, \psi \circ h}$ are equal to $f_{\phi,\psi}^2 \circ \bar{h}$, where $f_{\phi,\psi}^2$ are the transition functions of $\pi_2$.

EXERCISE 47. (a) Show that the functions $f_{\alpha,\beta}$ from Definition 1.3 satisfy $f_{\alpha,\gamma} = f_{\beta,\gamma} \cdot f_{\alpha,\beta}^1$ and $f_{\alpha,\gamma} = f_{\beta,\gamma}^2 \cdot f_{\alpha,\beta}$, where $f_{\alpha,\beta}^1$ and $f_{\beta,\gamma}^2$ are transition functions for $\pi_1$ and $\pi_2$, respectively.

(b) Conversely, suppose $\pi_i : M_i \to B$ are two coordinate bundles over $B$ with fiber $F$ and group $G$. Show that if there is a collection of maps $f_{\alpha,\beta}$ as in Definition 1.3 satisfying the identities in (a), then the bundles are equivalent.

EXERCISE 48. Identify $S^{4n+3} \subset \mathbb{R}^{4n+4} = \mathbb{H}^{n+1}$ with the set of all $(n+1)$-tuples of quaternions $(q_1, \ldots, q_{n+1})$ such that $\sum |q_i|^2 = 1$ (see Chapter 1, Example 8.1(iii)). Replace complex numbers by quaternions in Examples and remarks 1.1 (iii) to construct *quaternionic projective space* $\mathbb{H}P^n$ and a fiber bundle $\pi : S^{4n+3} \to \mathbb{H}P^n$ over $\mathbb{H}P^n$ with fiber and group $S^3$. This bundle is often called a *generalized Hopf fibration*.

EXERCISE 49. (a) Show that $\mathbb{C}P^1$ is diffeomorphic to the 2-sphere via

$$\mathbb{C}P^1 \longrightarrow S^2,$$

$$[z_1, z_2] \longmapsto \frac{1}{|z_1|^2 + |z_2|^2}(2z_1\bar{z}_2, |z_1|^2 - |z_2|^2).$$

(b) Show that $\mathbb{H}P^1$ from Exercise 48 may similarly be identified with $S^4$. Thus, for $n = 1$, the Hopf fibrations become $S^3 \to S^2$ with fiber $S^1$, and $S^7 \to S^4$ with fiber $S^3$.

## 2. Principal and Associated Bundles

The Hopf fibration discussed in the previous section is a prime example of the following key concept:

DEFINITION 2.1. A fiber bundle $\pi : P \to B$ with fiber and group $G$ is called a *principal G-bundle* if there exists a free right action of $G$ on $P$ and an atlas such that for each bundle chart $(\pi^{-1}(U), (\pi, \phi))$, the map $\phi : \pi^{-1}(U) \to G$ is $G$-equivariant; i.e.,

$$(\pi, \phi)(pg) = (\pi(p), \phi(p)g), \qquad p \in \pi^{-1}(U), \quad g \in G.$$

It follows that $B$ is the quotient space $P/G$: Since $\pi(pg) = \pi(p)$, the orbit $G(p) = \{pg \mid g \in G\}$ of $p$ is contained in $\pi^{-1}(\pi(p))$; conversely, if $(\pi, \phi)$ is a bundle chart around $p$, then for $q \in \pi^{-1}(\pi(p))$,

$$q = (\pi, \phi)^{-1}(\pi(q), \phi(q)) = (\pi, \phi)^{-1}(\pi(p), \phi(p)\phi(p)^{-1}\phi(q))$$

$$= (\pi, \phi)^{-1}(\pi(p), \phi(p)g) = pg,$$

where $g = \phi(p)^{-1}\phi(q) \in G$. Furthermore, the structure group is $G$ acting on itself by left translations: for $p \in P$,

$$f_{\phi,\psi}(\pi(p)) = \psi(p)\phi(p)^{-1},$$

where the choice of the element $p \in \pi^{-1}(\pi(p))$ is irrelevant because

$$\psi(pg)\phi(pg)^{-1} = \psi(p)g(\phi(p)g)^{-1} = \psi(p)gg^{-1}\phi(p)^{-1} = \psi(p)\phi(p)^{-1}.$$

EXAMPLES AND REMARKS 2.1. (i) The Hopf fibrations $S^{2n+1} \to \mathbb{C}P^n$ and $S^{4n+3} \to \mathbb{H}P^n$ are principal $S^1$ and $S^3$ bundles.

(ii) The *trivial principal G-bundle over B* is the projection $B \times G \to B$ onto the first factor. The action of $G$ is by right multiplication $(b, g_1)g = (b, g_1g)$ on the second factor.

(iii) Let $G$ be a Lie group, $H$ a closed subgroup of $G$, and denote by $B$ the homogeneous space $G/H$. We first show that the quotient space $G^n/H^k$ admits a (unique) differentiable structure of dimension $n - k$ for which the projection $\pi : G \to G/H$ becomes a submersion. This actually follows from Theorem 14.2 in Chapter 1, but we provide an independent argument, since that theorem won't be proved until Chapter 5. Observe that $\pi$ is an open map for the quotient topology on $G/H$: If $U$ is open in $G$, then so is $\pi(U)$ (in $G/H$), because $\pi^{-1}(\pi(U)) = \cup_{h \in H} R_h(U)$ is open in $G$. Furthermore, the quotient space is Hausdorff: If $\pi(a) \neq \pi(b)$, so that $a^{-1}b \notin H$, there exists a neighborhood of $a^{-1}b$ that does not intersect $H$. Such a neighborhood always contains an open set of the form $U \cdot a^{-1}b \cdot U$, where $U$ is a neighborhood of the identity with $U = U^{-1}$. Then $Ua^{-1}bU \cap H = \emptyset$, which implies that $bUH \cap aUH = \emptyset$. Thus, $\pi \circ L_b(U)$ and $\pi \circ L_a(U)$ are disjoint open sets containing $\pi(b)$ and $\pi(a)$, respectively.

In order to exhibit a manifold structure on $G/H$, recall that Frobenius' theorem applied to the distribution $L_{g*}H_e$, $g \in G$, guarantees the existence of a chart $(U, x)$ around $e$, with $x(U) = (0, 1)^n$, such that each slice

$$\{g \in U \mid x^{k+1}(g) = a_1, \ldots, x^n(g) = a_{n-k}\}$$

is contained in a left coset of $H$. If $S$ denotes the slice containing $e$, there exists a neighborhood $V$ of $e$ such that $V \cap S = V \cap H$ (since $H$ is a submanifold of $G$), and $V = V^{-1}$, $V \cdot V \subset U$. For the sake of simplicity, denote $V$ by $U$ again. Let $N = (\pi_1 \circ x)^{-1}(a)$, where $\pi_1 : \mathbb{R}^n \to \mathbb{R}^k \times 0$ denotes projection, and $a := \pi_1 \circ x(e)$. We claim that $\pi$ is one-to-one when restricted to $N$: Indeed, if $\pi(a) = \pi(b)$, then $a^{-1}b \in U \cap H = U \cap S$, so that $b$ belongs to $L_a(U \cap S)$. The latter set, being connected, is contained in a single slice. Since it also contains $a$, $a$ and $b$ lie in the same slice, so that $x(a) = x(b)$; i.e, $a = b$.

It follows that $\pi_{|N} : N \to W := \pi(N)$ is an open, bijective map, hence a homeomorphism. So is $\tilde{x} := \pi_2 \circ x \circ (\pi_{|N})^{-1} : W \to \tilde{x}(W) \subset 0 \times \mathbb{R}^{n-k}$, where $\pi_2 : \mathbb{R}^n \to 0 \times \mathbb{R}^{n-k}$ denotes the projection onto the other factor. We may then take $(W, \tilde{x})$ as a chart around $\pi(e)$. In order to produce a chart around $\pi(a)$, consider the homeomorphism $\mathbb{L}_a$ of $G/H$ induced by left-multiplication by $a$ in $G$, $\mathbb{L}_a(\pi(g)) := \pi(ag)$. The desired chart is then given by $(\mathbb{L}_a(W), \tilde{x} \circ \mathbb{L}_{a^{-1}})$. Given $b \in G$, the corresponding transition function is $\pi_2 \circ x_{|N} \circ L_{a^{-1}b} \circ (\pi_2 \circ x_{|N})^{-1}$, so that the collection $\{(\mathbb{L}_a(W), \tilde{x} \circ \mathbb{L}_{a^{-1}}) \mid a \in G\}$ induces a differentiable structure on $G/H$.

It remains to check that $\pi$ is differentiable at $g \in G$. Using the charts $(L_g(U), x \circ L_{g^{-1}})$ around $g$ and $(\mathbb{L}_g(W), \tilde{x} \circ \mathbb{L}_{g^{-1}})$ around $\pi(g)$, we have

$$\tilde{x} \circ \mathbb{L}_{g^{-1}} \circ \pi \circ (x \circ L_{g^{-1}})^{-1} = \tilde{x} \circ \mathbb{L}_{g^{-1}} \circ \pi \circ L_g \circ x^{-1} = \tilde{x} \circ \pi \circ x^{-1} = \pi_2,$$

which establishes the claim.

Finally, we show that $\pi : G \to G/H$ is a principal $H$-bundle: Notice that for any $[g] := \pi(g)$, there exists a neighborhood $U = \mathbb{L}_g(W)$ of $[g]$ on which $\pi$ has a right inverse $s_U$. In fact, taking $s_U = L_g \circ (\pi_{|N})^{-1} \circ \mathbb{L}_{g^{-1}|U}$, we have $\pi \circ s_U = 1_U$. Then the map

$$U \times H \to \pi^{-1}(U),$$

$$([g], h) \mapsto (s_U[g]) \cdot h$$

is a diffeomorphism. Its inverse is of the form $(\pi, \phi_U)$, where $\phi_U : \pi^{-1}(U) \to H$ is $H$-equivariant, since $\phi_U(g) = s_U(\pi(g))^{-1}g$, so that

$$\phi_U(gh) = s_U(\pi(gh))^{-1}gh = (s_U(\pi(g))^{-1} \cdot g)h = \phi_U(g)h.$$

Thus, the collection of such maps $(\pi, \phi_U)$ forms a principal bundle atlas on $G$ over $B$.

We have seen that given a fiber bundle $\pi : M \to B$ with fiber $F$ and group $G$, a bundle atlas $\{(\pi^{-1}(U_\alpha), (\pi, \phi_\alpha))\}$ determines a family of transition functions $f_{\alpha,\beta} : U_\alpha \cap U_\beta \to G$ which satisfy $f_{\alpha,\gamma} = f_{\beta,\gamma} \cdot f_{\alpha,\beta}$. It turns out that the bundle may be reconstructed from these transition functions. More generally, one has the following:

PROPOSITION 2.1. *Let $\{U_\alpha\}_{\alpha \in A}$ be an open cover of a manifold $B$, and $G$ a Lie group acting effectively on a manifold $F$. Suppose there is a collection of maps $f_{\alpha,\beta} : U_\alpha \cap U_\beta \to G$ such that*

$$(2.1) \qquad f_{\alpha,\gamma}(p) = f_{\beta,\gamma}(p) \cdot f_{\alpha,\beta}(p), \qquad p \in U_\alpha \cap U_\beta \cap U_\gamma, \quad \alpha, \beta, \gamma \in A.$$

*Then there exists a fiber bundle $\pi : M \to B$ with fiber $F$, structure group $G$, and a bundle atlas whose transition functions are the given collection $\{f_{\alpha,\beta}\}$. Furthermore, if $F = G$ and $G$ acts on itself by left translations, then the atlas is a principal bundle atlas.*

Notice that taking $\alpha = \beta = \gamma$ in (2.1) implies that $f_{\alpha,\alpha} \equiv e$. Taking $\alpha = \gamma$ then yields $f_{\alpha,\beta}^{-1} = f_{\beta,\alpha}$.

PROOF. Consider the disjoint union $\cup_{\alpha \in A}(U_\alpha \times F)$, and the quotient space $M$ under the equivalence relation:

$$(p, q_1) \sim (p, q_2) \text{ iff } q_2 = f_{\alpha,\beta}(p) q_1 \text{ for some } \alpha, \beta \in A.$$

If $\rho : \cup_\alpha(U_\alpha \times F) \to M$ denotes the projection, then each restriction $\rho : U_\alpha \times F \to \rho(U_\alpha \times F)$ is a homeomorphism, and its inverse $(\pi, \phi_\alpha)$ may be taken as a bundle chart. By construction, the transition functions of this atlas are the $f_{\alpha,\beta}$.                                                                              □

As a simple application, consider the group $G = \{\pm 1\}$ acting on $\mathbb{R}$ by multiplication. The circle $B = S^1$ of unit complex numbers admits $U_1 = S^1 \setminus \{-i\}$ and $U_2 = S^1 \setminus \{i\}$ as open cover. Then the map

$$f_{1,2} : U_1 \cap U_2 \to G,$$

$$z \mapsto \begin{cases} 1, & \text{if } \operatorname{Re} z > 0, \\ -1, & \text{otherwise,} \end{cases}$$

determines a rank 1 vector bundle over the circle, called a *Moebius band*.

DEFINITION 2.2. Let $\pi : M \to B$ be a fiber bundle with fiber $F$ and group $G$. The principal $G$-bundle obtained as in Proposition 2.1 from the transition functions of $\pi$ is called the *principal bundle associated to $\pi$*.

Thus, a fiber bundle with group $G$ induces an associated principal $G$-bundle. One can recover the original bundle from the principal one: More generally, let $\pi_P : P \to B$ be a principal $G$-bundle, $F$ a manifold on which $G$ acts effectively on the left. Define an equivalence relation $\sim$ on the space $P \times F$ by setting $(p, m) \sim (pg, g^{-1}m)$, and denote the quotient space $(P \times F)/\sim$ by $P \times_G F$. There is a well-defined map $\pi : P \times_G F \to B$ given by $\pi[p, m] = \pi_P(p)$.

THEOREM 2.1. *Let $\pi_P : P \to B$ be a principal $G$-bundle, $F$ a manifold on which $G$ acts on the left. Then the map $\pi : P \times_G F \to B$ constructed above is a fiber bundle over $B$ with fiber $F$ and structure group $G$, called the fiber*

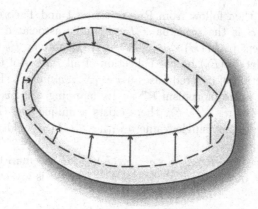

FIGURE 2. A Moebius band.

bundle with fiber $F$ associated to the principal bundle $\pi_P : P \to B$ and the given action of $G$. *Furthermore, the principal $G$-bundle associated to $\pi$ is $\pi_P$.*

PROOF. If $(\pi_P, \phi) : \pi_P^{-1}(U) \to U \times G$ is a principal bundle chart, define $\bar{\phi} : \pi^{-1}(U) \to U \times F$ by $\bar{\phi}[p,m] = \phi(p)m$. We claim that $(\pi, \bar{\phi})$ is a candidate for a bundle chart; i.e., it is invertible. Indeed, define $s : U \to \pi_P^{-1}(U)$ by $s(b) = (\pi_P, \phi)^{-1}(b, e)$; then $f : U \times F \to \pi^{-1}(U)$, where $f(b,m) = [s(b), m]$, is the inverse of $(\pi, \bar{\phi})$ : On the one hand,

$$(\pi, \bar{\phi}) \circ f(b,m) = (\pi, \bar{\phi})[s(b), m] = (\pi_P(s(b)), \phi(s(b))m) = (b, m);$$

on the other, given $p \in \pi_P^{-1}(U)$, we have $s(\pi_P(p)) = p\phi(p)^{-1}$, so that

$$f \circ (\pi, \bar{\phi})[p,m] = f(\pi_P(p), \phi(p)m) = [s(\pi_P(p)), \phi(p)m]$$
$$= [p\phi(p)^{-1}, \phi(p)m] = [p, m].$$

This establishes the claim. Since both $(\pi, \bar{\phi})$ and $f$ are continuous, they are homeomorphisms. Now, let $(\pi_P, \phi)$ and $(\pi_P, \psi)$ be two principal bundle charts with overlapping domains. Given $b$ in the projection of their intersection and $m \in F$, the transition function of the (candidates for) associated bundle charts at $b$ is given by

$$f_{\bar{\phi}, \bar{\psi}}(b)m = \bar{\psi} \circ (\pi, \bar{\phi})^{-1}(b, m) = \bar{\psi}[(\pi_P, \phi)^{-1}(b, e), m]$$
$$= (\psi \circ (\pi_P, \phi)^{-1}(b, e))m = (f_{\phi, \psi}(b)e)m$$
$$= f_{\phi, \psi}(b)m.$$

The collection of charts therefore induces a differentiable structure on $P \times_G F$ and satisfies the requirements for a bundle atlas. Since the transition functions of the bundle coincide with those of $\pi_P$, $\pi_P$ is the principal $G$-bundle associated to $\pi$. $\square$

EXAMPLE 2.1 (The Frame Bundle of a Vector Bundle). Let $\pi : E \to B$ denote a rank $n$ vector bundle over $B$. We shall construct a principal $GL(n)$-bundle $\pi_P : Fr(E) \to B$, called the *frame bundle of $E$*, with the same transition functions. It will then follow from Proposition 2.1 and Theorem 2.1 that the frame bundle of $E$ is the principal $GL(n)$-bundle associated to $\pi$, and that $E \to B$ is equivalent to $Fr(E) \times_{GL(n)} \mathbb{R}^n \to B$. Denote by $E_b$ the fiber $\pi^{-1}(b)$ over $b \in B$, and let $Fr(E_b)$ be the collection of all *frames* of the vector space $E_b$; i.e., the collection of ordered bases $p = (v_1, \ldots, v_n)$ of $E_b$. Each such frame can be viewed as an isomorphism $\mathbb{R}^n \to E_b$ mapping $\mathbf{e}_i$ to $v_i$ for $1 \le i \le n$. Given two frames $p_i : \mathbb{R}^n \to E_b$, there exists a unique $g \in GL(n)$ such that $p_1 = p_2 g$. Identifying any single frame $p$ with $e \in GL(n)$ yields a bijective map $Fr(E_b) \leftrightarrow GL(n)$.

Let $Fr(E) := \cup_{b \in B} Fr(E_b)$, $\pi_P : Fr(E) \to B$ the map that assigns the point $b$ to a frame of $E_b$. If $(\pi, \bar{\phi}) : \pi^{-1}(U) \to U \times \mathbb{R}^n$ is a vector bundle chart for $E$, define

$$\phi : \pi_P^{-1}(U) \to G = GL(n),$$

$$p \mapsto \bar{\phi}_{|E_{\pi_P(p)}} \circ p.$$

$\phi$ is $G$-equivariant by construction, and $(\pi_P, \phi) : \pi_P^{-1}(U) \to U \times G$ is therefore bijective. Given another vector bundle chart $(\pi, \bar{\psi})$ over $U$, we have

$$(\pi_P, \psi) \circ (\pi_P, \phi)^{-1}(b, g) = (\pi_P, \psi)(\bar{\phi}_{|E_b}^{-1} g) = (b, \bar{\psi} \circ (\bar{\phi}_{|E_b}^{-1})g).$$

The collection of maps $(\pi_P, \phi)$ therefore induces a differentiable structure on $Fr(E)$ and forms a principal bundle atlas with transition functions $f_{\phi, \psi} = f_{\bar{\phi}, \bar{\psi}}$. This establishes the claim.

Notice that there is an explicit equivalence between $Fr(E) \times_{GL(n)} \mathbb{R}^n \to B$ and $\pi$: if $p = (v_1, \ldots, v_n)$ is a frame of $E_b$, the equivalence maps $[p, (\alpha_1, \ldots, \alpha_n)] \in Fr(E) \times_G \mathbb{R}^n$ to $\sum \alpha_i v_i \in E_b$.

EXERCISE 50. Consider a principal $G$-bundle $\pi_P : P \to B$ and an associated bundle $\pi : P \times_G F \to B$.

(a) Show that $\rho : P \times F \to P \times_G F$, where $\rho(p, q) = [p, q]$, is a principal $G$-bundle, and that the projection $\pi_1 : P \times F \to P$ onto the first factor is a $G$-equivariant map inducing $\pi$ on the base spaces.

$$
\begin{array}{ccc}
P \times F & \xrightarrow{\pi_1} & P \\
\rho \downarrow & & \downarrow \pi_P \\
P \times_G F & \xrightarrow{\pi} & B
\end{array}
$$

(b) Show that for any $p \in P$, the map $F \to \pi^{-1}(\pi(p))$ given by $q \mapsto \rho(p, q)$ is a diffeomorphism.

(c) If $F$ is a vector space and $G$ acts linearly on $F$, show that $\pi$ is a vector bundle.

EXERCISE 51. Let $H$ be a Lie subgroup of $G$, $\pi_G : P_G \to B$, $\pi_H : P_H \to B$ two principal $G$ and $H$-bundles over $B$ respectively with $P_H \subset P_G$. $P_H \to B$ is said to be a *principal subbundle* of $P_G \to B$ if for any $b \in B$, there exists a neighborhood $U$ of $b$ and principal bundle charts $(\pi_H, \phi)$, $(\pi_G, \psi)$ of $\pi_H$, $\pi_G$ over $U$ such that

$$(\pi_G, \psi) \circ (\pi_H, \phi)^{-1} : U \times H \to U \times G$$

is the inclusion map. Show that in this case, given an action of $G$ on a manifold $F$, the total spaces $P_H \times_H F$ and $P_G \times_G F$ of the associated $F$-bundles are diffeomorphic via a fiber preserving map. We say the structure group $G$ of $P_G \times_G F \to B$ is *reducible* to $H$.

EXERCISE 52. Prove that the structure group $G$ of a bundle is reducible to $H$ (see Exercise 51) iff the bundle admits an atlas with $H$-valued transition functions.

## 3. The Tangent Bundle of $S^n$

In this section, we apply some of the concepts introduced above to discuss a basic example, that of the tangent bundle of the $n$-sphere.

The standard action of $SO(n+1)$ on $S^n$ yields a map $SO(n+1) \to S^n$ that sends $g \in SO(n+1)$ to $g(\mathbf{e}_1)$. The subgroup of $SO(n+1)$ acting trivially on $\mathbf{e}_1$ may be identified with $SO(n)$, and one has an induced map

$$SO(n+1)/SO(n) \to S^n,$$
$$[g] \mapsto g\mathbf{e}_1,$$

which is one-to-one by construction. It is also onto since $SO(n+1)$ acts transitively on the unit sphere. This map is a homeomorphism ($SO(n+1)/SO(n)$ being compact) which is easily checked to be a diffeomorphism. By Examples and Remarks 2.1(iii), $SO(n+1) \to S^n$ is a principal $SO(n)$-bundle.

On the other hand, the tangent bundle of $S^n$ is a vector bundle with group $GL(n)$. For $p \in S^n$, the derivative of the inclusion map $S^n \hookrightarrow \mathbb{R}^{n+1}$ induces an inner product on the tangent space of $S^n$ at $p$. By requiring the second component $\phi$ of each bundle chart $(\pi, \phi)$ to be a linear isometry $\phi_{|S_p^n} : S_p^n \to \mathbb{R}^n$, we obtain a reduction of the structure group to $O(n)$; cf. Exercise 52. Since the sphere is orientable, the group may further be reduced to $SO(n)$. (More generally, we will see in the next section that any vector bundle admits a reduction of its structure group $GL(n)$ to $O(n)$. The bundle is said to be *orientable* if its structure group is further reducible to $SO(n)$. The Moebius band from the preceding section is an example of a nonorientable bundle.)

In terms of principal bundles, we are reducing the frame bundle $Fr(TS^n)$ of Example 2.1 to the $SO(n)$-subbundle $SO(TS^n) \to S^n$ of *oriented orthonormal frames* whose fiber over $p \in S^n$ consists of all positively oriented orthonormal frames of $S_p^n$.

We claim that $SO(TS^n) \to S^n$ is equivalent to $SO(n+1) \to S^n$: In fact, the map $f : SO(n+1) \to SO(TS^n)$ which sends $g \in SO(n+1)$ to the ordered orthonormal frame $(\mathcal{J}_{g\mathbf{e}_1} g\mathbf{e}_2, \ldots, \mathcal{J}_{g\mathbf{e}_1} g\mathbf{e}_{n+1})$ of $S_{g\mathbf{e}_1}^n$ induces the identity on $S^n$. Its inverse maps an orthonormal frame $v_1, \ldots, v_n$ of $S_p^n$ to the element $g \in SO(n+1)$ defined by $g\mathbf{e}_1 = p$, $g\mathbf{e}_{i+1} = \mathcal{J}_p^{-1} v_i$, $1 \le i \le n$. Since $f$ is $SO(n)$-equivariant, the claim now follows from the following theorem:

THEOREM 3.1. *Let $\pi_i : P_i \to B$, $i = 1, 2$, be two principal $G$-bundles over $B$. If $h : P_1 \to P_2$ is a $G$-equivariant map inducing the identity on $B$, then the two bundles are equivalent.*

PROOF. If $(\pi_1^{-1}(U_\alpha), (\pi_1, \phi_\alpha))$ and $(\pi_2^{-1}(V_\beta), (\pi_2, \psi_\beta))$ are bundle charts of $\pi_1$ and $\pi_2$, there are smooth maps $\pi_1^{-1}(U_\alpha) \to G$ and $\pi_1^{-1}(V_\beta) \to G$ given by $p \mapsto \phi_\alpha(p)$ and $q \mapsto (\psi_\beta \circ h)(q)$ respectively. Thus, the assignment

$$f_{\alpha,\beta} : U_\alpha \cap V_\beta \to G,$$

$$b \mapsto (\psi_\beta \circ h)(p)\phi_\alpha(p)^{-1},$$

where $p$ is any element of the fiber over $b$, is a well-defined smooth map. The bundles are then equivalent by Definition 1.3, since $f_{\alpha,\beta}(b) = \psi_\beta \circ h \circ (\phi_{\alpha | \pi_1^{-1}(b)})$: Indeed, let $g \in G$, $a := \phi_\alpha(p)$; then $g = aa^{-1}g = \phi_\alpha(p)a^{-1}g = \phi_\alpha(pa^{-1}g)$, so that

$$\psi_\beta \circ h \circ (\phi_{\alpha | \pi_1^{-1}(b)})^{-1}(g) = \psi_\beta \circ h(pa^{-1}g) = \psi_\beta(h(p)) \cdot a^{-1}g$$

$$= (\psi_\beta \circ h)(p) \cdot \phi_\alpha(p)^{-1} \cdot g.$$

$\square$

COROLLARY 3.1. *Let $\pi_i : E_i \to B$, $i = 1, 2$, be two rank $n$ vector bundles over $B$. If $h : E_1 \to E_2$ is a diffeomorphism mapping each fiber $\pi_1^{-1}(b)$ linearly onto $\pi_2^{-1}(b)$, then the bundles are equivalent.*

PROOF. Define $f : Fr(E_1) \to Fr(E_2)$ by $f(p) = h \circ p$, where $p : \mathbb{R}^n \to \pi_1^{-1}(b)$ is a frame of $E_1$. By Theorem 3.1, the two frame bundles are equivalent, and therefore so are the associated vector bundles $E_i \to B$. $\square$

Corollary 3.1 provides another approach to the tangent bundle of the sphere, or more generally, to the tangent bundle $TM$ of any homogeneous space $M = G/H$: Let $p = eH \in M$, so that $H$ is the *isotropy group* at $p$ of the action; i.e., $H = \{g \in G \mid gp = p\}$. The *linear isotropy representation* at $p$ is the homomorphism $\rho : H \to GL(M_p)$ given by $\rho(h) = h_{*p}$. It is not difficult to show that if $M$ is connected and $G$ acts effectively on $M$, then $\rho$ is one-to-one; in this case, $\rho$ induces an effective linear action of $H$ on $M_p$.

PROPOSITION 3.1. *If $G$ acts effectively on the homogeneous space $M = G/H$, then the tangent bundle of $M$ is equivalent to the bundle $G \times_H M_p \to M$, where $H$ acts on $M_p$ via the linear isotropy representation at $p$.*

PROOF. Consider the map $f : G \times_H M_p \to TM$ defined by $f[g, u] = g_* u$, which is clearly smooth, and linear on each fiber. Its inverse is given as follows: if $v \in M_q$, then by transitivity of the action of $G$, there exists some $g \in G$ such that $gp = q$. Then $f^{-1}(v) = [g, g_{*p}^{-1}v]$. This is well-defined, for if $q = \bar{g}p = gp$, then $g^{-1}\bar{g} \in H$, so that $\bar{g} = gh$ for some $h \in H$, and

$$[g, g_{*p}^{-1}v] = [gh, \rho(h)^{-1}g_{*p}^{-1}v] = [gh, h_{*p}^{-1}g_{*p}^{-1}v] = [gh, (gh)_{*p}^{-1}v] = [\bar{g}, \bar{g}_{*p}^{-1}v].$$

$\square$

The hypothesis that $G$ act effectively on $M$ in Proposition 3.1 is not restrictive: Exercise 54 shows that $M$ can always be realized as $\bar{G}/\bar{H}$, where $\bar{G}$ acts effectively on $M$.

EXERCISE 53. Let $S^3$ denote the group of quaternions of norm 1. Identify $SO(3)$ with the special orthogonal group of span$\{i, j, k\} = \mathbb{R}^3$, and define $\rho : S^3 \to SO(3)$ by $\rho(p)q = pqp^{-1}$ (quaternion multiplication).

(a) Show that $\rho$ is a homomorphism with kernel $\{\pm 1\}$. It is not hard to see that $\rho$ is onto, so that $SO(3)$ is diffeomorphic to $\mathbb{R}P^3$ and $\rho$ is the standard double covering.

(b) Consider the principal $SO(2)$-bundle $\pi : SO(3) \to S^2$. Prove that $\pi \circ \rho : S^3 \to S^2$ is equivalent to the Hopf fibration.

EXERCISE 54. Let $M = G/H$ be a homogeneous space.

(a) Show that the subgroup of $G$ which acts trivially on $M$ is the largest normal subgroup $N(H)$ of $G$ which lies in $H$.

(b) Show that $\bar{G} = G/N(H)$ acts effectively on $M$, and that $M = \bar{G}/\bar{H}$, where $\bar{H} = H/N(H)$.

## 4. Cross-Sections of Bundles

A trivial bundle $B \times F \to B$ has the property that through any point $(b, m) \in B \times F$, there is a copy $B \times \{m\}$ of $B$; alternatively, the map $s : B \to B \times F$ given by $s(b) = (b, m)$ is a lift of the identity $1_B$ (in the sense that $\pi \circ s = 1_B$) through $(b, m)$. It is by no means clear that such lifts exist in general, and they have a special name:

DEFINITION 4.1. Let $\xi = \pi : M \to B$ be a fiber bundle. A map $s : B \to M$ is said to be a *cross-section of* $\xi$ if $\pi \circ s = 1_B$.

For example, a vector field on a manifold $M$ is a cross-section of the tangent bundle of $M$; a differential $k$-form on $M$ is a cross-section of the bundle $\Lambda_k^*(M) \to M$. It is common practice to abbreviate cross-section by *section*. Before looking at further examples, we point out that one can construct from a given vector bundle $\xi$ many other vector bundles whose structure is induced by that of $\xi$. We illustrate the procedure in detail for the dual $\xi^*$ of a vector bundle $\xi$. It is convenient to denote the fiber $\pi^{-1}(b)$ of a vector bundle $\pi : E \to B$ over $b$ by $E_b$, and we will often do so.

PROPOSITION 4.1. *Let $\xi = \pi : E \to B$ be a rank $n$ vector bundle, and define $E^* = \cup_{b \in B} E_b^*$. For $\alpha \in E_b^*$, let $\pi^*(\alpha) = b$. There exists a natural rank $n$ vector bundle structure on $\xi^* = \pi^* : E^* \to B$ induced by $\xi$. $\xi^*$ is called the dual bundle of $\xi$.*

PROOF. Let $(\pi, \phi)$ be a bundle chart of $\xi$ over $U \subset B$. Since $\phi_{|E_b} : E_b \to \mathbb{R}^n$ is an isomorphism for each $b \in U$, so is $\bar{\phi}_{|E_b^*} : E_b^* \to \mathbb{R}^{n*} \cong \mathbb{R}^n$, where $\bar{\phi}_{|E_b^*} := (\phi_{|E_b})^{-1*}$ (recall that the transpose of a linear transformation $L : V \to W$ is the linear map $L^* : W^* \to V^*$ given by $(L^*\alpha)v = \alpha(Lv)$ for $\alpha \in W^*$, $v \in V$). Then $(\pi^*, \bar{\phi}) : \pi^{*-1}(U) \to U \times \mathbb{R}^n$ is one-to-one, onto, and its restriction to each $E_b^*$ is linear; if $(\pi, \psi)$ is another bundle chart, then

$$(\pi^*, \bar{\psi}) \circ (\pi^*, \bar{\phi})^{-1}(b, \alpha) = (b, \bar{\psi} \circ (\bar{\phi}_{|E_b^*})^{-1}\alpha) = (p, (\phi \circ \psi_{|E_b}^{-1})^*\alpha).$$

Thus, there exist unique topological and differentiable structures on $E^*$ for which the maps $(\pi^*, \bar{\phi})$ become local diffeomorphisms. These maps form a bundle atlas, since the transition functions are given by $f_{\bar{\phi}, \bar{\psi}}(b) = f_{\psi, \phi}(p)^*$. $\square$

Given two vector bundles $\xi_i = \pi_i : E_i \to B$, one defines in a similar fashion the *tensor product bundle* $\xi_1 \otimes \xi_2$ with fiber $E_{1b} \otimes E_{2b}$ over $b$, the *homomorphism bundle* $\mathrm{Hom}(\xi_1, \xi_2)$ whose fiber over $b$ consists of all linear transformations $E_{1b} \to E_{2b}$, etc. The isomorphism $\mathrm{Hom}(E_{1b}, E_{2b}) \cong E_{1b}^* \otimes E_{2b}$ induces an equivalence $\mathrm{Hom}(\xi_1, \xi_2) \cong \xi_1^* \otimes \xi_2$.

DEFINITION 4.2. A *Euclidean metric* on a vector bundle $\xi = \pi : E \to B$ is a section $s$ of the bundle $(\xi \otimes \xi)^*$ such that $s(b)$ is an inner product on $E_b$ for each $b \in B$. A Euclidean metric on the tangent bundle of a manifold $M$ is called a *Riemannian metric* on $M$.

Loosely translated, a Euclidean metric on $\xi$ is just an inner product on the fibers that varies smoothly with the base point.

THEOREM 4.1. *Every vector bundle $\xi = \pi : E \to B$ admits a Euclidean metric.*

PROOF. Consider a locally finite cover of $B$ by sets $\{U_\alpha\}$ whose preimages are the domains of bundle charts $\{(\pi, \phi_\alpha)\}$. Define a Euclidean metric $s_\alpha$ on each $\pi^{-1}(U_\alpha)$ so that $\phi_\alpha$ becomes a linear isometry: $s_\alpha(u, v) = \langle \phi_\alpha u, \phi_\alpha v \rangle$, where $\langle , \rangle$ denotes the standard inner product on $\mathbb{R}^n$. Let $\{\psi_\alpha\}$ be a partition of unity subordinate to $\{U_\alpha\}$, and extend $s_\alpha$ to all of $B$ by setting $\bar{s}_\alpha(b) = \psi_\alpha(b)s_\alpha(b)$ if $b \in U_\alpha$ and $\bar{s}_\alpha(b) = 0$ otherwise. Then $s = \sum_\alpha \bar{s}_\alpha$ is a Euclidean metric on $\xi$. $\qquad\square$

Theorem 4.1 implies that every rank $n$ vector bundle admits a reduction of its structure group to $O(n)$, by requiring that charts be linear isometries when restricted to each fiber.

Notice that a vector bundle always admits a section, namely *the zero section* given by $s(b) = 0 \in E_b$. Principal bundles, on the other hand, do not, in general, admit sections:

THEOREM 4.2. *A principal $G$-bundle $\pi : P \to B$ admits a section iff it is trivial.*

PROOF. If $\pi : B \times G \to B$ is trivial, then for any fixed $g \in G$, the map $s(b) := (b, g)$ defines a section of $\pi$. Conversely, suppose $s : B \to P$ is a section. Since $p \in P$ and $s(\pi(p))$ belong to the same fiber, there is a well-defined equivariant map $\phi : P \to G$ such that $p = s(\pi(p))\phi(p)$. $(\pi, \phi) : P \to B \times G$ is then an equivalence by Theorem 3.1. $\qquad\square$

EXAMPLE 4.1. Recall from Section 3 that the principal $SO(n)$-bundle over $S^n$ associated to the tangent bundle of $S^n$ is $\pi : SO(n+1) \to SO(n+1)/SO(n) = S^n$. When $n = 3$, $S^3$ is identified with the group of quaternions of norm 1, and $\mathbf{e}_1 = 1 \in \mathbb{H}$.

Consider the map $s : S^3 \to SO(4)$ given by $s(q)u = qu$, for $q \in S^3 \subset \mathbb{H}$, $u \in \mathbb{H} = \mathbb{R}^4$. Then $(\pi \circ s)(q) = s(q)1 = q$; i.e., $s$ is a section of $\pi : SO(4) \to S^3$, and $SO(4)$ is diffeomorphic to $S^3 \times SO(3)$ (although not isomorphic, as a group, to the direct product $S^3 \times SO(3)$). Since $\pi$ is trivial, so is the associated tangent bundle $TS^3 \to S^3$. We saw in Chapter 1 that even-dimensional spheres do not admit a nowhere-zero vector field; i.e., their tangent bundle does not admit a nowhere-zero section, and is therefore nontrivial. Thus, none of the bundles $SO(n+1) \to S^n$ admit sections when $n$ is even.

EXERCISE 55. Consider the map
$$\phi : S^3 \times S^3 \to SO(4), \qquad \phi(q_1, q_2)u = q_1 u q_2^{-1}, \qquad q_i \in S^3, \quad u \in \mathbb{H} = \mathbb{R}^4.$$
Show that $\phi$ is a Lie group homomorphism, and determine its kernel. It is not hard to see that $\phi$ is onto, so that $S^3 \times S^3$ is the two-fold covering group of $SO(4)$, denoted $Spin(4)$. Notice that if $\imath, \Delta : S^3 \to S^3 \times S^3$ are the imbedding-homomorphisms given by $\imath(q) = (q, e)$ and $\Delta(q) = (q, q)$, then $\phi \circ \imath$ is the section from Example 4.1, and $\phi \circ \Delta$ is the two-fold covering from Exercise 53.

EXERCISE 56. Let $\xi_i = \pi_i : E_i \to B$ be vector bundles over $B$, $i = 1, 2$, and denote by $\Gamma \xi_i$ the collection of sections of $\xi_i$.
 (a) Show that $\Gamma \xi_i$ is a module over the ring of smooth functions $B \to \mathbb{R}$.
 (b) Show that $\Gamma \operatorname{Hom}(\xi_1, \xi_2)$ and $\operatorname{Hom}(\Gamma \xi_1, \Gamma \xi_2)$ are naturally isomorphic as modules.

EXERCISE 57. A *complex vector bundle* is a bundle with fiber $\mathbb{C}^n$ whose transition functions are complex linear. Show that a real rank $2n$ vector bundle $\xi$ admits a complex vector bundle structure iff there exists a section $J$ of the bundle $\operatorname{Hom}(\xi, \xi)$ such that $J^2$ equals minus the identity on the total space. $J$ is called a *complex structure* on $\xi$.

## 5. Pullback and Normal Bundles

Let $\xi = \pi : M \to B$ denote a fiber bundle with fiber $F$ and group $G$. Given a manifold $\bar{B}$ and a map $f : \bar{B} \to B$, one can construct in a natural way a bundle over $\bar{B}$ with the same fiber and group: Consider the subset
$$f^*M = \{(b, m) \in \bar{B} \times M \mid \pi(m) = f(b)\}$$
together with the subspace topology from $\bar{B} \times M$, and denote by $\pi_1 : f^*M \to \bar{B}$, $\pi_2 : f^*M \to M$ the projections.

PROPOSITION 5.1. $f^*\xi = \pi_1 : f^*M \to \bar{B}$ *is a fiber bundle with fiber $F$ and group $G$, called the* pullback bundle *of $\xi$ via $f$, and $\pi_2 : f^*M \to M$ is a bundle map covering $f$. Furthermore, $f^*\xi$ is uniquely characterized by the property that $\pi \circ \pi_2 = f \circ \pi_1$;*

$$
\begin{array}{ccc}
f^*M & \xrightarrow{\ \pi_2\ } & M \\
{\scriptstyle \pi_1} \downarrow & & \downarrow {\scriptstyle \pi} \\
\bar{B} & \xrightarrow[\ f\ ]{} & B
\end{array}
$$

*i.e., if $\bar{\xi} = \bar{\pi} : \bar{M} \to \bar{B}$ is a fiber bundle with fiber $F$ and group $G$, and there exists a bundle map $\bar{f} : \bar{\xi} \to \xi$ covering $f : \bar{B} \to B$, then $\bar{\xi} \cong f^*\xi$.*

PROOF. A bundle chart $(\pi, \phi)$ of $\xi$ over $U \subset B$ induces a chart $(\pi_1, \bar{\phi})$ of $f^*\xi$ over $f^{-1}(U)$, where $\bar{\phi} = \phi \circ \pi_2$. It is easily checked that the transition functions satisfy $f_{\bar{\phi}, \bar{\psi}} = f_{\phi, \psi} \circ f$, so that $f^*\xi$ is a bundle as claimed, and $\pi_2$ is a bundle map by definition. For the uniqueness part, let $\bar{\xi}$ be a bundle as in the statement. By the remark following Definition 1.3, the coordinate bundle over $\bar{B}$ with bundle charts of the form $(\bar{\pi}^{-1}(f^{-1}(U)), (\bar{\pi}, \phi \circ \bar{f}))$, where $(\pi^{-1}(U), (\pi, \phi))$ is a bundle chart of $\pi$, is equivalent to $\bar{\xi}$. Since it has the same transition functions as $f^*\xi$, $f^*\xi$ is equivalent to $\bar{\xi}$.                                        □

Observe that the structure group of $f^*\xi$ may very well be smaller than $G$, since its transition functions are those of $\xi$ composed with $f$.

If $\xi_i = \pi_i : E_i \to B$ are two vector bundles of rank $n_i$ over B, then $\xi_1 \times \xi_2 = \pi_1 \times \pi_2 : E_1 \times E_2 \to B \times B$ is a vector bundle of rank $n_1 + n_2$. Consider the diagonal imbedding $\Delta : B \to B \times B$, $\Delta(b) = (b, b)$.

DEFINITION 5.1. The *Whitney sum* $\xi_1 \oplus \xi_2$ is the rank $(n_1 + n_2)$ vector bundle $\Delta^*(\xi_1 \times \xi_2)$.

The fiber of $\xi_1 \oplus \xi_2$ over $b \in B$ is $E_{1b} \oplus E_{2b}$.

DEFINITION 5.2. Let $\xi_i = \pi_i : E_i \to B$ be two vector bundles over $B$. A map $h : E_1 \to E_2$ is said to be a *homomorphism* if it maps each fiber $E_{1b}$ linearly into $E_{2b}$.

Thus, a homomorphism $h : E_1 \to E_2$ is just another word for a section $s$ of the bundle $\mathrm{Hom}(\xi_1, \xi_2)$: We can go from one to the other via $s(b) = h_{|E_{1b}}$. By Corollary 3.1, if $h$ is an isomorphism on each fiber, then $h$ is an equivalence. Conversely, an equivalence is a homomorphism (and a bundle map). More generally:

PROPOSITION 5.2. *Let* $\xi_i = \pi_i : E_i \to B_i$ *be vector bundles over* $B_i$, $i = 1, 2$. *If* $h : E_1 \to E_2$ *maps each fiber* $\pi_1^{-1}(b_1)$ *linearly into a fiber* $\pi_2^{-1}(b_2)$, *then* $h = f \circ g$, *where* $g$ *is a homomorphism and* $f$ *a bundle map.*

PROOF. Consider the pullback bundle $\bar{h}^*\xi_2$, where $\bar{h} : B_1 \to B_2$ is the map induced by $h$. If $pr_2 : \bar{h}^*E_2 \to E_2$ is the bundle map given by projection onto the second factor, then $h = pr_2 \circ g$, where $g : E_1 \to \bar{h}^*E_2$ is the homomorphism $g(u) = (\pi_1(u), h(u))$.

$$
\begin{array}{ccccc}
E_1 & \xrightarrow{\ g\ } & \bar{h}^*E_2 & \xrightarrow{\ pr_2\ } & E_2 \\
{\scriptstyle \pi_1}\downarrow & & {\scriptstyle pr_1}\downarrow & & \downarrow{\scriptstyle \pi_2} \\
B_1 & \xrightarrow[\ 1_{B_1}\ ]{} & B_1 & \xrightarrow[\ \bar{h}\ ]{} & B_2
\end{array}
$$

□

THEOREM 5.1. *Let* $\xi_i = \pi_i : E_i \to B$ *denote vector bundles over* $B$, $h : E_1 \to E_2$ *a homomorphism.*

(1) *If* $h$ *is one-to-one (on each fiber), then* $\mathrm{coker}\, h = \xi_2/h(\xi_1)$ *is a vector bundle over* $B$.

(2) *If* $h$ *is onto, then* $\ker h$ *is a vector bundle over* $B$.

PROOF. (1) Suppose that for each $b \in B$, the restriction $h : E_{1b} \to E_{2b}$ is injective. If $\xi_1$ has rank $n$ and $\xi_2$ rank $n + k$, then the vector space $E_{2b}/h(E_{1b})$ has dimension $k$. We construct a bundle atlas for $\mathrm{coker}\, h = \cup_{b \in B} E_{2b}/h(E_{1b})$:

Let $b \in B$, $(\pi_1, \phi)$ and $(\pi_2, \psi)$ be bundle charts on $\pi_i^{-1}(U)$, where $U$ is a neighborhood of $b$. Consider the map $g : U \to \operatorname{Hom}(\mathbb{R}^n, \mathbb{R}^{n+k})$ given by

$$g(p) = \psi \circ h \circ (\phi_{|E_{1p}})^{-1}.$$

$g(p)$ has rank $n$ for all $p \in U$, and we may assume, by reordering coordinates if necessary, that the first $n$ rows of the matrix $M(b)$ of $g(b)$ with respect to the standard bases are linearly independent; i.e., that $pr_1 \circ g(b) : \mathbb{R}^n \to \mathbb{R}^n$ is an isomorphism, where $pr_1 : \mathbb{R}^{n+k} \to \mathbb{R}^n$ is the projection. By continuity, this holds for all $p$ in a neighborhood (which we also call $U$) of $b$. It follows that for each $p \in U$, the map

$$\mathbb{R}^{n+k} = \mathbb{R}^n \times \mathbb{R}^k \to \mathbb{R}^{n+k},$$
$$(u, v) \mapsto g(p)u + (0, v)$$

is an isomorphism, and $f : U \times \mathbb{R}^n \times \mathbb{R}^k \to U \times \mathbb{R}^{n+k}$, where $f(p, u, v) = (p, g(p)u + (0, v))$, is an equivalence of trivial bundles. Thus, $(\pi_2, \Psi) := f^{-1} \circ (\pi_2, \psi)$ is a bundle chart for $\xi_2$. By construction, $v \in E_{2p}$ belongs to $h(E_{1p})$ iff $(\pi_2, \Psi)(v) \in p \times \mathbb{R}^n \times 0 \subset U \times \mathbb{R}^n \times \mathbb{R}^k$. Therefore, if $\pi : \operatorname{coker} h \to B$ is the natural projection and $pr_2 : \mathbb{R}^n \times \mathbb{R}^k \to \mathbb{R}^k$ the projection onto the second factor, then the bundle chart $(\pi_2, \Psi)$ of $\xi_2$ induces a diffeomorphism

$$\pi^{-1}(U) \to U \times \mathbb{R}^k,$$
$$w + h(E_{1p}) \mapsto (p, (pr_2 \circ \Psi)w),$$

which is linear on each fiber. This yields a bundle atlas on $\operatorname{coker} h \to B$: smoothness of the transition functions follows from smoothness of $h$ and of the transition functions of $\xi_i$.

(2) Suppose $h : E_{1b} \to E_{2b}$ is onto for each $b \in B$, with $n+k$ and $n$ denoting the ranks of $\xi_1$ and $\xi_2$ respectively. Let $U$, $g : U \to \operatorname{Hom}(\mathbb{R}^{n+k}, \mathbb{R}^n)$ be as in (1). We may assume that $g(p)e_1, \ldots, g(p)e_n$ are independent for each $p$ in $U$. Define a bundle equivalence

$$f : U \times \mathbb{R}^{n+k} \to U \times \mathbb{R}^n \times \mathbb{R}^k,$$
$$(p, a) \mapsto (p, g(p)a, a_{n+1}, \ldots, a_{n+k}),$$

so that $(\pi_1, \Phi) := f \circ (\pi_1, \phi)$ is a bundle chart for $\xi_1$ over $U$. By construction, $h(v) = 0$ for $v \in \pi_1^{-1}(U)$ iff $(\pi_1, \Phi)(v) \in U \times 0 \times \mathbb{R}^k$. $(\pi_1, pr_2 \circ \Phi)$ is therefore a bundle chart for $\ker h$.    $\square$

If $h : \xi_1 \to \xi_2$ is a one-to-one homomorphism, then $h(\xi_1)$ is a subbundle of $\xi_2$ equivalent to $\xi_1$. An *exact sequence of bundle homomorphisms* is a sequence of homomorphisms

$$0 \longrightarrow \xi_1 \overset{h}{\longrightarrow} \xi_2 \overset{f}{\longrightarrow} \xi_3 \longrightarrow 0$$

such that the kernel of each map equals the image of the preceding one; thus, $h$ is one-to-one, $f$ is onto, and $h(\xi_1) = \ker f$.

PROPOSITION 5.3. *If* $0 \longrightarrow \xi_1 \overset{h}{\longrightarrow} \xi_2 \overset{f}{\longrightarrow} \xi_3 \longrightarrow 0$ *is an exact sequence of homomorphisms, then there exists an equivalence* $g : \xi_2 \to \xi_1 \oplus \xi_3$ *with* $g \circ h : \xi_1 \to \xi_1 \oplus \xi_3$ *being the inclusion, and* $f \circ g^{-1} : \xi_1 \oplus \xi_3 \to \xi_3$ *the projection.*

PROOF. Consider a Euclidean metric $\langle , \rangle$ on $\xi_2$ (cf. Theorem 4.1). Since the metric is smooth as a section, the orthogonal projection $\pi : \xi_2 \to h(\xi_1)$ is a bundle homomorphism. Being onto, its kernel $h(\xi_1)^\perp$ is a bundle, and the map

$$L : h(\xi_1) \oplus h(\xi_1)^\perp \to \xi_2,$$
$$(v, w) \mapsto v + w$$

is an equivalence. The restriction $\bar{h} : \xi_1 \to h(\xi_1)$ of $h$ is also an equivalence. Furthermore, the restriction $\bar{f} : h(\xi_1)^\perp \to \xi_3$ of $f$ is a one-to-one homomorphism because $\ker f = h(\xi_1)$, so that by rank considerations, it is an equivalence. Thus, $g := (\bar{h}^{-1} \oplus \bar{f}) \circ L^{-1} : \xi_2 \to \xi_1 \oplus \xi_3$ has the required properties.    $\square$

EXAMPLE 5.1. Let $\xi = \pi : E \to B$ be a vector bundle over $B$, and denote by $\tau_E$, $\tau_B$ the tangent bundles of $E$ and $B$. Since $\pi_* : TE \to TB$ maps the fiber over $u \in E$ linearly onto the fiber over $\pi(u) \in B$, $\pi_*$ induces an epimorphism $h : \tau_E \to \pi^* \tau_B$ by Proposition 5.2. Its kernel $\ker h = \ker \pi_*$ is therefore the total space of a bundle $\mathcal{V}\xi = \pi_\mathcal{V} : VE \to E$ over $E$, called the *vertical bundle of* $\xi$. By Proposition 5.3,

$$\tau_E \cong \mathcal{V}\xi \oplus \pi^* \tau_B.$$

The fiber $VE_u$ of $\mathcal{V}\xi$ over $u \in E$ can be described as follows: If $b = \pi(u)$, and $\imath : E_b = \pi^{-1}(b) \to E$ denotes inclusion, then $VE_u = \imath_*(E_b)_u$ as an immediate consequence of Proposition 6.2 in Chapter 1 (here, $(E_b)_u$ is the tangent space of $E_b$ at $u$).

Let $f : M \to N$ be an immersion. Since $f_* : TM \to TN$ is linear and one-to-one, it induces a monomorphism $h : \tau_M \to f^* \tau_N$.

DEFINITION 5.3. Let $f : M \to N$ be an immersion. The *normal bundle of* $f$ is the bundle $\nu(f) = f^* \tau_N / h(\tau_M)$ over $M$.

Since $0 \to \tau_M \to f^* \tau_N \to \nu(f) \to 0$ is an exact sequence of homomorphisms, Proposition 5.3 implies that $f^* \tau_N \cong \tau_M \oplus \nu(f)$. In fact, given a Euclidean metric on $f^* \tau_N$ (for instance one induced by a Riemannian metric on $N$), $\nu(f)$ is equivalent to the orthogonal complement of $h(\tau_M)$.

EXAMPLE 5.2. Consider the inclusion $\imath : S^n \to \mathbb{R}^{n+1}$. By the remark following Proposition 5.1, the pullback of the trivial tangent bundle of $\mathbb{R}^{n+1}$ via $\imath$ is the trivial rank $(n + 1)$ bundle $\epsilon^{n+1}$ over $S^n$. The normal bundle of $\imath$ is also the trivial rank 1 bundle $\epsilon^1$ over $S^n$: Indeed, the restriction of the position vector field $p \mapsto \mathcal{J}_p p$ to the sphere is a section of the frame bundle of $\tau_{S^n}^\perp$. Thus,

$$\tau_{S^n} \oplus \epsilon^1 \cong \epsilon^{n+1},$$

even though $\tau_{S^n}$ is not, in general, trivial.

EXERCISE 58. Show that if $\xi$ is a vector bundle, then $\xi \oplus \xi$ admits a complex structure, see Exercise 57. *Hint:* Let $J(u, v) = (-v, u)$.

EXERCISE 59. If $\xi = \pi : E \to B$ is a vector bundle, show that the vertical bundle of $\xi$ is equivalent to the pullback $\pi^* \xi$. *Hint:* Recall the canonical isomorphism $\mathcal{J}_u$ of the vector space $E_b$ with its tangent space $(E_b)_u$ at $u$. Show that $f : \pi^* E \to VE$ is an equivalence, where $f(u, v) = \mathcal{J}_u v$.

EXERCISE 60. If $\xi = \pi : E \to B$ is a vector bundle, then by Example 5.1 and Exercise 59, $\tau_E \cong \pi^*\xi \oplus \pi^*\tau_B$. Prove that if $s$ is the zero section of $\xi$, then

$$s^*\tau_E \cong \xi \oplus \tau_B.$$

Thus, the normal bundle of the zero section in $\xi$ is $\xi$ itself.

## 6. Fibrations and the Homotopy Lifting/Covering Properties

Although we have so far only considered bundles over manifolds, the definition used also makes sense for manifolds with boundary (and even for topological spaces—the traditional type of base in bundle theory— if we replace diffeomorphisms by homeomorphisms). Let $B$ be a manifold, $I = [0,1]$, and for $t \in I$, denote by $\imath_t : B \to B \times I$ the imbedding $\imath_t(b) = (b,t)$. Recall that two maps $f, g : \bar{B} \to B$ are said to be *homotopic* if there exists $H : \bar{B} \times I \to B$ with $H \circ \imath_0 = f$ and $H \circ \imath_1 = g$. $H$ is called a *homotopy of $f$ into $g$*.

Homotopies play an essential role in the classification of bundles: In this section, we will see that if $\xi$ is a bundle over $B$, then for any two homotopic maps $f, g : \bar{B} \to B$, the induced bundles $f^*\xi$ and $g^*\xi$ are equivalent.

We begin by introducing the notion of fibration, which is weaker than that of fiber bundle:

DEFINITION 6.1. A surjective map $\pi : M \to B$ is said to be a *fibration* if it has the *homotopy lifting property*: namely, given $f : \bar{B} \to M$, any homotopy $H : \bar{B} \times I \to B$ of $\pi \circ f$ can be lifted to a homotopy $\tilde{H} : \bar{B} \times I \to M$ of $f$; i.e., $\pi \circ \tilde{H} = H$, $\tilde{H} \circ \imath_0 = f$.

In order to show that a fiber bundle $\xi = \pi : M \to B$ is a fibration, we first rephrase the problem: Notice that a homotopy $H : \bar{B} \times I \to B$ can be lifted to $\tilde{H} : \bar{B} \times I \to M$ iff the pullback bundle $H^*M \to \bar{B} \times I$ admits a section. Indeed, if $\tilde{H}$ is a lift of $H$, then $(b,t) \mapsto (b,t,\tilde{H}(b,t))$ is a section. Conversely, if $s$ is a section of $H^*M \to \bar{B} \times I$, then $\pi_2 \circ s$ is a lift of $H$, where $\pi_2 : H^*M \to M$ is the second factor projection. In other words, the homotopy lifting property may be paraphrased as saying that if $\xi$ is a fiber bundle over $B \times I$, then any section of $\xi_{|B \times 0}$ can be extended to a section of $\xi$.

We begin with the following:

LEMMA 6.1. *Let $\xi$ be a principal bundle over $B \times I$. Then any $b \in B$ has a neighborhood $U$ such that the restriction $\xi_{|U \times I}$ is trivial.*

PROOF. By compactness of $b \times I$, there exist neighborhoods $V_1, \ldots, V_k$ of $b$, and intervals $I_1, \ldots, I_k$ such that $\{V_i \times I_i\}$ is a cover of $b \times I$, and each restriction $\xi_{|V_i \times I_i}$ is trivial. We claim that $U$ may be taken to be $V_1 \cap \cdots \cap V_k$. The proof will be by induction on $k$.

The case $k = 1$ being trivial, assume the statement holds for $k - 1$. Order the intervals $I_j$ by their left endpoints, so that if $I_j = (t_j^0, t_j^1)$, then $t_j^0 < t_{j+1}^0$ (if $t_j^0 = t_{j+1}^0$, then either $I_j$ or $I_{j+1}$ can be discarded). We may also assume that $t_1^1 < t_2^1$ since otherwise $I_2$ may be discarded. If $t_0 \in (t_2^0, t_1^1)$, then by the induction hypothesis, $\xi$ is trivial over $U_1 \times [0, t_0)$, and over $U_2 \times (t_0, 1]$, where $U_1 = V_1$ and $U_2 = V_2 \cap \cdots \cap V_k$. Let $s_1$ and $s_2$ be sections over these two sets. For each $(q,t) \in (U_1 \cap U_2) \times (t_2^0, t_1^1)$, there exists a unique $g(q,t) \in G$ such that

$s_1(q,t) = s_2(q,t)g(q,t)$, and $g : (U_1 \cap U_2) \times (t_2^0, t_1^1) \to G$ is smooth because the sections are. Extend $g$ to a differentiable map $g : (U_1 \cap U_2) \times (t_2^0, 1] \to G$. We then obtain a section $s$ of $\xi$ restricted to $(U_1 \cap U_2) \times I$ by defining

$$s(q,t) = \begin{cases} s_1(q,t), & \text{for } t \leq t_0, \\ s_2(q,t)g(q,t), & \text{for } t \geq t_0. \end{cases}$$

$\square$

THEOREM 6.1. *Let* $\xi = \pi : P \to B \times I$ *be a principal $G$-bundle, and consider the maps* $p : B \times I \to B \times 1$, $p(b,t) = (b,1)$, *and* $j : B \times 1 \to B \times I$, $j(b,1) = (b,1)$. *Then*

$$\xi \cong (j \circ p)^* \xi = p^* \xi_{|B \times 1}.$$

PROOF. Denote by $\pi_B : P \to B$ and $u : P \to I$ the maps obtained by composing $\pi$ with the projections of $B \times I$ onto its two factors. We will construct a $G$-equivariant bundle map $f : P \to \pi^{-1}(B \times 1)$ covering $p$; the theorem will then follow from Theorem 3.1 and Proposition 5.1.

By Lemma 6.1, there exists a countable cover $\{U_n\}$ of $B$ such that $\xi$ is trivial over each $U_n \times I$. Let $s_n$ denote a section of $\xi_{|U_n \times I}$, and $\{\phi_n\}$ a partition of unity subordinate to $\{U_n\}$. Since any element in $\pi^{-1}(b,t)$ with $b \in U_n$ can be written as $s_n(b,t)g$ for a unique $g \in G$, the assignment

$$f_n : \pi^{-1}(U_n \times I) \to \pi^{-1}(U_n \times I),$$
$$s_n(b,t)g \mapsto s_n(b, \min\{t + \phi_n(b), 1\})g$$

is a $G$-equivariant bundle map. Furthermore, $f_n$ is the identity on an open set containing $\pi^{-1}(\partial U_n \times I)$, and may therefore be continuously extended to all of $P$ by defining $f_n(q) = q$ for $q \notin \pi^{-1}(U_n \times I)$. Finally, set $f = f_1 \circ f_2 \circ \cdots$. The composition makes sense because all but finitely many $f_n$ are the identity on a neighborhood of any point. $f$ is $G$-equivariant since each $f_n$ is, and $u \circ f = \min\{u + (\sum \phi_n) \circ \pi_B, 1\}$, so that $u \circ f \equiv 1$. Thus, $f$ maps into $\pi^{-1}(B \times 1)$, and furthermore, $f$ is differentiable, because although $u \circ f_n$ is in general only continuous, $u \circ f$ is differentiable. This completes the proof. $\square$

COROLLARY 6.1 (Homotopy Lifting Property). *A fiber bundle is a fibration.*

PROOF. As noted at the beginning of this section, what needs to be shown is that if $\xi$ is a fiber bundle over $B \times I$ with group $G$ and fiber $F$, then any section $s$ of $\xi_{|B \times 1}$ can be extended to the whole bundle. With the notation of Theorem 6.1, if $\pi : P \to B \times I$ denotes the principal $G$-bundle associated to $\xi$, then there exists a bundle map $f : P \to \pi^{-1}(B \times 1)$ covering $p$. $f$ then induces a bundle map $f : \xi \to \xi_{|B \times 1}$ between the associated bundles with fiber $F$. Thus, $f^{-1} \circ s \circ p$ is a section of $\xi$. Furthermore, the restriction of $f$ to $\pi^{-1}(B \times 1)$ is the identity, so $f^{-1} \circ s \circ p$ is an extension of $s$. $\square$

Recall that $\iota_t : B \to B \times I$ denotes the imbedding $\iota_t(b) = (b,t)$.

COROLLARY 6.2. *Let $\xi$ be a fiber bundle over $B \times I$. Then* $\iota_0^* \xi \cong \iota_1^* \xi$.

PROOF. We may assume that $\xi$ is a principal bundle. Let $p_1 : B \times I \to B$ denote the projection onto the first factor. With notation as in Theorem 6.1, $j \circ p = \imath_1 \circ p_1$, and $\xi \cong (j \circ p)^* \xi = p_1^* \imath_1^* \xi$. Thus,

$$\imath_0^* \xi \cong \imath_0^* p_1^* \imath_1^* \xi = (p_1 \circ \imath_0)^* \imath_1^* \xi = \imath_1^* \xi,$$

since $p_1 \circ \imath_0 = 1_B$. $\qquad\square$

COROLLARY 6.3 (The Homotopy Covering Property). *Let $\xi$ denote a fiber bundle over $B$. If $f, g : \bar{B} \to B$ are homotopic, then $f^* \xi \cong g^* \xi$.*

PROOF. Let $H : \bar{B} \times I \to B$ be a homotopy with $H \circ \imath_0 = f$ and $H \circ \imath_1 = g$. By Corollary 6.2,

$$f^* \xi = \imath_0^*(H^* \xi) \cong \imath_1^*(H^* \xi) = g^* \xi.$$

$\qquad\square$

EXERCISE 61. Show that a bundle over a contractible space (one for which the identity map is homotopic to a constant map) is trivial.

EXERCISE 62. Suppose $M \to B$ is a fiber bundle, and let $\tilde{H} : \bar{B} \times I \to M$ be a lift (the existence of which is guaranteed by Corollary 6.1) of some homotopy $H : \bar{B} \times I \to B$. Show that $\tilde{H}$ may be chosen to be *stationary* with respect to $H$; i.e., if $H(b, t)$ is constant in $t$ for some $b \in \bar{B}$, then so is $\tilde{H}(b, t)$.

EXERCISE 63. Let $\pi : M \to B$ be a fibration, $b \in B$, $p \in \pi^{-1}(B)$.
(a) Show that any curve $c : I \to B$ with $c(0) = b$ may be lifted to a curve $\bar{c}$ in $M$ with $\bar{c}(0) = p$.
(b) Let $c_i$, $i = 1, 2$, denote two curves in $B$ from $b$ to $\tilde{b}$, and $H$ a homotopy of $c_1$ into $c_2$ with $H(0, s) = b$, $H(1, s) = \tilde{b}$ for all $s \in I$. Prove that, if $\bar{c}_i$ is a lift of $c_i$ to $M$ with $\bar{c}_i(0) = p$, then $\bar{c}_1$ is homotopic to $\bar{c}_2$, and the two curves have the same endpoint.
(c) Prove that the lift of $c$ in (a) is unique.

## 7. Grassmannians and Universal Bundles

The collection $G_{n,k}$ of all $n$-dimensional subspaces (or $n$-*planes*) of $\mathbb{R}^{n+k}$ is called the *Grassmannian manifold* of $n$-planes of $\mathbb{R}^{n+k}$. Consider the map $\pi : O(n+k) \to G_{n,k}$ given by $\pi(L) = L(\mathbb{R}^n)$, where $\mathbb{R}^n$ denotes the subspace $\mathbb{R}^n \times 0 \subset \mathbb{R}^{n+k}$. $\pi$ is onto, and $\pi(L) = \pi(T)$ iff $L^{-1} \circ T(\mathbb{R}^n) = \mathbb{R}^n$; i.e., iff $L^{-1} \circ T \in O(n) \times O(k) \subset O(n+k)$. $\pi$ therefore induces a bijective correspondence

$$O(n+k)/O(n) \times O(k) \longleftrightarrow G_{n,k},$$

and we endow $G_{n,k}$ with the differentiable structure for which this correspondence becomes a diffeomorphism. $G_{n,k}$ is then a compact homogeneous space of dimension $\binom{n+k}{2} - \binom{n}{2} - \binom{k}{2} = nk$. $G_{1,k}$, for example, is just $\mathbb{R}P^k$.

One can explicitly describe a differentiable atlas for $G_{n,k}$: Given an $n$-plane $P$ in $G_{n,k}$, decompose $\mathbb{R}^{n+k} = P \oplus P^\perp$, and denote by $\pi_1, \pi_2$ the projections of $\mathbb{R}^{n+k}$ onto $P$ and $P^\perp$ respectively. Let $U$ be the open neighborhood of $P$ consisting of all $n$-planes $V$ such that $\pi_{1|V} : V \to P$ is an isomorphism, and define $x : U \to \mathrm{Hom}(P, P^\perp)$ by $x(V) = \pi_2 \circ \pi_{1|V}^{-1}$. $x$ is a homeomorphism with inverse $x^{-1}(L) = \{u + Lu \mid u \in P\}$. Since $\mathrm{Hom}(P, P^\perp) \cong P^* \otimes P^\perp$ is a vector space of dimension $nk$, $x$ may be considered as a coordinate map $x : U \to \mathbb{R}^{nk}$. It is straightforward to check that the transition maps for the collection of all such charts are differentiable.

There is a canonical rank $n$ vector bundle $\gamma_{n,k}$ over $G_{n,k}$: its total space is the subset $E(\gamma_{n,k})$ of $G_{n,k} \times \mathbb{R}^{n+k}$ consisting of all pairs $(P, u)$ such that $u \in P$, and $\pi : E(\gamma_{n,k}) \to G_{n,k}$ is given by $\pi(P, u) = P$. Thus, the fiber over $P \in G_{n,k}$ is $P$ itself. The differentiable atlas of $G_{n,k}$ described above induces a bundle atlas on $\gamma_{n,k}$: given $P \in G_{n,k}$, the orthogonal projection $p : \mathbb{R}^{n+k} \to P$, and $U = \{V \in G_{n,k} \mid p_{|V} \text{ is an isomorphism}\}$, let $\phi : \pi^{-1}(U) \to P \cong \mathbb{R}^n$ be given by $\phi(v) = p(v)$. Then $(\pi, \phi) : \pi^{-1}(U) \to U \times \mathbb{R}^n$ is a diffeomorphism which maps each fiber isomorphically onto $\mathbb{R}^n$.

$\gamma_{n,k}$ is called the *universal rank $n$ bundle* over $G_{n,k}$, the reason being that *any* rank $n$ vector bundle over a manifold $B$ is equivalent to $f^*\gamma_{n,k}$ for sufficiently large $k$ and some map $f : B \to G_{n,k}$. Recall that the pullback of a bundle is less twisted than the original, since the transition functions of the former equal those of the latter composed with the pullback map. Roughly speaking, the universal bundle is so twisted that any other bundle is a diluted version of it. Some more work is needed before we are in a position to prove this, but it can already be established in the case of a tangent bundle:

EXAMPLE 7.1. A classical theorem in topology states that any $n$-manifold $M$ can be immersed in Euclidean space $\mathbb{R}^{n+k}$, provided $k$ is large enough. If $f$ is such an immersion, then $f_* M_p$ is an $n$-dimensional subspace of $\mathbb{R}^{n+k}_{f(p)}$ for each $p$ in $M$, and $\mathcal{J}^{-1}_{f(p)} f_* M_p$ is an element of $G_{n,k}$. The map

$$\bar{h} : TM \to E(\gamma_{n,k}),$$
$$v \mapsto (\mathcal{J}^{-1}_{f(p)} f_* M_p, \mathcal{J}^{-1}_{f(p)} f_* v),$$

for $v \in M_p$, is a bundle map covering $h : M \to G_{n,k}$, where $h(p) = \mathcal{J}^{-1}_{f(p)} f_* M_p$. Thus, the tangent bundle of $M$ is equivalent to $h^*\gamma_{n,k}$ by Proposition 5.1.

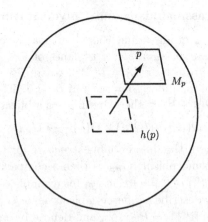

FIGURE 3. A classifying map $h : S^2 \to G_{2,1}$ for $\tau S^2$.

In order to deal with the general case, we will need the following:

LEMMA 7.1. *Let $\xi$ denote a vector bundle over an n-dimensional manifold B. Then B can be covered by $n+1$ sets $U_0, \ldots, U_n$, where each restriction $\xi_{|U_i}$ is trivial.*

PROOF. Choose an open cover of $B$ such that $\xi$ is trivial over each element. It is a well-known theorem in topology that this (and in fact any) cover of an $n$-dimensional manifold $B$ admits a refinement $\{V_\alpha\}_{\alpha \in A}$ with the property that any point in $B$ belongs to at most $n+1$ $V_\alpha$'s. Let $\{\phi_\alpha\}$ be a partition of unity subordinate to this cover, and denote by $A_i$ the collection of all subsets of $A$ with $i+1$ elements. Given $a = \{\alpha_0, \ldots, \alpha_i\} \in A_i$, denote by $W_a$ the set consisting of those $b \in B$ such that $\phi_\alpha(b) < \phi_{\alpha_0}(b), \ldots, \phi_{\alpha_i}(b)$ for all $\alpha \neq \alpha_0, \ldots, \alpha_i$. Then

(1) each $W_a$ is open,
(2) $W_a \cap W_{a'} = \emptyset$ if $a \neq a'$, and
(3) $\xi_{|W_a}$ is trivial.

Statements (1) and (2) follow immediately from the definition of these sets; (3) holds because $W_a \subset \cap_{j=0}^i \operatorname{supp} \phi_{\alpha_j} \subset \cap_{j=0}^i V_{\alpha_j}$, and $\xi$ is trivial over each $V_\alpha$. Define $U_i = \cup_{a \in A_i} W_a$. By (1), $U_i$ is open, and by (2) and (3), $\xi$ is trivial over $U_i$.

It remains to show that $U_0, \ldots, U_n$ cover $B$. For any fixed $b \in B$, consider the set $a = \{\alpha \in A \mid \phi_\alpha(b) > 0\}$. $a$ is nonempty because $\phi_\alpha(b) > 0$ for some $\alpha$, and $a \in A_j$ for some $j \leq n$ because at most $n+1$ of the sets $V_\alpha$ contain $b$, so that at most $n+1$ of the functions $\phi_\alpha$ are positive at $b$. Then $b \in W_a \subset U_j$. $\square$

THEOREM 7.1. *Let $\xi$ be a rank n vector bundle over B. For large enough $l$, there is a map $f : B \to G_{n,l}$ such that $\xi \cong f^* \gamma_{n,l}$.*

$G_{n,l}$ is then called a *classifying space* and $f$ a *classifying map* for $\xi$.

PROOF OF THEOREM 7.1. By Lemma 7.1, there is an open cover $U_1, \ldots, U_k$ of $B$ with the restriction of $\xi$ over each $U_i$ being trivial, so that there exist bundle charts $(\pi, \phi_i) : \pi^{-1}(U_i) \to U_i \times \mathbb{R}^n$. Let $\psi_1, \ldots, \psi_k$ be a partition of unity subordinate to $U_1, \ldots, U_k$, and define $\Phi_i : E(\xi) \to \mathbb{R}^n$ for each $i = 1, \ldots, k$ by

$$\Phi_i(u) = \begin{cases} (\psi_i \circ \pi)(u)\phi_i(u), & u \in \pi^{-1}(U_i), \\ 0, & u \notin \pi^{-1}(U_i). \end{cases}$$

$\Phi_i$ is linear on each fiber of $\xi$, but not one-to-one in general. However, $\Phi = (\Phi_1, \ldots, \Phi_k) : E(\xi) \to \mathbb{R}^{nk}$ is one-to-one: suppose $\Phi(u) = 0$; if $b = \pi(u)$, then $\psi_j(b) > 0$ for some $j$, and $b \in U_j$. Since $\Phi(u) = 0$, $\Phi_j(u)$ must also vanish. But $\Phi_j$ is an isomorphism on $E_b$, so that $u = 0$, and $\Phi$ is one-to-one. Then

$$\bar{f} : E(\xi) \to E(\gamma_{n,n(k-1)}),$$
$$u \mapsto (\Phi(\pi^{-1}(\pi(u))), \Phi(u))$$

is a bundle map covering

$$f : B \to G_{n,n(k-1)},$$
$$b \mapsto \Phi(\pi^{-1}(b)),$$

and $\xi \cong f^* \gamma_{n,n(k-1)}$ as claimed. $\square$

It follows from Lemma 7.1 and the proof of Theorem 7.1 that the integer $l$ in that theorem may always be chosen to equal $nk$, where $k$ is the dimension of $B$, and $n$ the rank of the bundle. In some cases, it may be taken to be much smaller: When $B$ is a $k$-sphere, one can choose $l = n$, since $S^k$ may be covered by two contractible open sets, and the restriction of the bundle to each of these is trivial by Exercise 61.

A classifying map is not unique, since any homotopic map will induce the same bundle by Corollary 6.3. Furthermore, if $k < k'$, then the inclusion $\mathbb{R}^{n+k} \subset \mathbb{R}^{n+k'}$ induces an inclusion $\imath : G_{n,k} \hookrightarrow G_{n,k'}$ and a bundle map $\gamma_{n,k} \to \gamma_{n,k'}$ covering $\imath$. Thus, $\gamma_{n,k} \cong \imath^* \gamma_{n,k'}$, and if $f : B \to G_{n,k}$ is a classifying map for $\xi$, then so is $\imath \circ f$. We claim, however, that for large enough $k'$, the homotopy class of $\imath \circ f$ is uniquely determined by $\xi$. To be more precise, let $B$ denote a $k$-dimensional manifold. Then $G_{n,nk}$ is a classifying space for rank $n$ bundles over $B$, and by the above remark, so is $G_{n,n(2k+1)}$. Let $[B, G_{n,n(2k+1)}]$ denote the collection of homotopy classes of maps $B \to G_{n,n(2k+1)}$, and $\mathrm{Vect}_n(B)$ the collection of equivalence classes of rank $n$ vector bundles over $B$.

THEOREM 7.2. *If $B$ is a $k$-dimensional manifold, then there exists a bijective correspondence between $\mathrm{Vect}_n(B)$ and $[B, G_{n,n(2k+1)}]$, which maps the equivalence class of the vector bundle $\xi$ over $B$ to the homotopy class of $\imath \circ f$, where $f : B \to G_{n,nk}$ is a classifying map for $\xi$, and $\imath : G_{n,k} \to G_{n,n(2k+1)}$ is the inclusion.*

PROOF. We only need to check that the correspondence is well-defined, since it will then be onto by Proposition 5.1, and one-to-one by the homotopy covering property.

So suppose $\xi \cong f^* \gamma_{n,nk} \cong g^* \gamma_{n,nk}$; we must show that $\imath \circ f, \imath \circ g : B \to G_{n,n(2k+1)}$ are homotopic. This will be done in two steps:

*Step 1:* Let $L : \mathbb{R}^{2n(k+1)} = \mathbb{R}^{n(k+1)} \times \mathbb{R}^{n(k+1)} \to \mathbb{R}^{2n(k+1)}$ be the isomorphism given by $L(u, v) = (-v, u)$, and denote by $\tilde{L}$ the induced map from $G_{n,n(2k+1)}$ into itself. Then $\tilde{L} \circ \imath \circ g \cong \imath \circ g$.

*Step 2:* $\imath \circ f \cong \tilde{L} \circ \imath \circ g$.

For the first step, observe that $L$ actually lies in $SO(2n(k + 1))$, and the latter is connected. Any smooth curve $c$ in $SO(2n(k + 1))$ joining the identity to $L$ induces a homotopy between the identity map of $G_{n,n(2k+1)}$ and $\tilde{L}$, so that $\tilde{L} \circ \imath \circ g \cong \imath \circ g$. (One example of such a $c : I \to SO(2n(k+1))$ can actually be explicitly described: Let $c(t)$ be the direct sum of rotations by angle $\pi t/2$ in each 2-plane $P_i$ spanned by $\mathbf{e}_i, \mathbf{e}_{i+n(k+1)}$.)

For the second step, the equivalence $f^* \gamma_{n,nk} \cong g^* \gamma_{n,nk}$ induces an equivalence $(\imath \circ f)^* \gamma_{n,n(2k+1)} \cong (\imath \circ g)^* \gamma_{n,n(2k+1)}$. By step 1, $(\imath \circ f)^* \gamma_{n,n(2k+1)} \cong (\tilde{L} \circ \imath \circ g)^* \gamma_{n,n(2k+1)}$. The latter equivalence is just a smooth section $h$ of the homomorphism bundle $\mathrm{Hom}((\imath \circ f)^* \gamma, (\tilde{L} \circ \imath \circ g)^* \gamma)$ (where we have dropped the subscripts for brevity) such that $h(b) : (\imath \circ f)(b) \to (\tilde{L} \circ \imath \circ g)(b)$ is an isomorphism for each $b \in B$. Let $H(b, t)$ be the subspace of $\mathbb{R}^{2n(k+1)}$ given by

$$H(b, t) = ((1 - t)1_{\mathbb{R}^{2n(k+1)}} + th(b))((\imath \circ f)(b)).$$

Then $H(b,0) = (\imath \circ f)(b)$, and $H(b,1) = (\tilde{L} \circ \imath \circ g)(b)$. It remains to show that $H(b,t)$ is $n$-dimensional for each $t$, so that $H$ maps into $G_{n,n(2k+1)}$ (this is the reason why we had to go to a higher-dimensional Grassmannian). Notice that $(\imath \circ f)(b)$ is a subspace of $\mathbb{R}^{n(k+1)} \times 0 \subset \mathbb{R}^{n(k+1)} \times \mathbb{R}^{n(k+1)}$, whereas $h(b)((\imath \circ f)(b)) = (\tilde{L} \circ \imath \circ g)(b)$ is a subspace of $0 \times \mathbb{R}^{n(k+1)} \subset \mathbb{R}^{n(k+1)} \times \mathbb{R}^{n(k+1)}$ because of the way $L$ was defined. Thus, for $u \in (\imath \circ f)(b)$,

$$((1-t)1_{\mathbb{R}^{2n(k+1)}} + th(b))u = ((1-t)u, tv) \in \mathbb{R}^{n(k+1)} \times \mathbb{R}^{n(k+1)},$$

where $(0,v) = h(b)u$. This expression can only vanish when $u = 0$, thereby completing the proof. $\qquad\square$

There is a way of avoiding having to consider different classifying spaces in the previous discussion: define $BGL(n)$ to be the union of the increasing sequence $G_{n,1} \subset G_{n,2} \subset \cdots$ with the weak topology; i.e., a subset of $BGL(n)$ is open iff its intersection with each $G_{n,k}$ is open. Let $\mathbb{R}^\infty$ denote the union of the increasing sequence $\mathbb{R} \subset \mathbb{R}^2 \subset \cdots$ with the corresponding topology, and set

$$E(\gamma_n) = \{(P,u) \in BGL(n) \times \mathbb{R}^\infty \mid u \in P\}.$$

Endow $E(\gamma_n)$ with the subspace topology, and define $\pi : E(\gamma_n) \to BGL(n)$ by $\pi(P,u) = P$. If we relax the conditions in the definition of fiber bundle by allowing bundle charts to be homeomorphisms instead of diffeomorphisms, and transition functions to be continuous rather than differentiable, then $\gamma_n = \pi : E(\gamma_n) \to BGL(n)$ is a rank $n$ vector bundle. Furthermore, if $\imath : G_{n,k} \to BGL(n)$ denotes inclusion, then $\imath^* \gamma_n$ is (continuously) equivalent to $\gamma_{n,k}$. The arguments in the previous theorem can be adapted to yield a bijective correspondence

$$\mathrm{Vect}_n(B) \longleftrightarrow [B, BGL(n)]$$

between equivalence classes of rank $n$ vector bundles over a manifold $B$ and homotopy classes of maps $B \to BGL(n)$. $BGL(n)$ is called the *classifying space* for rank $n$ bundles, and $\gamma_n$ the *universal rank $n$ bundle*.

The work in this section carries over to principal $GL(n)$-bundles: Indeed, let $P_\xi$, $P_\eta$ denote the principal frame bundles of $\xi$, $\eta$ respectively. If $\eta \cong f^*\xi$, then $P_\eta \cong f^*P_\xi$. Thus, if $G_{n,k}$ is a classifying space for rank $n$ vector bundles over $B$, then any principal $GL(n)$ bundle over $B$ is the pullback of the principal frame bundle of $\gamma_{n,k}$. The total space of the latter is called the *Stiefel manifold* $V_{n,k}$. By definition of the frame bundle, it is the open subset of $\mathbb{R}^{n(n+k)}$ consisting of all $n$-tuples $(v_1, \ldots, v_n) \in \mathbb{R}^{n+k} \times \cdots \times \mathbb{R}^{n+k}$ with $v_1, \ldots, v_n$ linearly independent. It admits a principal $O(n)$ subbundle with total space $V_{n,k}^0$ consisting of those $(v_1, \ldots, v_n) \in V_{n,k}$ for which $\langle v_i, v_j \rangle = \delta_{ij}$. By Exercise 51, $\gamma_{n,k}$ admits a reduction of its structure group to $O(n)$. By Theorem 7.1, *every* rank $n$ bundle admits a reduction of its structure group to $O(n)$. This is just another way of proving that a vector bundle admits a Euclidean metric.

Although $\gamma_{n,k}$ is not orientable and therefore does not admit a reduction of its structure group to $SO(n)$, there is a classifying space for bundles with group $SO(n)$: Let $\tilde{G}_{n,k}$ denote the collection of *oriented* $n$-planes in $\mathbb{R}^{n+k}$. $SO(n+k)$ acts transitively on $\tilde{G}_{n,k}$, and the map $\pi : SO(n+k) \to$

$\tilde{G_{n,k}}$ which assigns to $L \in SO(n + k)$ the $n$-plane $L(\mathbb{R}^n \times 0)$ oriented by the ordered basis $(Le_1, \ldots, Le_n)$ identifies $\tilde{G_{n,k}}$ with the homogeneous space $SO(n+k)/SO(n) \times SO(k)$. Since a given $n$-plane has exactly two orientations, the projection $\tilde{G_{n,k}} \to G_{n,k}$ (which forgets orientation) is a 2-fold covering. As before, one defines the *universal oriented $n$-plane bundle* $\gamma_{\tilde{n,k}}$ over the *oriented Grassmannian* $\tilde{G_{n,k}}$ with total space $E(\gamma_{\tilde{n,k}}) = \{(P, u) \in \tilde{G_{n,k}} \times \mathbb{R}^{n+k} \mid u \in P\}$. The total space of the associated oriented orthonormal frame bundle is $V_{n,k}^0$: The projection $\pi : V_{n,k}^0 \to \tilde{G_{n,k}}$ assigns to an $n$-tuple of orthonormal vectors the plane spanned by them together with the orientation determined by the ordering of the vectors. The group of this bundle is $SO(n)$, so that $\gamma_{\tilde{n,k}}$ admits $SO(n)$ as structural group.

We say two oriented bundles are equivalent if there exists an equivalence which is orientation-preserving on each fiber.

THEOREM 7.3. *Let $\xi$ be an oriented rank $n$ bundle over $B$. Then for sufficiently large $k$, there is a map $f : B \to \tilde{G_{n,k}}$ with $f^*\gamma_{\tilde{n,k}} \cong \xi$.*

The proof is left as an exercise. The procedure followed in constructing $BGL(n)$ and $\gamma_n$ can be applied to obtain the *classifying space* $BSO(n)$ for vector bundles with group $SO(n)$, and the *universal bundle* $\tilde{\gamma}_n$ over $BSO(n)$.

EXERCISE 64. Let $\xi$ denote a rank $n$ vector bundle over $B$ together with a Euclidean metric. The *unit sphere bundle* of $\xi$ is the bundle $\xi^1$ with fiber $S^{n-1}$ over $B$ and total space $E(\xi^1) = \{u \in E(\xi) \mid \langle u, u \rangle = 1\}$. Show that the total space of the unit sphere bundle of $\tau S^n$ is diffeomorphic to the Stiefel manifold $V_{2,n-1}^0$.

EXERCISE 65. Show that $G_{n,k}$ is diffeomorphic to $G_{k,n}$.

EXERCISE 66. Define the *complex Grassmannian* $G_{n,k}(\mathbb{C})$ to be the set of all complex $n$-dimensional subspaces of $\mathbb{C}^{n+k}$.

(a) Show that there exists a bijective correspondence $G_{n,k}(\mathbb{C}) \leftrightarrow U(n + k)/U(n) \times U(k)$, where $U(n)$ denotes the unitary group of $n \times n$ complex matrices $A$ such that $A\bar{A}^t = I$, see Chapter 6. Under this identification, $G_{n,k}(\mathbb{C})$ becomes a compact homogeneous space.

(b) Construct a universal complex rank $n$ vector bundle $\gamma_{n,k}(\mathbb{C})$ over $G_{n,k}(\mathbb{C})$ as we did for $G_{n,k}$.

(c) When $n = k = 1$, $G_{1,1}(\mathbb{C}) = \mathbb{C}P^1 \sim S^2$ by Exercise 49. Show that the unit sphere bundle of $\gamma_{1,1}(\mathbb{C})$ is equivalent to the Hopf fibration $S^3 \to S^2$ from Examples and Remarks 1.1(iii).

EXERCISE 67. Prove Theorem 7.3.

EXERCISE 68. Let $\xi$ be a rank $n$ vector bundle over a manifold $B$, so that $\xi \cong f^*\gamma_{n,k}$ for some $k$ and $f : B \to G_{n,k}$.

(a) Show that there is a one-to-one homomorphism $\xi \to \epsilon^{n+k}$ of $\xi$ into the trivial rank $n + k$ bundle $\epsilon^{n+k}$ over $B$.

(b) Conclude that there exists a rank $k$ bundle $\eta$ over $B$ such that $\xi \oplus \eta$ is trivial.

# CHAPTER 3

# Homotopy Groups and Bundles Over Spheres

## 1. Differentiable Approximations

We saw in Chapter 2 that the concept of homotopy plays a central role in bundle theory. Although we have so far dealt only with differentiable maps, it is more convenient, when working with homotopies, to consider continuous maps, and we will do so in this chapter. One purpose of this section is to try and convince the reader that we are not introducing new objects when, for example, we consider the pullback $f^*\xi$ of a bundle via a continuous map $f$: Explicitly, we will show that any continuous map between manifolds is homotopic to a differentiable one, and the latter can be chosen to be arbitrarily close to the original one.

We begin with the following:

THEOREM 1.1 (Tubular Neighborhood Theorem). *Let $h : M^n \to \mathbb{R}^{n+k}$ be a differentiable imbedding. Then there is a neighborhood of the zero section in the normal bundle $E(\nu_h)$ of $h$ which is mapped diffeomorphically onto a neighborhood of $h(M)$ in $\mathbb{R}^{n+k}$.*

PROOF. Recall that $E(\nu_h) = \{(p,u) \in M \times T\mathbb{R}^{n+k} \mid u \in (h_*M_p)^\perp\}$, using the canonical Euclidean metric on $T\mathbb{R}^{n+k}$. Define $f : E(\nu_h) \to \mathbb{R}^{n+k}$ by $f(p,u) = h(p) + \mathcal{J}_{h(p)}^{-1}u$. If $s : M \to E(\nu_h)$ denotes the zero section, we claim that $f_*$ is an isomorphism at each $s(p)$, $p \in M$; equivalently, $f_* \circ \pi_2 : s^*TE(\nu_h) \to T\mathbb{R}^{n+k}$ is an isomorphism on each fiber, where $\pi_2 : s^*TE(\nu_h) \to TE(\nu_h)$ is projection onto the second factor.

Now, by Exercise 60, $s^*\tau_{E(\nu_h)} \cong \tau_M \oplus \nu_h$, and under this equivalence, $f_* \circ \pi_2$ maps $(u,v) \in E(\tau_M \oplus \nu_h)$ to $h_*u + v$; since $v \perp h_*u$, the assertion is clear. Thus, there is an open neighborhood of $s(M)$ in $E(\nu_h)$ on which $f$ is a local diffeomorphism. Since the restriction of $f$ to $s(M)$ is a homeomorphism onto its image $h(M)$, the statement of the theorem is now a consequence of the next lemma. $\square$

LEMMA 1.1. *Let $M$, $N$ be second countable Hausdorff spaces, with $M$ locally compact, and $f : M \to N$ a local homeomorphism. If the restriction of $f$ to some closed subset $A$ of $M$ is a homeomorphism, then $f$ is a homeomorphism on some neighborhood $V$ of $A$.*

PROOF. We construct $V$ inductively. Let $W_i$ be a sequence of nested compact sets $W_1 \subset W_2 \subset \cdots$ whose union equals $M$, and set $A_i = A \cap W_i$. Notice that if $f$ is one-to-one on a compact set $C$, then it remains so on some compact neighborhood of $C$: Otherwise, there would exist sequences $p_n \to p \in C$,

$q_n \to q \in C$, such that $p_n \neq q_n$ but $f(p_n) = f(q_n)$. By continuity, $f(p) = f(q)$, so that $p = q$. But since any neighborhood of $p$ contains $p_n$ and $q_n$ for large enough $n$, $f$ would not be a local homeomorphism at $p$.

A similar argument shows that if $C$ is a compact subset of $A$, then $C$ has a neighborhood $U$ with compact closure such that $f$ is one-to-one on $\bar{U} \cup A$. Otherwise, we could choose a sequence $\{p_n\}$ in $\bar{U} \setminus A$ converging to some $p \in C$, and a sequence $\{q_n\}$ in $A$ with $f(p_n) = f(q_n)$. Since $f(q_n) \to f(p)$ and $f$ is a homeomorphism on $A$, $q_n \to p$, contradicting the fact that $f$ is a local homeomorphism at $p$.

By the above, there exists a neighborhood $V_1$ of $A_1$ such that $\bar{V}_1$ is compact and $f$ is a homeomorphism on $\bar{V}_1 \cup A_1$. Inductively, if $V_i$ is a neighborhood of $A_i$ satisfying these conditions, then $\bar{V}_i \cup A_{i+1}$ is a compact subset of $\bar{V}_i \cup A$, and since $f$ is a homeomorphism on the latter, there exists a neighborhood $V_{i+1}$ of $\bar{V}_i \cup A_{i+1}$ such that $f$ is one-to-one on $\bar{V}_{i+1} \cup A$. Then $f$ is one-to-one on the neighborhood $V := \cup V_i$ of $A$. □

Let $(N, d)$ be a metric space as defined in Chapter 5, Section 7, and let $\epsilon > 0$. A map $g : M \to N$ is said to be an $\epsilon$-*approximation* of $f : M \to N$ if $d(g(p), f(p)) \leq \epsilon$ for all $p \in M$. A homotopy $H : M \times I \to N$ is said to be $\epsilon$-*small* if $d(H(p, t_0), H(p, t_1)) \leq \epsilon$ for all $t_0, t_1 \in I$, $p \in M$.

LEMMA 1.2. *Let $M$ be a manifold, $f : M \to \mathbb{R}^k$ be a continuous map. For any $\epsilon > 0$, there exists a differentiable $\epsilon$-approximation $g : M \to \mathbb{R}^k$ of $f$ which is homotopic to $f$ via an $\epsilon$-small homotopy.*

PROOF. Denote by $|a|$ the norm of $a \in \mathbb{R}^k$, and by $B_\epsilon(a)$ the set of all $b \in \mathbb{R}^k$ with $|a - b| < \epsilon$. For each $p \in M$, let $V_p = f^{-1}(B_\epsilon(f(p)))$, and define $h_p : V_p \to \mathbb{R}^k$ by $h_p(q) = f(p)$. Consider a locally finite refinement $\{U_\alpha\}_{\alpha \in A}$ of $\{V_p\}_{p \in M}$, so that for each $\alpha \in A$, there exists some $p(\alpha) \in M$ with $U_\alpha \subset V_{p(\alpha)}$. Let $\{\phi_\alpha\}$ be a partition of unity subordinate to $\{U_\alpha\}$, and define for each $\alpha$ a differentiable map $g_\alpha : M \to \mathbb{R}^k$ by

$$g_\alpha(q) = \begin{cases} \phi_\alpha(q) h_{p(\alpha)}(q), & \text{for } q \in U_\alpha, \\ 0, & \text{otherwise.} \end{cases}$$

We claim that $g := \sum_\alpha g_\alpha$ satisfies the conclusion of the lemma: g is differentiable, since it is a finite sum of smooth maps in a neighborhood of any point. Furthermore,

$$|g(q) - f(q)| = \left| \sum_{\{\alpha | q \in U_\alpha\}} \phi_\alpha(q) h_{p(\alpha)}(q) - f(q) \right| = \left| \sum \phi_\alpha(q) f(p(\alpha)) - f(q) \right|$$

$$= \sum \phi_\alpha(q) |f(p(\alpha)) - f(q)| < \epsilon \sum \phi_\alpha(q) = \epsilon.$$

The homotopy $H : M \times I \to \mathbb{R}^k$ can be taken to be given by $H(q, t) = (1 - t)f(q) + tg(q)$. □

THEOREM 1.2. *Let $f : M \to N$ be a continuous map between differentiable manifolds, where $N$ is compact. Endow $N$ with a metric $d$. Given any $\epsilon > 0$, there exists a differentiable $\epsilon$-approximation $g$ of $f$, which is homotopic to $f$ via an $\epsilon$-small homotopy.*

PROOF. Choose some imbedding $h : N \to \mathbb{R}^k$, cf. Example 7.1 in Chapter 2. Since $h : N \to h(N)$ is a homeomorphism between compact spaces, there exists some $\delta > 0$ with the property that $|h(p) - h(q)| < \epsilon$ whenever $d(p, q) < \delta$. Again by compactness, we may assume that the tubular neighborhood of $h(N)$ guaranteed by Theorem 1.1 is $B_{\delta/2}(h(N))$. Let $r : B_{\delta/2}(h(N)) \to h(N)$ be the smooth retraction that corresponds to the normal bundle projection under the diffeomorphism from 1.1. Choose some differentiable $\delta/2$- approximation $\tilde{f}$ of $h \circ f$. We claim that $g := h^{-1} \circ r \circ \tilde{f} : M \to N$ satisfies the conclusion of the theorem: This map is by definition differentiable, and for any $p \in M$,

$$|(r \circ \tilde{f})(p) - (h \circ f)(p)| \leq |(r \circ \tilde{f})(q) - \tilde{f}(p)| + |\tilde{f}(p) - (h \circ f)(p)| < \frac{\delta}{2} + \frac{\delta}{2} = \delta,$$

so that $d(f(p), g(p)) = d(h^{-1}(h(f(p))), h^{-1}((r \circ \tilde{f})(p))) < \epsilon$. The homotopy is given by $H(p, t) = (h^{-1} \circ r)(t\tilde{f}(p) + (1 - t)(h \circ f)(p))$. $\square$

EXERCISE 69. Modify the proofs of Lemma 1.2 and Theorem 1.2 to show that if the restriction of the continuous map $f : M \to N$ to a closed subset $A$ of $M$ is differentiable, then the differentiable $\epsilon$-approximation of $f$ may be chosen to agree with $f$ on $A$.

EXERCISE 70. Let $f : M \to N$ be a continuous map between differentiable manifolds, where $N$ is no longer assumed to be compact. Show that $f$ is homotopic to a differentiable map $g : M \to N$. Given $\epsilon > 0$, can $g$ always be chosen to be an $\epsilon$-approximation of $f$?

## 2. Homotopy Groups

The boundary $\partial I^n$ of the $n$-cube $I^n$ consists of the $(n - 1)$-faces $\{a \in I^n \mid a_i = \alpha\}$, $1 \leq i \leq n$, $\alpha = 0, 1$. The initial $(n - 1)$-face $I^{n-1} \times 0$ will be identified with $I^{n-1}$, and the union of the remaining faces will be denoted $J^{n-1}$.

Let $X$ be a topological space, $A$ a subset of $X$, and $p \in A$. We will be dealing in this section with the set $C^n(X, A, p)$ (sometimes denoted $C^n$ when there is no risk of confusion) of all continuous maps $f : (I^n, I^{n-1}, J^{n-1}) \to (X, A, p)$ from $I^n$ to $X$ that map $I^{n-1}$ into $A$ and $J^{n-1}$ into $p$.

Two maps $f, g \in C^n$ are said to be *homotopic in* $C^n$ if there exists a homotopy $H$ between $f$ and $g$ with $H \circ \imath_t$ in $C^n$ for all $t \in I$. The relation $f \sim g$ if $f$ and $g$ are $C^n$-homotopic is an equivalence relation, and partitions $C^n$ into equivalence classes. The collection of these classes is denoted $\pi_n(X, A, p)$, and $\pi_n(X, p)$ in the case $A = \{p\}$.

Given $f, g \in C^n$, define their *sum* $f + g \in C^n$ to be

$$(f + g)(a_1, \ldots, a_n) = \begin{cases} f(2a_1, a_2, \ldots, a_n), & \text{if } 0 \leq a_1 \leq 1/2, \\ g(2a_1 - 1, a_2, \ldots, a_n), & \text{if } 1/2 \leq a_1 \leq 1. \end{cases}$$

If $H_f$ and $H_g$ are $C^n$-homotopies between $f, f_1$, and $g, g_1$ respectively, then $f + g$ is $C^n$-homotopic to $f_1 + g_1$ via $H$, where

$$H(a, t) = \begin{cases} H_f(2a_1, a_2, \ldots, a_n, t), & \text{if } 0 \leq a_1 \leq 1/2, \\ H_g(2a_1 - 1, a_2, \ldots, a_n, t), & \text{if } 1/2 \leq a_1 \leq 1. \end{cases}$$

There is therefore a well-defined addition in $\pi_n(X, A, p)$ given by $[f]+[g] := [f+g]$, where $[f]$ denotes the element in $\pi_n$ determined by $f$. It is straightforward to check that $\pi_n(X, A, p)$ together with this operation is a group, called *the n-th homotopy group of X relative to A, based at p*; cf. the exercises below.

THEOREM 2.1. $\pi_2(X, p)$ *is abelian, and* $\pi_n(X, A, p)$ *is abelian when* $n > 2$.

PROOF. Consider a homeomorphism $F$ of $I^2$ with the unit disk $D$ mapping the line segment $a_1 = 1/2$ to the diameter $0 \times [-1, 1]$ in $D$, and let $H : D \times I \to D$ be given by $H(a, t) = e^{i\pi t}a$. Then $\tilde{H} : I^2 \times I \to I^2$, where $\tilde{H}(a, t) = F^{-1}(H(F(a), t))$, can be viewed as a 'rotation' of $I^2$ by angle $\pi t$, $0 \leq t \leq 1$. Now, any $f \in C^2(X, p)$ is homotopic to $\bar{f} := f \circ \tilde{H} \circ \imath_1$; since $a \mapsto \tilde{H}(a, 1)$ interchanges the two half-squares $0 \leq a_1 \leq 1/2$ and $1/2 \leq a_1 \leq 1$, we have that

$$f + g \sim \overline{f + g} = \bar{g} + \bar{f} \sim g + f.$$

When $n > 2$, the homeomorphism $F$ can be extended to $I^n = I^2 \times I^{n-2}$ by keeping the last $n - 2$ variables fixed. The rotation $\tilde{H} \circ \imath_t$ defined above then belongs to $C^n(I^n, I^{n-1}, J^{n-1})$, so that any $f \in C^n(X, A, p)$ is $C^n$-homotopic to $\bar{f}$, with $\bar{f}$ as before; the rest of the argument then goes through to show that $\pi_n(X, A, p)$ is abelian. $\qquad\square$

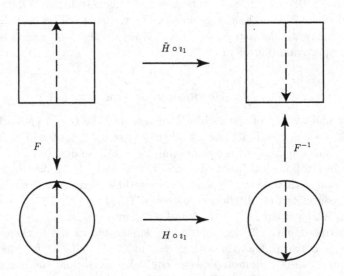

FIGURE 1. The "rotation" $\tilde{H} \circ \imath_1$ of $I^2$.

Notice that if $f$ and $g$ are homotopic in $C^n(X, A, p)$, then their restrictions to $I^{n-1}$ are homotopic in $C^{n-1}(A, p)$, and the assignment $f \mapsto f_{|I^{n-1}}$ is a homotopy group mapping.

DEFINITION 2.1. For $n \geq 2$, the *boundary operator* is the map $\partial : \pi_n(X, A, p) \to \pi_{n-1}(A, p)$ given by $\partial[f] = [f_{|I^{n-1}}]$.

The boundary operator is a group homomorphism, since $(f + g)_{|I^{n-1}} = f_{|I^{n-1}} + g_{|I^{n-1}}$.

Another important group homomorphism is the one induced by a continuous map $h : (X, A, p) \to (Y, B, q)$. When composed with $f \in C^n(X, A, p)$, $h$ yields an element $h \circ f \in C^n(Y, B, q)$. Moreover, $h \circ f$ and $h \circ g$ are $C^n$-homotopic whenever $f$ and $g$ are, so that $h$ induces a map

$$h_\# : \pi_n(X, A, p) \to \pi_n(Y, B, q),$$
$$[f] \mapsto [h \circ f],$$

which is a homomorphism because $h \circ (f + g) = h \circ f + h \circ g$.

DEFINITION 2.2. The *homotopy sequence* of $(X, A, p)$ is the sequence of homomorphisms

$$\cdots \longrightarrow \pi_n(A, p) \xrightarrow{\imath_\#} \pi_n(X, p) \xrightarrow{\jmath_\#} \pi_n(X, A, p) \xrightarrow{\partial} \pi_{n-1}(A, p) \longrightarrow \cdots$$
$$\cdots \xrightarrow{\partial} \pi_1(A, p) \xrightarrow{\imath_\#} \pi_1(X, p),$$

where $\imath : (A, p) \to (X, p)$ and $\jmath : (X, p, p) \to (X, A, p)$ are the inclusion maps.

THEOREM 2.2.     (1) *The homotopy sequence of $(X, A, p)$ is exact.*
(2) *If $h : (X, A, p) \to (Y, B, q)$ is continuous, then the diagram*

$$
\begin{array}{ccccccc}
\pi_n(A, p) & \xrightarrow{\imath_\#} & \pi_n(X, p) & \xrightarrow{\jmath_\#} & \pi_n(X, A, p) & \xrightarrow{\partial} & \pi_{n-1}(A, p) \\
h_\# \downarrow & & h_\# \downarrow & & h_\# \downarrow & & h_\# \downarrow \\
\pi_n(B, q) & \xrightarrow{\imath_\#} & \pi_n(Y, q) & \xrightarrow{\jmath_\#} & \pi_n(Y, B, q) & \xrightarrow{\partial} & \pi_{n-1}(B, q)
\end{array}
$$

*commutes.*

PROOF. Statement (1) is fairly straightforward, and we illustrate the procedure by showing exactness of the portion

$$\pi_n(X, p) \xrightarrow{\jmath_\#} \pi_n(X, A, p) \xrightarrow{\partial} \pi_{n-1}(A, p),$$

which is perhaps also the more instructive part. If $[f] \in \pi_n(X, p)$, then $(\jmath \circ f)(\partial I^n) = \{p\}$, and $(\jmath \circ f)_{|I^{n-1}}$ is a constant map. Then $\partial \jmath_\#[f] = [(\jmath \circ f)_{|I^{n-1}}]$ is 0 by Exercise 71, so that $\operatorname{im} \jmath_\# \subset \ker \partial$. Conversely, if $[f] \in \ker \partial$, then $f_{|I^{n-1}}$ is homotopic in $C^{n-1}(A, p)$ to the constant map sending $I^{n-1}$ to $p$ via some $H : (I^{n-1} \times I, \partial I^{n-1} \times I) \to (A, p)$ with $H \circ \imath_0 = f$. Consider the subset $E = I^n \times 0 \cup \partial I^n \times I$ of the boundary $\partial I^{n+1}$ of $I^{n+1} = I^n \times I$, and extend $H$ to $\tilde{H} : (E, I^{n-1} \times I, J^{n-1} \times I) \to (X, A, p)$ by setting $\tilde{H}(a, 0) = f(a)$ for $a \in I^n$, $\tilde{H}(a, t) = H(a, t)$ for $(a, t) \in I^{n-1} \times I$, and $\tilde{H}(a, t) = f(a) = p$ for $(a, t) \in J^{n-1} \times I$. Now, there exists a retraction $r$ (see Exercise 74) from $I^n \times I$ onto $E$: For example, if $p_0 = (1/2, \ldots, 1/2, 2) \in \mathbb{R}^{n+1}$, let $r(q)$ be the point where the line through $p_0$ and $q$ intersects $E$. Then $\tilde{H} \circ r : (I^n \times I, I^{n-1} \times I, J^{n-1} \times I) \to (X, A, p)$ is a homotopy of $f$ into a map $g : (I^n, \partial I^n) \to (X, p)$, and $[f] = \jmath_\#[g]$. Thus, $\ker \partial = \operatorname{im} \jmath_\#$.

Statement (2) follows from Exercise 73 and the fact that $(k \circ h)_\# = k_\# \circ h_\#$ for maps $h : (X, A, p) \to (Y, B, q)$ and $k : (Y, B, q) \to (Z, C, r)$.  $\square$

EXAMPLES AND REMARKS 2.1. (i) If $h : (X, A, p) \to (Y, B, q)$ is a homotopy equivalence (i.e., there exists $k : (Y, B, q) \to (X, A, p)$ such that $k \circ h \sim 1_X$ and $h \circ k \sim 1_Y$), then $h_\# : \pi_n(X, A, p) \to \pi_n(Y, B, q)$ is an isomorphism for each $n$. Indeed, $k_\# \circ h_\# = (k \circ h)_\# = (1_X)_\# = 1_{\pi_n(X,A,p)}$, and similarly for $h_\# \circ k_\#$. Notice that for the second equality, we used the fact that if $H : (X \times I, A \times I, p \times I) \to (Y, B, q)$ is a homotopy, then $(H \circ \imath_0)_\# = (H \circ \imath_1)_\#$. In particular, if $X$ is contractible, then all its homotopy groups vanish.

(ii) $\pi_k(S^n, p) = 0$ for all $k < n$: To see this, notice first that given $q \in S^k$, any map $(I^k, \partial I^k) \to (X, p)$ can be considered as a map $(S^k, q) \to (X, p)$, so that $\pi_k(X, p)$ consists of homotopy classes of such maps. The above assertion then says that any $f : (S^k, q) \to (S^n, p)$ is homotopic to a constant map if $k < n$. By Theorem 1.2, we may assume that $f$ is differentiable. By Sard's theorem, $f$ cannot be onto (otherwise all values would be critical because $k < n$), so that $f$ maps into a sphere with a point deleted. The latter is contractible, and by (i), $f$ is homotopic to a constant map.

(iii) $\pi_n(S^n, p) \cong \mathbb{Z}$: As observed in (ii), $\pi_n(S^n, p)$ may be identified with homotopy classes of maps $(S^n, p) \to (S^n, p)$. Recall from Chapter 1, Theorem 15.3 and Theorem 1.2 that each class contains a differentiable representative $f$ to which we may assign an integer $\deg f \in \mathbb{Z}$. Furthermore, two homotopic maps have the same degree, and it can be shown that conversely, maps with the same degree are homotopic. Thus, there is a one-to-one function $\deg : \pi_n(S^n, p) \to \mathbb{Z}$. Since the degree is computed by adding the local degrees in the preimage of a regular value, $\deg(f + g) = \deg f + \deg g$, and $\deg$ is a group homomorphism. In particular, $\deg(k \cdot 1_{S^n}) = k$, and $\deg$ is an isomorphism.

We next examine the dependence of $\pi_n$ on the base point. Suppose $X$ is path-connected. We claim that for any two points $p_0$ and $p_1$ of $X$, $\pi_n(X, p_0) \cong \pi_n(X, p_1)$. To see this, consider a curve $c : I \to X$ from $p_0$ to $p_1$. Given $f \in C^n(X, p_0)$, define

$$h_{c,f} : E = I^n \times 0 \cup \partial I^n \times I \to X,$$

$$(q, t) \mapsto \begin{cases} f(q), & \text{if } t = 0, \\ c(t), & \text{if } q \in \partial I^n. \end{cases}$$

If $r : I^n \times I \to E$ denotes the retraction used in the proof of Theorem 2.2, then $H_{c,f} := h_{c,f} \circ r : I^n \times I \to X$ is a homotopy of $f$ into a map $c(f) := H_{c,f} \circ \imath_1 : (I^n, \partial I^n) \to (X, p_1)$, and $c(f)$ represents an element of $\pi_n(X, p_1)$. Furthermore, given a homotopy $F$ in $C^n(X, p_0)$ of $f$ into $g$, the map

$$G : E \times I \to X,$$

$$(q, t, s) \mapsto \begin{cases} F(q, s), & \text{if } (q, t, s) = (q, 0, s) \in I^n \times 0 \times I, \\ c(t), & \text{if } (q, t, s) \in \partial I^n \times I \times I, \end{cases}$$

homotopes $h_{c,f}$ into $h_{c,g}$. Then $H_{c,f}$ and $H_{c,g}$ are homotopic, so that $[c(f)] = [c(g)]$, and $c$ induces a map $c_\# : \pi_n(X, p_0) \to \pi_n(X, p_1)$. Similarly, if $F$ is a homotopy of $c$ into another curve $\bar{c}$ from $p_0$ to $p_1$, then $G : E \times I \to X$ is a homotopy of $h_{c,f}$ into $h_{\bar{c},f}$, where $G(q, 0, s) = f(q)$ when $q \in I^n$, and $G(q, t, s) = F(t, s)$ when $q \in \partial I^n$. Thus, $c_\#$ only depends on the homotopy class of $c$.

Finally, it is clear from the above construction that $c_\#$ is a homomorphism. It follows that $c_\#$ is an isomorphism with inverse $c_\#^{-1}$, where $c^{-1}(t) = c(1 - t)$.

If we now restrict ourselves to the case when $p_0 = p_1$, we obtain the following:

PROPOSITION 2.1. *The fundamental group $\pi_1(X,p)$ acts on $\pi_n(X,p)$ by means of $[c]([f]) = c_\#([f])$.*

Although we used a particular homotopy $H$ of $f$ to construct $c(f)$, the only condition required of $H$ is that $H(\partial I^n, t) = c(t)$:

LEMMA 2.1. *Let $c$ be a curve in $X$ from $p_0$ to $p_1$. If $f \in C^n(X,p_0)$ and $H$ is a homotopy of $f$ such that $H(\partial I^n,t) = c(t)$ for all $t$, then $H \circ \imath_1$ represents the same element as $c(f)$ in $\pi_n(X,p_1)$.*

PROOF. Let

$$h(p,t) = \begin{cases} H_{c,f}(p, 1-2t), & t \leq 1/2, \\ H(p, 2t-1), & t \geq 1/2. \end{cases}$$

The map $h$ homotopes $c(f)$ into $H \circ \imath_1$, and $h(\partial I^n, t) = c^{-1}c(t)$. Consider a homotopy $F : (I^2, J^1) \to (X, p_1)$ of $F \circ \imath_0 = c^{-1}c$ into $F \circ \imath_1 = p_1$, and define a map $\bar{h}$ on the subset $E = I^n \times I \times 0 \cup \partial(I^n \times I) \times I$ of $I^{n+2}$ by

$$\bar{h}(p,t,s) = \begin{cases} h(p,t), & \text{if } s = 0, \\ c(f)(p), & \text{if } t = 0, \\ (H \circ \imath_1)(p), & \text{if } t = 1, \\ F(h(p,t),s), & \text{if } p \in \partial I^n. \end{cases}$$

Let $r : I^n \times I \times I = I^{n+2} \to E$ be the retraction onto $E$ from Theorem 2.2, and extend $\bar{h}_{|E}$ to all of $I^{n+2}$ by $\bar{h}_{|E} \circ r$. Finally, define $k : I^n \times I \to X$ by $k(p,t) = \bar{h}(p,t,1)$. Then $k \circ \imath_0 = c(f)$, $k \circ \imath_1 = H \circ \imath_1$, and $k(\partial I^n, t) = p_1$, so that $k$ is a homotopy in $C^n(X,p_1)$ of $c(f)$ into $H \circ \imath_1$.                    □

A space $X$ is said to be *simply connected* if $\pi_1(X,p) = \{e\}$ for all $p \in X$, and *$n$-simple* if the action of the fundamental group on $\pi_n$ from Proposition 2.1 is trivial. Of course, any simply connected $X$ is $n$-trivial for all $n$, but another large class of such spaces is given in the following:

EXAMPLE 2.1. Any Lie group $G$ is $n$-simple for every $n$: Let $f \in C^n(G, e)$, and consider a curve from $e$ to $g_0$. Then $H(a,t) = c(t)f(a)$ is a homotopy of $f$ satisfying the hypotheses of Lemma 2.1, and $[c(f)] = [H \circ \imath_1]$. But $(H \circ \imath_1)(a) = g_0 \cdot f(a)$, so that the isomorphism $c_\# : \pi_n(G,e) \to \pi_n(G, g_0)$ is induced by left translation $L_{g_0}$ (it is also induced by right translation, if we take $H(a,t) = f(a)c(t)$). The special case $g_0 = e$ then yields that the action of $\pi_1$ on $\pi_n$ is trivial, and $G$ is $n$-simple.

Another interesting fact about $\pi_n(G, e)$ is that addition is given by group multiplication; i.e., if $f_1, f_2 \in C^n(G, e)$, then

$$[f_1] + [f_2] = [f_1 \cdot f_2].$$

To see this, let $e$ denote the constant map; then $f_1 + f_2 = (f_1 + e) \cdot (e + f_2)$ by definition of addition, and $(f_1 + e) \cdot (e + f_2)$ is homotopic to $f_1 \cdot f_2$ by multiplying the homotopies $f_1 + e \sim f_1$ and $e + f_2 \sim f_2$.

When $X$ is not $n$-simple, one can consider the set $\pi_n(X,p)/\pi_1(X,p)$ of orbits under the action of $\pi_1$:

PROPOSITION 2.2. *If $X$ is path-connected, then there is a bijection between the set of orbits $\pi_n(X,p)/\pi_1(X,p)$ and the set $[S^n, X]$ of (free) homotopy classes of maps $S^n \to X$.*

PROOF. We have already seen that any $f : (I^n, \partial I^n) \to (X,p)$ can be viewed as $f : (S^n, q) \to (X,p)$, yielding a map $h : \pi_n(X,p) \to [S^n, X]$. Furthermore, $c(f)$ is homotopic to $f$, so that $h$ factors through $h : \pi_n(X,p)/\pi_1(X,p) \to [S^n, X]$.

To see that $h$ is onto, let $f : S^n \to X$, and set $p_0 = f(q)$. View $f'$: $(I^n, \partial I^n) \to (X, p_0)$. If $c$ is a curve in $X$ from $p_0$ to $p$, then $c(f) : (I^n, \partial I^n) \to (X,p)$ is homotopic to $f$, so that $h$ is onto.

Finally, suppose two orbits $O_0$ and $O_1$ are mapped to the same homotopy class in $[S^n, X]$. Choose $f_i \in C^n(X,p)$ with $[f_i] \in O_i$, $i = 0, 1$. Now, $f_0$ homotopes into $f_1$ via some $H : I^n \times I \to X$ with $H$ constant on each $\partial I^n \times t$. If $c(t) = H(\partial I^n \times t)$, then $[c] \in \pi_1(X,p)$, and by Lemma 2.1, $[f_1] = [H \circ \iota_1] = c_{\#}[f_0]$, so that $O_0 = O_1$. Thus, $h$ is bijective. $\square$

EXERCISE 71. Show that if $f \in C^n(X, A, p)$ maps $I^n$ into $A$, then $[f] = 0 \in \pi_n(X, A, p)$.

EXERCISE 72. Show that if $f \in C^n(X, A, p)$, then $-[f] = [-f]$, where $-f$ is given by $-f(a_1, \ldots, a_n) = f(1 - a_1, a_2, \ldots, a_n)$.

EXERCISE 73. Let $h : (X, A, p) \to (Y, B, q)$. Prove that if $\bar{h} : (A, p) \to (B, q)$ denotes the restriction of $h$ to $A$, then $\partial \circ h_{\#} = \bar{h}_{\#} \circ \partial$.

EXERCISE 74. Let $A \subset X$, and suppose there is a retraction $r : X \to A$ of $X$ onto $A$; i.e., $r_{|A} = 1_A$. Let $\iota : (A, p) \to (X, p)$, $\jmath : (X, p) \to (X, A, p)$ denote inclusions.
  (a) Show that $\iota_{\#} : \pi_n(A, p) \to \pi_n(X, p)$ is one-to-one, and $\jmath_{\#} : \pi_n(X, p) \to \pi_n(X, A, p)$ is onto.
  (b) Show that $\pi_n(X, p) \cong \pi_n(A, p) \oplus \pi_n(X, A, p)$.

## 3. The Homotopy Sequence of a Fibration

Our goal in this section is to develop a powerful tool for investigating and classifying bundles, and more generally fibrations. It is based on the homotopy exact sequence of $(M, F, m)$ for a fibration $\pi : M \to B$ with fiber $F$ based at a point $m \in M$, and consists of replacing the triple $\pi_n(M, F, m)$ in that sequence by an isomorphic homotopy group of the base space $B$.

LEMMA 3.1. *Let $\pi : M \to B$ be a fibration, $X$ a space which admits a subspace $A$ as a strong deformation retract; i.e., there is a homotopy $H$ between $1_X$ and some retraction $r : X \to A$, with $H(a, t) = a$ for all $a \in A$, $t \in I$. Then given a map $f : X \to B$, any lift $g : A \to M$ of $f_{|A}$ can be extended to a lift of $f$.*

PROOF. Let $\bar{g} : X \to M$ denote $g \circ r$. Now, $f \circ H$ is a homotopy of $f$ into $f \circ r$. Since $f \circ r = \pi \circ g \circ r = \pi \circ \bar{g}$, there exists, by the homotopy lifting property together with Exercise 62, a homotopy $h : X \times I \to M$ covering $f \circ H$ and stationary with respect to it, such that $h \circ \imath_1 = \bar{g}$. Then $h \circ \imath_0$ is a lift of $f$ (because $\pi \circ (h \circ \imath_0) = f \circ H \circ \imath_0 = f \circ 1_X = f$) which extends $g$: Given $a \in A$, $(f \circ H)(a, t) = f(a)$, so that $(h \circ \imath_0)(a) = (h \circ \imath_1)(a) = \bar{g}(a) = g(a)$. $\qquad\square$

THEOREM 3.1. *Let $\pi : M \to B$ denote a fibration. If $b \in B$, $F = \pi^{-1}(b)$, and $m \in F$, then $\pi_\# : \pi_n(M, F, m) \to \pi_n(B, b)$ is an isomorphism for $n \geq 2$.*

PROOF. We first show $\pi_\#$ is onto: Given $f \in C^n(B, b)$, the constant map $g : J^{n-1} \to M$ sending everything to $m$ is a partial lift of $f$. An extension $\bar{g}$ of $g$ to a lift of $f$ is then an element of $C^n(M, F, m)$, and $\pi_\#[\bar{g}] = [f]$. Next, suppose that $\pi_\#[f] = 0$ for $f \in C^n(M, F, m)$, so that there is a $C^n(B, b)$-homotopy $h$ of $\pi \circ f$ into the constant map $h \circ \imath_1 = b$. Define $g : E = I^n \times 0 \cup I^n \times 1 \cup J^{n-1} \times I \to M$ by $g(a, 0) = f(a)$ if $a \in I^n$, $g(a, 1) = m$ again if $a \in I^n$, and $g(a, t) = m$ for $a \in J^{n-1}$. Then $g$ is a partial lift of $h$, and since $E$ is a strong deformation retract of $I^n \times I$, $g$ can be extended to a lift $\bar{g} : I^n \times I \to M$ of $h$. $\bar{g}$ is then a homotopy in $C^n(M, F, m)$ of $f$ into the constant map $m$, so that $[f] = 0$. $\qquad\square$

Let $\pi : M \to B$ be a fibration, $b \in B$, $F = \pi^{-1}(b)$, $m \in F$. Recall from Theorem 2.2 that the homotopy sequence

$$\cdots \longrightarrow \pi_n(F, m) \xrightarrow{\imath_\#} \pi_n(M, m) \xrightarrow{\jmath_\#} \pi_n(M, F, m) \xrightarrow{\partial} \pi_{n-1}(F, m) \longrightarrow \cdots$$

of $(M, F, m)$ is exact. Using Theorem 3.1, we may replace the third term by the isomorphic group $\pi_n(B, b)$. Since $\pi \circ \jmath = \pi$, we obtain an exact sequence

$$\cdots \longrightarrow \pi_n(F, m) \xrightarrow{\imath_\#} \pi_n(M, m) \xrightarrow{\pi_\#} \pi_n(B, b) \xrightarrow{\Delta} \pi_{n-1}(F, m) \longrightarrow \cdots,$$

where $\Delta := \partial \circ \pi_\#^{-1}$ may be described as follows: Given $f : (I^n, \partial I^n) \to (B, b)$, the partial lift $g : J^{n-1} \to M$ sending $J^{n-1}$ to $m$ may be extended to a lift $g : (I^n, I^{n-1}, J^{n-1}) \to (M, F, m)$ of $f$. Then $g_{|I^{n-1}}$ represents $\Delta[f]$.

The above sequence terminates in $\pi_1(M, m)$. When we add the map $\pi_\# : \pi_1(M, m) \to \pi_1(B, b)$, the resulting sequence is called the *homotopy sequence of the fibration* based at $m$.

THEOREM 3.2. *The homotopy sequence of a fibration is exact.*

PROOF. Exactness of the last portion $\pi_1(F, m) \xrightarrow{\imath_\#} \pi_1(M, m) \xrightarrow{\pi_\#} \pi_1(B, b)$ is what remains to be established. If $[c] \in \pi_1(F, m)$, then $\pi \circ \imath \circ c$ is the constant curve $b$, so that $\pi_\# \imath_\#$ is the zero homomorphism, and $\operatorname{im} \imath_\# \subset \ker \pi_\#$. Conversely, given $[c] \in \pi_1(M, m)$ with $\pi_\#[c] = e$, there exists a homotopy $h : (I^2, I^1, J^1) \to (B, (\pi \circ c)(I), b)$. If $\bar{h}$ is a covering homotopy of $h$ stationary with respect to it, and such that $\bar{h} \circ \imath_0 = c$, then $\bar{h} \circ \imath_1$ is a curve lying in $F$ which is $C^1(M, m)$- homotopic to $c$; i.e., $[c] = \imath_\#[\bar{h} \circ \imath_1] \in \imath_\#(\pi_1(F, m))$. $\qquad\square$

EXAMPLES AND REMARKS 3.1. (i) Suppose that $\pi : M \to B$ is a covering map, so that it can be considered a bundle with discrete fiber $F$ over $M$. Then $\pi_n(F, m) = 0$, so that $\pi_\# : \pi_n(M, m) \to \pi_n(B, b)$ is an isomorphism for $n \geq 2$, and a monomorphism for $n = 1$.

(ii) Since $\pi_k(S^1) \cong \pi_k(\mathbb{R}) = 0$ when $k > 1$, Theorem 3.2 applied to the Hopf fibration $S^{2n+1} \to \mathbb{C}P^n$ with fiber $S^1$ implies that $\pi_k(S^{2n+1}) \cong \pi_k(\mathbb{C}P^n)$ for $k > 2$. In particular, $\pi_3(S^2) \cong \pi_3(S^3) \cong \mathbb{Z}$, and the Hopf fibration $\pi : S^3 \to S^2$ is a generator of $\pi_3(S^2)$.

(iii) Consider the principal fibration $SO(n+1) \to S^n = SO(n+1)/SO(n)$. Since $\pi_k(S^n) = 0$ when $k < n$, the homotopy sequence of this fibration yields isomorphisms $\pi_k(SO(n)) \cong \pi_k(SO(n+1))$ for $n > k+1 \geq 2$. Thus, $\pi_k(SO(n))$ is independent of $n$ provided $n$ is large enough; it is called the $k$-th *stable homotopy group* $\pi_k(SO)$.

(iv) Let $M = SO(n+k)/SO(k)$. If $k > q+1 > 2$, then by (ii),

$$\pi_q(SO(k)) \xrightarrow{\cong} \pi_q(SO(n+k)) \longrightarrow \pi_q(M) \longrightarrow \pi_{q-1}(SO(k)) \xrightarrow{\cong}$$
$$\xrightarrow{\cong} \pi_{q-1}(SO(n+k)).$$

By exactness, the last isomorphism implies that $\pi_q(M) \to \pi_{q-1}(SO(k))$ is the zero map. The first isomorphism implies that $\pi_q(M) \to \pi_{q-1}(SO(k))$ is one-to-one. Thus, $\pi_q(M) = 0$ whenever $k > q+1 > 2$.

In order to compute $\pi_1(M)$, we use the following:

LEMMA 3.2. *Let $\xi = \pi : M \to B$ be a fiber bundle with connected fiber $F$. Then the homotopy sequence of $\xi$ terminates in $\pi_1(M, m) \to \pi_1(B, b) \to 0$; equivalently, $\pi_1(M, m) \to \pi_1(B, b)$ is onto.*

PROOF. Let $[c] \in \pi_1(B, b)$, and consider the map $f$ from the one-point space $\{0\}$ into $M$ given by $f(0) = m$. Then $c$ is a homotopy of $\pi \circ f$, so there is a lift $\tilde{c}$ of $c$ with $\tilde{c}(0) = m$. Since $c(1) \in F$ and the latter is connected, there exists a curve $\bar{c}$ in $F$ from $\tilde{c}(1)$ to $m$. Then $[\bar{c}\tilde{c}] \in \pi_1(M, m)$, and $\pi_\#[\bar{c}\tilde{c}] = [ce] = [c]$. $\square$

(iv) (continued) $\pi_1(SO(k)) \cong \pi_1(SO(n+k))$ when $k > 2$ by (iii). Since $SO(k)$ is connected, the homotopy sequence for $SO(n+k) \to M$ terminates in

$$\pi_1(SO(k)) \xrightarrow{\cong} \pi_1(SO(n+k)) \longrightarrow \pi_1(M) \longrightarrow 0$$

by Lemma 3.2, and $\pi_1(M) = e$ by exactness.

(v) Recall that the Grassmannian of oriented $n$-planes in $\mathbb{R}^{n+k}$ is $\tilde{G}_{n,k} = SO(n+k)/SO(n) \times SO(k)$. Thus, $M = SO(n+k)/SO(k)$ from (iii) is the total space of a principal $SO(n)$ bundle over the Grassmannian; it is in fact the orthonormal frame bundle of the universal bundle $\tilde{\gamma}_{n,k}$, see Exercise 75 below. Applying (iv) and Lemma 3.2 to the homotopy sequence of this bundle implies that

$$\pi_q(\tilde{G}_{n,k}) \cong \begin{cases} \pi_{q-1}(SO(n)), & \text{if } k > q+1 > 2, \\ 0, & \text{if } q = 1 \text{ and } k > 2. \end{cases}$$

The above example actually provides us with a way of classifying vector bundles over spheres (this problem will be approached in a geometric and more general manner in the next section). But we must first establish the following:

LEMMA 3.3. *Any vector bundle over a simply connected base is orientable.*

PROOF. Let $\xi = \pi : E \to B$ denote a rank $n$ coordinate vector bundle over $B$ with transition functions $f_{\alpha,\beta} : U_\alpha \cap U_\beta \to O(n)$. Since $det : O(n) \to O(1) = \{\pm 1\}$ is a homomorphism, there exists, by Proposition 2.1 in Chapter 2, a principal $O(1)$-bundle $det\,\xi$ over $B$ with transition functions $det\,f_{\alpha,\beta} : U_\alpha \cap U_\beta \to O(1)$. It is called the *determinant bundle* of $\xi$. Thus, $\xi$ is orientable if and only if its determinant bundle is trivial. But the total space of $det\,\xi$ is either $B$ or a two-fold cover of it, so if $B$ is simply connected, this bundle must be trivial. $\square$

THEOREM 3.3. *The collection* $\mathrm{Vect}_n(S^k)$ *of equivalence classes of rank $n$ vector bundles over $S^k$, $k > 1$, is in bijective correspondence with the collection* $[S^{k-1}, SO(n)]$ *of homotopy classes of maps* $S^{k-1} \to SO(n)$. *When $k = 1$,* $\mathrm{Vect}_n(S^1)$ *consists of two elements.*

PROOF. Any vector bundle over $S^k$ admits an atlas with two charts, since $S^k$ can be covered by two contractible open sets $S^k \setminus \{p\}$, $S^k \setminus \{q\}$, with $p \neq q$, and a bundle over a contractible space is trivial by Exercise 61. Lemma 3.3 together with the proof of Theorem 7.1 in Chapter 2 shows that $\tilde{G}_{n,n}$ is a classifying space for rank $n$ bundles over $S^k$ when $k > 1$, and the proof of Theorem 7.2 that $\mathrm{Vect}_n(S^k)$ is in bijective correspondence with homotopy classes of maps $S^k \to \tilde{G}_{n,3n}$. By Proposition 2.2 and Examples and remarks 3.1(v),

$$[S^k, \tilde{G}_{n,3n}] \leftrightarrow \pi_k(\tilde{G}_{n,3n})/\pi_1(\tilde{G}_{n,3n}) = \pi_k(\tilde{G}_{n,3n}) \cong \pi_{k-1}(SO(n)).$$

Finally, $\pi_{k-1}(SO(n))$ is in bijective correspondence with $[S^{k-1}, SO(n)]$ by Example 2.1.

When $k = 1$, the bundle need no longer be orientable. In this case, the classifying space is $G_{n,3n}$, and the collection of rank $n$ bundles over $S^1$ is identified with $[S^1, G_{n,3n}]$. Since the oriented Grassmannian is a simply connected 2-fold cover of the nonoriented one, there is exactly one nontrivial rank $n$ bundle over $S^1$. $\square$

EXAMPLE 3.1. Any vector bundle over $S^3$ is trivial: In view of Theorem 3.3, this amounts to showing that $\pi_2(SO(k)) = 0$ for all $k$. This is clear for $k = 1, 2$. By Exercise 53, $S^3 \to SO(3)$ is a 2-fold cover, so $\pi_2(SO(3)) \cong \pi_2(S^3) = 0$. The portion

$$\pi_2(SO(3)) \longrightarrow \pi_2(SO(4)) \longrightarrow \pi_2(S^3)$$

of the homotopy sequence of $SO(4) \to SO(4)/SO(3) = S^3$ implies that $\pi_2(SO(4))$ is trivial. $\pi_2(SO(k))$ then vanishes for all $k$ by Examples and Remarks 3.1(iii).

The observant reader has no doubt noticed that we have sometimes replaced the fiber $F_b$ over $b \in B$ in the homotopy sequence of a bundle by the fiber $F$ of the bundle. Laziness notwithstanding, this may be justified as follows: Let $\xi = \pi : M \to B$ denote a fiber bundle with fiber $F$ and group $G$. Consider the trivial bundle $\epsilon : F \to *$ over a point $*$ with group $G$. Let $b \in B$, $\imath : F_b \to M$ denote inclusion. A map $f : F \to F_b$ is called *admissible* if $\imath \circ f : \epsilon \to \xi$ is a bundle map covering $* \to b$. Given $m \in F_b$, $f_\# : \pi_k(F, f^{-1}(m)) \to \pi_k(F_b, m)$ is an isomorphism for all $k$, and we may replace $\pi_k(F_b)$ by $\pi_k(F)$ in the homotopy sequence of $\xi$. If $h : F \to F_b$ is another admissible map, then $h^{-1} \circ f : F \to F$ coincides with the operation of an element of $G$. Thus, the identifications of homotopy groups induced by two admissible maps differ by an isomorphism $g_\#$ induced by the action of some $g \in G$. In the case when the bundle is principal, $F = G$, and if one considers only base-point preserving admissible maps $(G, e) \to (G_b, m)$, then the element $g$ above must equal $e$ (because $h^{-1} \circ f :$ $(G, e) \to (G, e)$), so that the identification is unique.

More generally, suppose $f : \xi_1 \to \xi_2$ is a bundle map. By Theorem 2.2(2), $f$ induces a homomorphism of the homotopy sequences of $\xi_1$ and $\xi_2$, which according to the preceding paragraph may be represented by the commutative diagram

$$
\begin{array}{ccccccccc}
\cdots & \longrightarrow & \pi_k(M_1) & \longrightarrow & \pi_k(B_1) & \xrightarrow{\Delta} & \pi_{k-1}(F) & \longrightarrow & \cdots \\
& & \downarrow{\scriptstyle f_\#} & & \downarrow{\scriptstyle f_\#} & & \downarrow{\scriptstyle \cong} & & \downarrow \\
\cdots & \longrightarrow & \pi_k(M_2) & \longrightarrow & \pi_k(B_2) & \xrightarrow{\Delta} & \pi_{k-1}(F) & \longrightarrow & \cdots
\end{array}
$$

If $\xi_i$ are principal bundles, so that $F = G$, and one considers only base-point preserving admissible maps, then the isomorphism $g_\# : \pi_{k-1}(G, e) \to \pi_{k-1}(G, e)$ must be the identity, and the diagram becomes

$$
\begin{array}{ccccccccc}
\cdots & \longrightarrow & \pi_k(M_1) & \longrightarrow & \pi_k(B_1) & \xrightarrow{\Delta} & \pi_{k-1}(F) & \longrightarrow & \cdots \\
& & \Vert & & \downarrow{\scriptstyle f_\#} & & \Vert & & \downarrow \\
\cdots & \longrightarrow & \pi_k(M_2) & \longrightarrow & \pi_k(B_2) & \xrightarrow{\Delta} & \pi_{k-1}(F) & \longrightarrow & \cdots
\end{array}
$$

Now suppose $\xi$ is a rank $n$ vector bundle over $S^k$. By Theorem 3.3, the equivalence class of $\xi$ is determined by the homotopy class of a classifying map $f : S^k \to \tilde{G}_{n,3n}$, and $[f]$ may be considered an element of $\pi_{k-1}(SO(n))$. This identification may be explicitly seen as follows: Denote by $SO(\xi)$ and $SO(\tilde{\gamma}_{n,3n})$ the principal $SO(n)$-bundles associated to $\xi$ and $\tilde{\gamma}_{n,3n}$ (in other words, the bundles of oriented orthonormal frames). By Exercise 75(a) below, the total space $E(SO(\tilde{\gamma}_{n,3n}))$ has vanishing homotopy groups, so that the homomorphism of homotopy sequences induced by $f$ becomes

$$
\begin{array}{ccccccccc}
\cdots & \longrightarrow & \pi_k(E(SO(\xi)) & \longrightarrow & \pi_k(S^k) & \xrightarrow{\Delta} & \pi_{k-1}(SO(n)) & \longrightarrow & \cdots \\
& & \Vert & & \downarrow{\scriptstyle f_\#} & & \Vert & & \downarrow \\
\cdots & \longrightarrow & 0 & \longrightarrow & \pi_k(\tilde{G}_{n,3n}) & \xrightarrow{\cong} & \pi_{k-1}(SO(n)) & \longrightarrow & \cdots
\end{array}
$$

We have proved the following:

THEOREM 3.4. *Let $\xi$ denote a rank $n$ vector bundle over $S^k$. Then the element $\alpha \in \pi_{k-1}(SO(n))$ which classifies $\xi$ in Theorem 3.3 is $\alpha = \Delta[1_{S^k}]$, where $\Delta$ is the boundary operator in the homotopy sequence of the principal $SO(n)$-bundle $SO(\xi)$ associated to $\xi$.*

Exactness of the homotopy sequence yields direct sum theorems in many cases. We illustrate one such below, and leave some others as exercises.

THEOREM 3.5. *If the bundle $\xi = \pi : M \to B$ with fiber $F$ admits a cross-section $s$, then $s_\# : \pi_n(B) \to \pi_n(M)$ and $\imath_\# : \pi_n(F) \to \pi_n(M)$ are one-to-one,* *and*

$$\pi_n(M) \cong s_\# \pi_n(B) \oplus \iota_\# \pi_n(F), \qquad n \geq 2.$$

*The above relation also holds for $n = 1$, provided $s_\# \pi_1(B)$ is a normal subgroup of $\pi_1(M)$ (e.g., if the fundamental group of $M$ is abelian).*

PROOF. Since $\pi \circ s = 1_B$, $\pi_\# : \pi_n(M) \to \pi_n(B)$ is onto, and $s_\#$ is one-to-one. Applying this to $n + 1$ instead of $n$ in the homotopy sequence, we see that $\Delta : \pi_{n+1}(B) \to \pi_n(F)$ is zero, so that $\iota_\# : \pi_n(F) \to \pi_n(M)$ is one-to-one. Thus, we have a short exact sequence

$$0 \longrightarrow \pi_n(F) \xrightarrow{\iota_\#} \pi_n(M) \xrightarrow{\pi_\#} \pi_n(B) \longrightarrow 0.$$

It is a well-known and easy to verify fact that if a short exact sequence $0 \longrightarrow A \xrightarrow{i} B \xrightarrow{j} C \longrightarrow 0$ of groups admits a homomorphism $f : C \to B$ such that $j \circ f = 1_C$, then $B \cong i(A) \oplus f(C)$, provided $f(C)$ is normal in $B$. In our case, $f$ is provided by $s_\#$. □

EXAMPLES AND REMARKS 3.2. (i) $\pi_n(B \times C) \cong \pi_n(B) \oplus \pi_n(C)$. This follows from Theorem 3.5, since a trivial bundle $B \times C \to B$ admits a cross section. When $n = 1$, one of the summands is normal; the other one is then also normal by symmetry.

(ii) Let $\xi = \pi : E \to B$ be a vector bundle. Then $\pi_n(E) \cong \pi_n(B)$ for all $n$. This again follows from Theorem 3.5 applied to the zero section, together with the fact that $\pi_n(F) = 0$ (of course, it also follows from the fact that $B$ is a strong deformation retract of $E$).

EXERCISE 75. (a) Show that $SO(n + k)/SO(k) = V_{n,k}^0$, and is therefore the total space of the principal $SO(n)$-bundle associated to $\tilde{\gamma}_{n,k}$. Thus, by Examples and Remarks 3.1, $\pi_q(V_{n,k}^0) = 0$ for all $q < k - 1$ if $k > 2$.

(b) Let $ESO(n)$ denote the union of the increasing sequence $V_{n,1}^0 \subset V_{n,2}^0 \subset \cdots$ with the weak topology, so that $ESO(n) \to BSO(n)$ is the principal $SO(n)$-bundle associated to $\tilde{\gamma}_n$. Show that all the homotopy groups of $ESO(n)$ are trivial.

EXERCISE 76. The *unit sphere bundle* of a vector bundle $\xi = \pi : E(\xi) \to B$ with Euclidean metric is the bundle $\xi^1$ over $B$ with total space $E(\xi^1) = \{ u \in E(\xi) \mid |u| = 1 \}$. The bundle projection is the restriction of $\pi$.

(a) Show that $\xi^1$ is the bundle with fiber $S^{n-1}$ associated to the principal $O(n)$-bundle $O(\xi)$, so that its total space is $E(O(\xi)) \times_{O(n)} S^{n-1}$. If $\xi$ is orientable, then $O(\xi)$ may be replaced by $SO(\xi)$.

(b) Let $\tilde{\gamma}_n^1$ denote the unit sphere bundle of the universal bundle. Show that $E(\tilde{\gamma}_n^1)$ has the same homotopy groups as $BSO(n)$. *Hint:* Define for each $k$ a map $f : E(\tilde{\gamma}_{n,k}^1) \to \tilde{G}_{n-1,k+1}$ which assigns to $u \in P$ the plane $u^\perp \cap P$ with appropriate orientation. Show that this is a fiber bundle with fiber $S^{k-2}$, and use Theorem 3.2.

EXERCISE 77. Let $M \to B$ be a fiber bundle with fiber $F$. Show that if the fiber $F_b$ over $b \in B$ is a retract of $M$, then the conclusion of Theorem 3.5 holds.

## 4. Bundles Over Spheres

Bundles over contractible spaces are trivial. Since a sphere is the simplest example of a connected space which is not contractible, bundles over spheres provide the simplest examples of nontrivial bundles. In Theorem 3.3, we classified vector bundles over $S^n$ by means of classifying spaces. In this section, we adopt a geometric approach to the more general problem of classifying $G$-bundles over $S^n$.

LEMMA 4.1. *Let $\pi_i : M_i \to B$ denote two coordinate bundles over $B$ with the same fiber $F$, group $G$, and trivializing cover $\{U_\alpha\}$ of $B$. Let $f^i_{\alpha,\beta} : U_\alpha \cap U_\beta \to G$ denote the transition functions of $\pi_i$. Then the bundles are equivalent iff there exist maps $g_\alpha : U_\alpha \to G$ such that*

$$(4.1) \qquad f^2_{\alpha,\beta}(p) = g_\beta(p)^{-1} \cdot f^1_{\alpha,\beta}(p) \cdot g_\alpha(p), \qquad p \in U_\alpha \cap U_\beta.$$

PROOF. Suppose the bundles are equivalent. By Exercise 47(a), there exist maps $f_{\alpha,\beta} : U_\alpha \cap U_\beta \to G$ satisfying

$$f_{\alpha,\gamma} = f_{\beta,\gamma} \cdot f^1_{\alpha,\beta} = f^2_{\beta,\gamma} \cdot f_{\alpha,\beta}.$$

In particular, $f^2_{\alpha,\beta} = f_{\beta,\beta} \cdot f^{-1}_{\beta,\alpha}$, and $f^1_{\alpha,\beta} = f_{\alpha,\alpha} \cdot f^{-1}_{\beta,\alpha}$. Setting $g_\alpha = f^{-1}_{\alpha,\alpha}$, we obtain $f^2_{\alpha,\beta} = f_{\beta,\beta} \cdot f^1_{\alpha,\beta} \cdot f^{-1}_{\alpha,\alpha} = g^{-1}_\beta \cdot f^1_{\alpha,\beta} \cdot g_\alpha$. Conversely, suppose there exists a collection $g_\alpha : U_\alpha \to G$ satisfying (4.1). If we define $f_{\alpha,\beta} := g^{-1}_\beta \cdot f^1_{\alpha,\beta}$, then $f_{\alpha,\gamma} = g^{-1}_\gamma \cdot f^1_{\alpha,\gamma} = g^{-1}_\gamma \cdot f^1_{\beta,\gamma} \cdot f^1_{\alpha,\beta} = f_{\beta,\gamma} \cdot f^1_{\alpha,\beta}$, and similarly, $f_{\alpha,\gamma} = f^2_{\beta,\gamma} \cdot f_{\alpha,\beta}$. The bundles are then equivalent by Exercise 47(b). $\square$

Notice that if $\pi : M \to B$ is a coordinate bundle with transition functions $f^1_{\alpha,\beta} : U_\alpha \cap U_\beta \to G$, then, given any collection $g_\alpha : U_\alpha \to G$, there exists, by Proposition 2.1 in Chapter 2, a coordinate bundle over $B$ with transition functions $f^2_{\alpha,\beta}$ defined as in (4.1), and this bundle is equivalent to the original one.

Let $G$ be a connected Lie group. By Example 2.1 and Proposition 2.2, there is a bijective correspondence between $\pi_k(G)$ and $[S^k, G]$. If $\xi = \pi : M \to S^n$ is a bundle with group $G$, we define a *characteristic map* $T$ of $\xi$ as follows: Let $p_1 = (0, \ldots, 0, 1)$ and $p_2 = (0, \ldots, 0, -1)$ denote the north and south poles of $S^n$, $U_1 = S^n \setminus \{p_2\}$, and $U_2 = S^n \setminus \{p_1\}$. Then $\{U_1, U_2\}$ is an open cover of $S^n$, and since $U_i$ is contractible, there exists a bundle atlas $\{\pi^{-1}(U_i), (\pi, \phi_i)\}$ consisting of two charts. The transition function $f_{1,2} : U_1 \cap U_2 \to G$ is defined on a neighborhood of the equator $S^{n-1}$. The map $T = f_{1,2|S^{n-1}} : S^{n-1} \to G$ is called a *characteristic map* of $\xi$, and its homotopy class $C(\xi) = [T] \in \pi_{n-1}(G)$ the *characteristic class* of $\xi$.

THEOREM 4.1. *Let $G$ be a connected Lie group acting effectively on a manifold $F$. The operation which assigns to each bundle $\xi$ with group $G$ and fiber $F$ over $S^n$ the characteristic class of $\xi$ yields a bijective correspondence between the collection of equivalence classes of such bundles and the set $[S^{n-1}, G]$ of homotopy classes of maps $S^{n-1} \to G$.*

PROOF. We first check that equivalent bundles $\xi_1$ and $\xi_2$ have homotopic characteristic maps. By Lemma 4.1, there exist maps $g_i : U_i \to G$, $i = 1, 2$, satisfying (4.1). If $h_i$ denotes the restriction of $g_i$ to the equator, then

$T_2(p) = h_2(p)^{-1} \cdot T_1(p) \cdot h_1(p)$ for $p \in S^{n-1}$. Let $H_i : U_i \times I \to \{p_0\}$ be strong deformation retractions of $U_i$ onto some $p_0$ in the equator, so that $g_i \circ H_{i|S^{n-1} \times I}$ are homotopies of $h_i$ into constant maps $S^{n-1} \to a_i \in G$. Then

$$S^{n-1} \times I \to G,$$

$$(p,t) \mapsto (g_2 \circ H_2)(p,t)^{-1} \cdot T_1(p) \cdot (g_1 \circ H_1)(p,t)$$

is a homotopy of $T_2$ into $a_2^{-1}T_1 a_1$. Composing it with the homotopy given by $(p,t) \mapsto c_2(t)^{-1}T_1(p)c_1(t)$, where $c_i : I \to G$ are curves from $a_i$ to $e$, yields a homotopy of $T_2$ into $T_1$.

Conversely, suppose $\xi_i$ are bundles with characteristic maps $T_i$, and $H$ is a homotopy of $T_1$ into $T_2$. We will construct maps $g_i$ as in (4.1). If we set $g_1(p) = f_{1,2}^1(p)^{-1}f_{1,2}^2(p)$ and $g_2(p) = e$, then the $g_i$ formally satisfy (4.1). The problem is of course that $g_1$ is not defined at the north pole $p_1$. We can, however, use $H$ to extend $g_{1|S^{n-1}}$ to all of $U_1$: Indeed, $(p,t) \mapsto T_1(p)^{-1}H(p,t)$ is a homotopy of the constant map $S^{n-1} \to e$ into $T_1^{-1}T_2$. If $E_1$ denotes the closure of the northern hemisphere, then each point of $E_1$ can be represented as $\cos(\pi t/2)p_1 + \sin(\pi t/2)p$ with $t \in I$ and $p \in S^{n-1}$. Define $T : E_1 \to G$ by $T(\cos(\pi t/2)p_1 + \sin(\pi t/2)p) = T_1(p)^{-1}H(p,t)$. The formula makes sense at $p_1$ because the right side is then $e$. Furthermore, the restriction of $T$ to the equator equals $T_1^{-1}T_2$, and

$$g_1 : U_1 \to G,$$

$$p \mapsto \begin{cases} T(p), & p \in E_1, \\ f_{1,2}^1(p)^{-1}f_{1,2}^2(p), & p \in U_1 \setminus E_1, \end{cases}$$

is a continuous extension of $T$ to $U_1$. Now, we may in both bundles replace the charts over $U_2$ by their restrictions to the open southern hemisphere $V_2$. If we set $g_2$ to be the constant map $e$ on $V_2$, then (4.1) holds, and the bundles are therefore equivalent.

To complete the proof, it must be shown that any map $T : S^{n-1} \to G$ is the characteristic map of some bundle. Let $r : U_1 \cap U_2 \to S^{n-1}$ be the retraction that maps $p$ to the closest point $r(p)$ on the equator, and define $f_{i,j} : U_i \cap U_j \to G$ by $f_{i,i} = e$, $f_{1,2} = T \circ r$, and $f_{2,1} = f_{1,2}^{-1}$. By Proposition 2.1 from Chapter 2, there is a bundle $\xi$ with these transition functions, and $T$ is a characteristic map of $\xi$. $\qquad\qquad\qquad\qquad\qquad\qquad\qquad\square$

Since $G$ is connected, $[S^{n-1}, G]$ equals $\pi_{n-1}(G)$, and the characteristic class $C(\xi)$ of a principal $G$-bundle $\xi$ over $S^n$ belongs to $\pi_{n-1}(G)$. On the other hand, the boundary operator $\Delta$ from the homotopy sequence of $\xi$ maps into $\pi_{n-1}(G)$. In view of Theorem 3.4, the following result should come as no surprise:

PROPOSITION 4.1. $C(\xi) = \Delta(1)$, where 1 is the class in $\pi_n(S^n)$ containing $1_{S^n}$.

PROOF. As before, $p_2$ will denote the south pole of $S^n$, and $E_1$, $E_2$ the closed northern and southern hemispheres. Recall that the boundary homomorphism $\Delta$ maps $\pi_n(S^n, p_2)$ into $\pi_{n-1}(G_2, q_2)$, where $G_2$ is the fiber over $p_2$, $q_2 \in G_2$. The latter homotopy group is then identified with $\pi_{n-1}(G)$ via an admissible map $k : G_2 \to G$, which in this case will be taken to be $\phi_{2|G_2}$. The argument will be based on the following two steps:

(a) construction of a map $f : (D^n, S^{n-1}) \to (S^n, p_2)$ with $[f] = 1$, and

(b) construction of a lift $\bar{f} : (D^n, S^{n-1}) \to (E(\xi), G_2)$ of $f$ such that $k \circ \bar{f}_{|S^{n-1}} = T$.

This will suffice: Since $\pi \circ \bar{f} = f$, we have that $\pi_\#^{-1}[f] = [\bar{f}]$, and

$$k_\# \Delta(1) = k_\# \partial \pi_\#^{-1}[f] = k_\# \partial[\bar{f}] = [k \circ \bar{f}_{|S^{n-1}}] = [T]$$

as claimed. For (a), view $D^n$ as $E_1$. Given $p \in E_1 \setminus \{p_1\}$, let $h(p)$ denote the point on the boundary $S^{n-1}$ that is closest to $p$. Then any $p \in E_1$—even $p_1$—can be expressed as $p = (\cos t)p_1 + (\sin t)h(p)$ for some $0 \le t \le \pi/2$. Define $f(p) = (\cos 2t)p_1 + (\sin 2t)h(p)$. Then $f : (D^n, S^{n-1}) \to (S^n, p_2)$, and $f$ has degree $+1$ because $p_1$ is a regular value of $f$, and $f_{*p_1}$ equals two times the identity on the tangent space. Thus, $[f] = 1$. Next, define

$$\bar{f}(p) = \begin{cases} (\pi, \phi_1)^{-1}(f(p), e), & \text{if } f(p) \in E_1, \\ (\pi, \phi_2)^{-1}(f(p), Th(p)), & \text{if } f(p) \in E_2. \end{cases}$$

The two definitions coincide on the overlap: When $f(p) \in S^{n-1}$, $h(p) = f(p)$. If $m = (\pi, \phi_2)^{-1}(f(p), Tf(p))$, then $\phi_2(m) = Tf(p)$. But the bundle is principal, so that $Tf(p) = \phi_2(\bar{m})\phi_1(\bar{m})^{-1}$ for any $\bar{m}$ in the fiber over $Tf(p)$, and in particular for $\bar{m} = m$. Thus, $\phi_1(m) = e$, and both parts of the definition agree.

It remains to show that $k \circ \bar{f}_{|S^{n-1}} = T$. When $p \in S^{n-1}$, $f(p) = p_2$ and $h(p) = p$. Then $\bar{f}(p) = (\pi, \phi_2)^{-1}(p_2, T(p))$, so that $k \circ \bar{f}(p) = \phi_2 \circ \bar{f}(p) = T(p)$. $\qquad\square$

EXERCISE 78. (a) Prove that $\pi_1(SO(n)) = \mathbb{Z}_2$ if $n \ge 3$. It is not difficult to show that the universal cover $\tilde{G}$ of a Lie group $G$ admits a Lie group structure for which the covering map $\rho : \tilde{G} \to G$ becomes a group homomorphism. The simply connected 2-fold cover of $SO(n)$, $n \ge 3$, is called the *spinor group* $Spin(n)$.

(b) Show that $Spin(3) \cong S^3$, and $Spin(4) \cong S^3 \times S^3$.

EXERCISE 79. Let $\xi = \pi : E \to B$ denote an oriented rank $n$ vector bundle over $B$, and $P$ the total space of the oriented orthonormal frame bundle $SO(\xi)$ of $\xi$. Then $\xi$ is said to admit a *spin structure* if there exists a principal $Spin(n)$-bundle $Spin(\xi) = \tilde{P} \to B$ together with a principal bundle homomorphism $f$ over the identity: i.e., $f : \tilde{P} \to P$ satisfies

$$f(pg) = f(p)\rho(g), \qquad p \in \tilde{P}, \quad g \in Spin(n).$$

(a) Show that if $\xi$ admits a spin structure, then $f : \tilde{P} \to P$ is a 2-fold covering map, and $E$ is diffeomorphic to $\tilde{P} \times_{Spin(n)} \mathbb{R}^n$. Here $Spin(n)$ acts non-effectively on $\mathbb{R}^n$ via the covering homomorphism $\rho$.

(b) Prove that any vector bundle of rank $\ge 3$ over $S^n$ admits a spin structure if $n \ge 3$.

## 5. The Vector Bundles Over Low-Dimensional Spheres

The tools developed in previous sections enable us to analyze in more detail the vector bundles over spheres of dimension $\leq 4$. Bundles over $S^1$ were discussed in Section 3. For $n > 1$, rank $k$ bundles over $S^n$ are orientable, and are classified by $\pi_{n-1}(SO(k))$. Since $SO(1) = \{1\}$, any line bundle over $S^n$ is trivial.

*Bundles over $S^2$:* Since $\pi_1(SO(2)) \cong \mathbb{Z}$, there is a plane bundle $\xi_k$ with $C(\xi_k) = k$ for each integer $k$. Consider the portion

$$\pi_2(S^3) \longrightarrow \pi_2(S^2) \xrightarrow{\Delta} \pi_1(SO(2)) \longrightarrow \pi_1(S^3)$$

of the homotopy sequence of the Hopf fibration. The end groups are trivial, so that $\Delta(1) = \pm 1$. The sign here is merely a matter of orientation: Indeed, denote by $-\xi$ the bundle over $S^2$ obtained from $\xi$ by reversing the orientation. The identity map $E(\xi) \to E(-\xi)$ is an orientation-reversing equivalence, and if $(\pi, \phi)$ is a positively oriented chart of $\xi$, then $(\pi, L \circ \phi)$ is a positively oriented chart of $-\xi$, where $L$ is the automorphism of $\mathbb{R}^2$ given by $L(a_1, a_2) = (a_2, a_1)$. Thus, $-f_{1,2} = L f_{1,2} L^{-1}$, and $[-T] = [LTL^{-1}]$. By Lemma 4.1, $\xi$ and $-\xi$ are equivalent as bundles with group the full orthogonal group $O(2)$. However, viewing $SO(2)$ as the unit circle, conjugation by $L$ is a reflection in the $x$-axis and the induced map on the fundamental group sends each element of $\pi_1(S^1)$ into its inverse. It follows that $[-T] = -[T]$, and $\xi_{-k}$ is $\xi_k$ with the opposite orientation.

The standard orientation on the Hopf bundle—the one induced on each fiber of the total space $S^3 \times_{S^1} \mathbb{R}^2$ by the standard orientation of $\mathbb{R}^2$—corresponds to $\xi_{-1}$; this will be established below for the Hopf bundle $S^7 \times_{S^3} \mathbb{R}^4 \to S^4$, and the same argument carries over to the former bundle if one replaces quaternions by complex numbers.

Since the Hopf bundle is $\xi_{-1}$, it is a generator of all oriented plane bundles over $S^2$: If $f_k : S^2 \to S^2$ has degree $k$, then $\xi_{-k} \cong f_k^* \xi_{-1}$, because the transition functions of $f_k^* \xi_{-1}$ equal those of the Hopf bundle composed with $f_k$. Alternatively, the total space of $SO(\xi_{\pm k})$ can be viewed as a lens space $S^3/\mathbb{Z}_k$, where $\mathbb{Z}_k$ denotes a cyclic subgroup of $S^1$ of order $k$.

The tangent bundle $\tau S^2$ is $\xi_{\pm 2}$ by Exercise 53. In order to determine the sign, we construct a characteristic map for the associated principal bundle $\pi : SO(3) \to S^2$, with $\pi(A) = A e_1$. Let $s : S^2 \setminus \{-e_1\} \to SO(3)$ be given by

$$s(p) = \begin{pmatrix} p_1 & -p_2 & -p_3 \\ p_2 & 1 - \frac{p_2^2}{1+p_1} & -\frac{p_2 p_3}{1+p_1} \\ p_3 & -\frac{p_2 p_3}{1+p_1} & 1 - \frac{p_3^2}{1+p_1} \end{pmatrix}, \qquad p = (p_1, p_2, p_3).$$

When $p = e_1$, $s(p) = I_3$. Otherwise, $(0, -p_3, p_2)$ is left fixed by $s(p)$, and is orthogonal to $e_1$ and $p$. $s(p)$ is therefore a rotation in the plane spanned by $e_1$ and $p$, sending $e_1$ to $p$. In particular, $\pi \circ s(p) = p$, and $s$ is a section of the bundle over $S^2 \setminus \{-e_1\}$. Let $U_1 = S^2 \setminus \{-e_3\}$ and $U_2 = S^2 \setminus \{e_3\}$. If $A$ denotes the matrix

$$\begin{pmatrix} 0 & 0 & -1 \\ 0 & 1 & 0 \\ 1 & 0 & 0 \end{pmatrix}$$

in $SO(3)$, then $s_1(p) := A^t s(Ap)$ and $s_2(p) := As(A^t p)$ define sections over $U_1$ and $U_2$ respectively. Any $B \in SO(3)$ which lies in some fiber over $U_1 \cap U_2$ can therefore be written as

$$(5.1) \qquad B = s_1(\pi(B))\phi_1(B) = s_2(\pi(B))\phi_2(B),$$

where $(\pi, \phi_i)$ are bundle charts over $U_i$. The transition function $f_{1,2}$ at $\pi(B)$ is given by $f_{1,2}(\pi(B)) = \phi_2(B)\phi_1(B)^{-1}$. Taking $B = s_1(p)$ for $p$ in the equator and substituting in (5.1) implies $s_1(p) = s_2(p)T(p)$, so that $T(p) = s_2(p)^t s_1(p)$. A straightforward computation with $p = (\cos\theta, \sin\theta, 0) \in S^1 \times 0$ then yields

$T(\cos\theta, \sin\theta, 0)$

$$= \begin{pmatrix} \cos\theta & \sin\theta & 0 \\ -\sin\theta\cos\theta & \cos^2\theta & -\sin\theta \\ -\sin^2\theta & \sin\theta\cos\theta & \cos\theta \end{pmatrix} \begin{pmatrix} \cos\theta & -\sin\theta\cos\theta & \sin^2\theta \\ \sin\theta & \cos^2\theta & -\sin\theta\cos\theta \\ 0 & \sin\theta & \cos\theta \end{pmatrix}$$

$$= \begin{pmatrix} 1 & 0 & 0 \\ 0 & \cos2\theta & -\sin2\theta \\ 0 & \sin2\theta & \cos2\theta \end{pmatrix} \in 1 \times SO(2).$$

Thus, $\tau S^2 = \xi_2$.

Bundles of rank $k \geq 3$ over $S^2$ are classified by $\pi_1(SO(k)) = \mathbb{Z}_2$. Thus, there is exactly one nontrivial rank $k$ bundle over $S^2$. Consider for example the nontrivial rank 3 bundle $\xi$. Exactness of the portion

$$\pi_1(SO(2)) \longrightarrow \pi_1(SO(3)) \longrightarrow \pi_1(S^2) = 0$$

in the homotopy sequence of $SO(3) \to S^2$ shows that the inclusion $SO(2) \to SO(3)$ induces an epimorphism $\pi_1(SO(2)) \to \pi_1(SO(3))$ of fundamental groups. $\xi$ is therefore equivalent in $SO(3)$ to a bundle $\xi'$ with group $SO(2)$, and $E(\xi) = P \times_{SO(2)} \mathbb{R}^3$, where $P$ is the total space of the principal $SO(2)$-bundle associated to $\xi'$. Since $SO(2)$ acts trivially on $0 \times \mathbb{R} \subset \mathbb{R}^2 \times \mathbb{R}$, the map $s: S^2 \to P \times_{SO(2)} \mathbb{R}^3$, with $s(q) = [p, \mathbf{e}_3]$, $p \in P_q$, is a well-defined cross-section of $\xi$. Thus, $\xi$ splits as a Whitney sum of a rank 2 bundle with the trivial line bundle $\epsilon^1$. This rank 2 bundle is not unique: if $\xi_k$ denotes the plane bundle with $C(\xi_k) = k$, then $\xi \cong \xi_{2n+1} \oplus \epsilon^1$, since the homomorphism $\pi_1(SO(2)) \to \pi_1(SO(3))$ maps an even multiple of a generator to 0. This also implies that $\xi_{2n} \oplus \epsilon^1$ is the trivial bundle, a phenomenon already observed earlier for $\tau S^2 = \xi_2$.

Notice that $\xi$ does not admit a spin structure; cf. Exercise 79(b): If it did, the latter would be the trivial bundle $S^2 \times Spin(3) \to S^2$ since $\pi_1(Spin(3)) = \{e\}$, and by Exercise 79(a), $E(\xi) = (S^2 \times Spin(3)) \times_{Spin(3)} \mathbb{R}^3 = S^2 \times \mathbb{R}^3$, contradicting the fact that $\xi$ is not trivial.

*Bundles over $S^3$*: Any bundle over $S^3$ is trivial by Example 3.1.

*Bundles over $S^4$*: $\pi_3(SO(2)) = 0$, so any plane bundle over $S^4$ is trivial. Since $\pi_3(SO(3)) \cong \pi_3(S^3) \cong \mathbb{Z}$, there are infinitely many rank 3 bundles over $S^4$. View $S^7$ as the collection of all pairs $(u_1, u_2)$ of quaternions with $|u_1|^2 + |u_2|^2 = 1$. The quotient $S^7/\mathbb{Z}_2$ under the equivalence relation $(u_1, u_2) \sim -(u_1, u_2)$ is $\mathbb{R}P^7$. Recall the homomorphism $\rho: S^3 \to 1 \times SO(3) \subset SO(4)$ given by $\rho(z)u = zuz^{-1}$ for $u \in \mathbb{H} = \mathbb{R}^4$, and identify $1 \times SO(3)$ with $SO(3)$. If $\rho(z) = A \in SO(3)$, then $\rho^{-1}(A) = \pm z$. There is therefore a well-defined free right action of $SO(3)$ on $\mathbb{R}P^7$ given by

$$[u_1, u_2]A := [u_1\rho^{-1}(A), u_2\rho^{-1}(A)],$$

where $[u_1, u_2]$ denotes the equivalence class of $(u_1, u_2)$ in $\mathbb{R}P^7$. The quotient $\mathbb{R}P^7/SO(3)$ is diffeomorphic to $S^4$ by mapping the $SO(3)$-orbit of $[u_1, u_2]$ to the point $(2u_1\bar{u}_2, |u_1|^2 - |u_2|^2) \in S^4 \subset \mathbb{H} \times \mathbb{R}$. We therefore have a principal $SO(3)$-bundle $\mathbb{R}P^7 \to S^4$. Let $\xi$ denote the associated rank 3 vector bundle, with total space $E(\xi) = \mathbb{R}P^7 \times_{SO(3)} \mathbb{R}^3$. The portion

$$\pi_4(S^4) = \mathbb{Z} \xrightarrow{\Delta} \pi_3(SO(3)) = \mathbb{Z} \longrightarrow \pi_3(\mathbb{R}P^7) = 0$$

of the homotopy sequence of $\mathbb{R}P^7 \to S^4$ implies that $\Delta(1)$ is a generator of $\pi_3(SO(3))$. Thus, any rank 3 bundle over $S^4$ is equivalent to $f^*(\xi)$ for some $f : S^4 \to S^4$ of appropriate degree.

Rank 4 vector bundles over $S^4$ are classified by $\pi_3(SO(4)) \cong \pi_3(S^3 \times SO(3)) \cong \pi_3(S^3) \oplus \pi_3(SO(3)) \cong \mathbb{Z} \oplus \mathbb{Z}$, with generators $s : S^3 \to SO(4)$, $\rho : S^3 \to 1 \times SO(3) \subset SO(4)$, where $s(q)u = qu$, $u \in \mathbb{H}$, and $\rho$ is as above; cf. Exercise 55: Indeed, the map $S^3 \times SO(3) \to SO(4)$ which sends $(q, \rho(A))$ to $s(q)\rho(A)$ is a diffeomorphism with inverse $A \mapsto (q, s(q^{-1})A)$, where $q := Ae_1 \in S^3$. The restriction of this diffeomorphism to $S^3 \times e$ is $s$, and the restriction to $1 \times SO(3)$ is the identity. The induced isomorphism on the homotopy level therefore maps $(m[1_{S^n}], n[\rho]) \in \pi_3(S^3) \oplus \pi_3(SO(3))$ to $m[s] + n[\rho]$.

Let $\xi_{m,n}$ denote the vector bundle with $\Delta(1) = m[s] + n[\rho]$. Then the structure group of $\xi_{m,0}$ is reducible to $S^3$, and that of $\xi_{0,n}$ to $1 \times SO(3)$. Since $1 \times SO(3)$ leaves $\mathbb{R} \times 0 \subset \mathbb{R} \times \mathbb{R}^3$ fixed, $\xi_{0,n}$ admits a nowhere-zero cross-section and is equivalent to the Whitney sum $\epsilon^1 \oplus \xi_n$ of the trivial line bundle $\epsilon^1$ and the rank 3 bundle $\xi_n$ with $C(\xi_n) = n$ discussed earlier. Similarly, $\xi_{m,0} \cong f_m^*\xi_{1,0}$, where $f_m : S^4 \to S^4$ has degree $m$, and we only need to identify $\xi_{1,0}$. But the latter is, up to sign, the bundle associated to the Hopf fibration $S^7 \to S^4$, since exactness of

$$\pi_4(SO(4)) \cong \mathbb{Z} \xrightarrow{\Delta} \pi_3(S^3) \cong \mathbb{Z} \longrightarrow \pi_3(S^7) = 0$$

implies that $\Delta$ is an isomorphism.

These bundles can also be described by their characteristic maps $T : S^3 \to SO(4)$ from the discussion in Section 4. For instance, characteristic maps for $\xi_{1,0}$ and $\xi_{0,1}$ are $s$ and $\rho$ respectively. This actually enables us to determine the sign of the Hopf bundle: Recall that the Hopf fibration is the free right action of $S^3$ on $S^7 = \{(u_1, u_2) \in \mathbb{H}^2 \mid |u_1|^2 + |u_2|^2 = 1\}$ given by right multiplication $(u_1, u_2)q = (u_1q, u_2q)$. The quotient space $\mathbb{H}P^1 = S^7/S^3$ is diffeomorphic to $S^4$, and under this identification, the bundle projection $\pi : S^7 \to S^4$ is given by

$$\pi(u_1, u_2) = (2u_1\bar{u}_2, |u_1|^2 - |u_2|^2) \in S^4 \subset \mathbb{H} \times [-1, 1].$$

In order to compute a characteristic map for this bundle, consider $U_1 = S^4 \setminus \{-e_5\}$ and $U_2 = S^4 \setminus \{e_5\}$, so that $\pi^{-1}(U_i) = \{(u_1, u_2) \in S^7 \mid u_i \neq 0\}$. Then $(\pi, \phi_i) : \pi^{-1}(U_i) \to U_i \times S^3$, with $\phi_i(u_1, u_2) = u_i/|u_i|$, are principal bundle charts. Notice that $q = \pi(u_1, u_2)$ belongs to the equator of $S^4$ iff $|u_1| = |u_2| = 1/\sqrt{2}$, and in this case, $q = 2u_1\bar{u}_2 \in S^3$. By definition of a characteristic map,

$$T(q) = f_{1,2}(q) = \phi_2(u_1, u_2) \cdot \phi_1(u_1, u_2)^{-1} = \frac{u_2}{|u_2|} \frac{\bar{u}_1}{|u_1|} = 2u_2\bar{u}_1 = \bar{q} = \rho(q)^{-1}.$$

By the remark following Example 2.1, $[T] = -[\rho] \in \pi_3(SO(4))$, and the Hopf bundle $S^7 \times_{S^3} \mathbb{R}^4 \to S^4$ is $\xi_{-1,0}$.

It is also clear from the above discussion that $\xi_{n,0}$ is the bundle corresponding to the spin structure of $\xi_{0,n}$, cf. Exercise 85. In particular, if $P_n$ denotes the total space of the principal $Spin(3)$-bundle of $\xi_{n,0}$, then $E(\xi_{0,n})$ is diffeomorphic to $P_n \times_{Spin(3)} \mathbb{R}^4$, where $Spin(3)$ acts on $\mathbb{R}^4$ via $\rho$.

An argument similar to the one for $\tau S^2$ shows that a characteristic map for the bundle $SO(5) \to S^4$ is given by $T(q)p = qpq$. Thus, $T(q) = \rho(q)^{-1}s(q)^2$, and $[T] = 2[s] - [\rho]$; i.e., $\tau S^4 = \xi_{2,-1}$.

If we denote by $S_{m,n} = E(\xi_{m,n}) \times_{SO(4)} S^3$ the total space of the sphere bundle of $\xi_{m,n}$, then $S_{\pm 1,n}$ is known to always be homeomorphic to $S^7$. In general, however, $S_{1,n}$ is not diffeomorphic to $S^7$; namely, if $n(n+1) \not\equiv 0$ mod 56. Such manifolds are called *exotic* 7-spheres, and were first discovered by Milnor.

We have only discussed bundles over $S^n$ ($n \leq 4$) of rank $\leq n$. Higher rank bundles are accounted for via the following:

PROPOSITION 5.1. *If $\xi^{n+k}$ is a rank $n+k$ bundle over $S^n$, then there exists a rank $n$ bundle $\eta^n$ over $S^n$ such that $\xi^{n+k} \cong \eta^n \oplus \epsilon^k$, where $\epsilon^k$ denotes the trivial bundle of rank $k$.*

PROOF. The exact homotopy sequence of the bundle $SO(n+1) \to S^n$ implies that $\pi_{n-1}(SO(n)) \to \pi_{n-1}(SO(n+1))$ is surjective. By Examples and Remarks 3.1(iii), the inclusion homomorphism $\pi_{n-1}(SO(n)) \to \pi_{n-1}(SO(n+k))$ is then also onto, so that the structure group of $\xi^{n+k}$ reduces to $SO(n)$. □

The problems that follow provide an alternative approach to some rank 4 bundles over $S^4$: View $\mathbb{R}^{4n} = \mathbb{H}^n$ as a right vector space over $\mathbb{H}$, so that $(u_1, \ldots, u_n)q := (u_1q, \ldots, u_nq)$. Given $p = (p_1, \ldots, p_n)$, $q = (q_1, \ldots, q_n) \in \mathbb{H}^n$, the *symplectic inner product* of $p$ with $q$ is

$$\langle p, q \rangle := \sum_{i=1}^n p_i \bar{q}_i \in \mathbb{H},$$

where the conjugate $\bar{q}$ of $q = a_1 + ia_2 + ja_3 + ka_4$ is the quaternion $a_1 - ia_2 - ja_3 - ka_4$ obtained after reflection in the real axis. The *symplectic group* $Sp(n)$ is the group of $n \times n$ quaternion matrices $A$ with $A\bar{A}^t = \bar{A}^t A = I_n$.

EXERCISE 80. (a) Show that $\langle q, p \rangle = \overline{\langle p, q \rangle}$ for $p, q \in \mathbb{H}^n$. Thus, the relation $p \perp q$ iff $\langle p, q \rangle = 0$ is symmetric.

(b) Show that $Sp(n)$ is the group of linear transformations of $\mathbb{H}^n$ that preserve the symplectic inner product.

(c) Prove that $Sp(1) \cong S^3$.

EXERCISE 81. (a) Show that $Sp(n)/1 \times Sp(n-1) = S^{4n-1}$. In particular, $S^7 = Sp(2)/1 \times Sp(1)$.

(b) Prove that $f : S^3 \to Sp(2)$, $f(p) = \begin{pmatrix} 1 & 0 \\ 0 & p \end{pmatrix}$ is a generator of $\pi_3 Sp(2)$.

(c) Use the homotopy sequence of the bundle $Sp(2) \to Sp(2)/Sp(1) \times 1 = S^7$ to show that $h : S^3 \to Sp(2)$, $h(p) = \begin{pmatrix} p & 0 \\ 0 & 1 \end{pmatrix}$ is also a generator of $\pi_3(Sp(2))$.

EXERCISE 82. The free action by right multiplication of $Sp(1) \times 1$ on $Sp(2)$ descends to $S^7 = Sp(2)/1 \times Sp(1)$. Show that the resulting principal $Sp(1)$-bundle $S^7 \to Sp(2)/Sp(1) \times Sp(1)$ is the Hopf fibration.

EXERCISE 83. Let $\mathbb{Z}_2 = \{\pm 1\} \subset S^3$, so that the map $S^3 \times (S^3/\mathbb{Z}_2) \to SO(4)$ which sends $(p, \pm q)$ to $s(p)\rho(q)$ is a diffeomorphism. Prove that the group operation on $S^3 \times (S^3/\mathbb{Z}_2)$ for which this map becomes an isomorphism is the semi-direct product

$$(p_1, \pm q_1) \cdot (p_2, \pm q_2) = (p_1 q_1 q_2 \bar{q}_1, \pm q_1 q_2).$$

EXERCISE 84. In this exercise, we construct an exotic sphere, following an example due to Rigas [33]. Identify $SO(4)$ with $S^3 \times (S^3/\mathbb{Z}_2)$ as in Exercise 83. Let $\mathbb{Z}_2$ denote the subgroup $\{\begin{pmatrix} 1 & 0 \\ 0 & 1 \end{pmatrix}, \begin{pmatrix} 1 & 0 \\ 0 & -1 \end{pmatrix}\}$ of $Sp(2)$, and consider the right action of $SO(4)$ on $Sp(2)/\mathbb{Z}_2$ given by

$$\left[\begin{pmatrix} a & b \\ c & d \end{pmatrix}\right] (p, \pm q) = \left[\begin{pmatrix} \pm q & 0 \\ 0 & \pm q \end{pmatrix} \begin{pmatrix} a & b \\ c & d \end{pmatrix} \begin{pmatrix} p(\pm q) & 0 \\ 0 & 1 \end{pmatrix}\right],$$

where the bracket denotes the $\mathbb{Z}_2$-class.

(a) Show that this is a well-defined free action, and that the quotient is diffeomorphic to $S^4$ via the map that sends the $SO(4)$-orbit of $\begin{pmatrix} a & b \\ c & d \end{pmatrix}$ to $(2\bar{b}d, |b|^2 - |d|^2) \in S^3 \times [-1, 1]$.

(b) Identify the fiber of this principal bundle whith the $SO(4)$-orbit of the identity. Use Exercise 81 to show that the homomorphism $\imath_\# : \pi_3 SO(4) \to \pi_3(Sp(2)/\mathbb{Z}_2)$ in the homotopy sequence of the bundle sends the elements $[s]$ and $[\rho]$ to generators of $\pi_3(Sp(2)/\mathbb{Z}_2) \cong \mathbb{Z}$.

(c) Conclude that the associated vector bundle $\xi_{m,n}$ satisfies $|m| = |n| = 1$. The total space $S_{m,n} = E(\xi_{m,n}) \times_{SO(4)} S^3$ of the corresponding sphere bundle is then an exotic 7-sphere, according to the remarks at the end of the section.

EXERCISE 85. Prove that the principal $Spin(3)$-bundle associated to $\xi_{n,0}$ is a spin structure for $\xi_{0,n}$.

# CHAPTER 4

# Connections and Curvature

There are many more classical applications of homotopy theory to geometry, but the central theme here being differential geometry, we wish to return to the differentiable category. Our next endeavor is to try and understand how bundles fail to be products by parallel translating vectors around closed loops. This depends of course on what is meant by "parallel translation" (which is explained in the section below), but roughly speaking, if parallel translation always results in the original vector, then the bundle is said to be *flat*. Otherwise, it is *curved*, and the amount of curvature is measured by the *holonomy group*, which tallies the difference between the end product in parallel translation and the original one. In this chapter, all maps and bundles are once again assumed to be differentiable.

## 1. Connections on Vector Bundles

In Euclidean space $\mathbb{R}^n$, there is a natural way of identifying tangent spaces at different points: Given $p$, $q \in \mathbb{R}^n$, the isomorphisms $\mathcal{J}_p : \mathbb{R}^n \to \mathbb{R}^n_p$ and $\mathcal{J}_q : \mathbb{R}^n \to \mathbb{R}^n_q$ combine to yield an identification $\mathcal{J}_q \circ \mathcal{J}_p^{-1} : \mathbb{R}^n_p \to \mathbb{R}^n_q$. It is called *parallel translation* from $p$ to $q$, and for $u \in \mathbb{R}^n_p$, $\mathcal{J}_q \circ \mathcal{J}_p^{-1} u$ is called the *parallel translate* of $u$. Notice that if $u = \sum_i a_i D_i(p)$, then $\mathcal{J}_q \circ \mathcal{J}_p^{-1} u = \sum_i a_i D_i(q)$.

More generally, a manifold $M^n$ is said to be *parallelizable* if there exist vector fields $X_1, \ldots, X_n$ on $M$ with the property that $\{X_i(p)\}$ is a basis of $M_p$ for all $p \in M$. $\{X_i\}$ is then called a *parallelization* of $M$. For example, Euclidean space is parallelizable via $\{D_i\}$. So is any Lie group $G$, by taking a basis $\{u_i\}$ of $G_e$, and considering the left-invariant vector fields $X_i$ on $G$ with $X_i(e) = u_i$.

If $\{X_i\}$ is a parallelization of $M$, then one can define parallel translation (with respect to $\{X_i\}$) from $p$ to $q$ to be the isomorphism $M_p \to M_q$ which sends $X_i(p)$ to $X_i(q)$ for $i = 1, \ldots, n$. Most manifolds are not parallelizable, however. We have seen for example that $S^2$ does not even admit a single nowhere-zero vector field.

All this can be interpreted in the broader context of vector bundles over $M$, if the tangent space at $p$ is replaced by the fiber of the bundle over $p$. Thus, a vector bundle is parallelizable iff it is a trivial bundle.

If $\xi = \pi : E \to M$ is a vector bundle over $M$, there are ways of parallel translating vectors in $E_p$ to $E_q$, provided $p$ and $q$ can be joined by a curve. In general, though, the isomorphism $E_p \to E_q$ will depend on the chosen curve, so that parallel translation along a closed curve need not equal the identity map.

We will approach the problem from an infinitesimal point of view: Suppose some system of parallel translation has been defined on $\xi$, so that given $p \in M$, and a curve $c : I \to M$ emanating from $p$, there exists, for each $u \in E_p$, a unique map $X_u : I \to E$ with $\pi \circ X = c$, where $X_u(t)$ is the parallel translate of $u$ along $c$ from $c(0)$ to $c(t)$. Then $\dot{X}_u(0)$ is a vector in the tangent space of $E$ at $u$, and $\pi_* \dot{X}_u(0) = \dot{c}(0)$. It is reasonable to expect that iterating this procedure for all curves emanating from $p$ produces a subspace $\mathcal{H}_u$ of $T_u E$ such that $\pi_* : \mathcal{H}_u \to M_p$ is an isomorphism (here, as elsewhere, $T_u E$ denotes the tangent space of $E$ at $u$, since we have reserved the notation $E_p$ for the fiber of the bundle over $p \in M$). In order for parallel translation to be an isomorphism, the field $aX_u$ must also be parallel for $a \in \mathbb{R}$. This motivates the following:

DEFINITION 1.1. Let $\xi = \pi : E \to M$ be a vector bundle. A *connection* $\mathcal{H}$ on $\xi$ is a distribution on the total space $TE$ of the tangent bundle of $E$ such that

   (1) $\pi_{*u} : \mathcal{H}_u \to M_{\pi(u)}$ is an isomorphism for all $u \in E$, and
   (2) $\mu_{a*}\mathcal{H}_u = \mathcal{H}_{au}$, where $\mu_a(u) = au$ is multiplication by $a \in \mathbb{R}$.

If we denote by $\mathcal{V}$ the total space of the vertical bundle of $\xi$ (see Chapter 2, Example 5.1), then (1) implies that the assignment $v \mapsto v + \mathcal{V}_u$ is an isomorphism between $\mathcal{H}_u$ and $T_u E / \mathcal{V}_u$. Thus, $\mathcal{H}$ admits a vector bundle structure for which it becomes the normal bundle of $\mathcal{V}$ in $TE$, and $\mathcal{H} \cong \pi^* \tau M$.

Notice that by (2), $\mathcal{H}_{0_p} = s_{*p} M_p$, where $s$ denotes the zero section of $\xi$: In fact, if $c$ is a curve in $E$ with $\dot{c}(0) = v \in \mathcal{H}_u$, and $\pi(u) = p$, then the image of $\mu_0 \circ c$ lies in $s(M)$, so that $\mu_{0*}v \in s_{*p} M_p$. Thus, $\mathcal{H}_{0_p} = \mu_{0*}\mathcal{H}_u \subset s_{*p} M_p$. The two spaces then coincide by dimension considerations.

THEOREM 1.1. *Every vector bundle admits a connection.*

PROOF. The statement is clear for a trivial bundle $M \times \mathbb{R}^k \to M$: Given $u \in \mathbb{R}^k$, let $\iota_u : M \to M \times \mathbb{R}^k$ be given by $\iota_u(p) = (p, u)$, and define $\mathcal{H}_{(p,u)} = \iota_{u*} M_p$. Then $\pi_* \mathcal{H}_{(p,u)} = M_p$. Furthermore, given $a \in \mathbb{R}$, we have $\mu_a \iota_u = \iota_{au}$, so that $\mu_{a*} \mathcal{H}_{(p,u)} = \mu_{a*} \iota_{u*} M_p = \iota_{au*} M_p = \mathcal{H}_{(p,au)}$.

In general, let $\{U_\alpha\}$ be a locally finite open cover of $M$ such that the bundle is trivial over each $U_\alpha$, and $\{\phi_\alpha\}$ a subordinate partition of unity. Choose a connection $\mathcal{H}^\alpha$ on $\pi^{-1}(U_\alpha)$. Given $p \in M$, $u \in E_p$, define $L_u : M_p \to T_u E$ by

$$L_u(v) = \sum_{\{\alpha \mid p \in U_\alpha\}} \phi_\alpha(p) w_\alpha,$$

where $w_\alpha$ is the vector in $\mathcal{H}_u^\alpha$ such that $\pi_* w_\alpha = v$. $L_u$ is then a linear transformation satisfying $\pi_{*u} \circ L_u = 1_{M_p}$. The distribution $\mathcal{H}$, where $\mathcal{H}_u := L_u(M_p)$, therefore satisfies the conditions of Definition 1.1. $\square$

Given a connection on $\xi$, the splitting $\tau E = \mathcal{H} \oplus \mathcal{V}$ induces a decomposition $w = w^h + w^v \in \mathcal{H} \oplus \mathcal{V}$ for $w \in TE$. $w$ is said to be *horizontal* if $w^v = 0$, *vertical* if $w^h = 0$. If $u \in E_p$ and $v \in M_p$, the *horizontal lift* of $v$ to $u$ is the unique vector $\bar{v} \in \mathcal{H}_u$ such that $\pi_* \bar{v} = v$. A vector field $X$ on $M$ then determines a vector field $\bar{X}$ on $E$, where $\bar{X}(u)$ is the horizontal lift of $X(\pi(u))$, $u \in E$.

Recall that a map $f : N \to M$ induces a pullback bundle $f^*E \to N$ such that the diagram

$$
\begin{array}{ccc}
f^*E & \xrightarrow{\ \pi_2\ } & E \\
\pi_1 \downarrow & & \downarrow \pi \\
N & \xrightarrow[\ f\ ]{} & M
\end{array}
$$

commutes by Proposition 5.1 in Chapter 2. The following is a key property of connections:

**THEOREM 1.2.** *Let $\mathcal{H}$ be a connection on $E \to M$. Then for any $f : N \to M$, the distribution $f^*\mathcal{H} := (\pi_{2*})^{-1}\mathcal{H}$ defines a connection on the bundle $f^*E \to N$, called the* pullback *of $\mathcal{H}$ via $f$.*

It will be convenient to have a description of the tangent space of $f^*E$ at a point $(p, u)$ in order to prove Theorem 1.2.

**LEMMA 1.1.** *Given $(p, u) \in f^*E$, the map*

$$(\pi_{1*}, \pi_{2*}) : T_{(p,u)}f^*E \to \pi_{1*}T_{(p,u)}f^*E \times \pi_{2*}T_{(p,u)}f^*E \subset N_p \times T_uE$$

*is an isomorphism. Under this identification,*

$$T_{(p,u)}f^*E = \{(v, w) \in N_p \times T_uE \mid f_{*p}v = \pi_{*u}w\}.$$

**PROOF OF LEMMA 1.1.** In order to show that $(\pi_{1*}, \pi_{2*})$ is an isomorphism onto its image, it suffices to check that it is one-to-one. So assume $(\pi_{1*}, \pi_{2*})v = 0$. Since $\pi_{1*}v = 0$, $v \in \mathcal{V}(f^*E) = \imath_*T_{(p,u)}(p \times E_{f(p)})$, where $\imath : p \times E_{f(p)} \hookrightarrow f^*E$ is inclusion. Thus, $v = \imath_*w$ for some $w \in T_{(p,u)}(p \times E_{f(p)})$. But $\pi_2 \circ \imath : p \times E_{f(p)} \to E_{f(p)}$ is the isomorphism $(p, v) \mapsto v$, so that $0 = \pi_{2*}v = (\pi_2 \circ \imath)_*w$ implies $w = 0$, and therefore also $v = 0$.

For the second statement, recall that $f_* \circ \pi_{1*} = \pi_* \circ \pi_{2*}$, so that $T_{(p,u)}f^*E \subset \{(v, w) \in N_p \times T_uE \mid f_{*p}v = \pi_{*u}w\}$. It remains to show that both spaces have the same dimension. Now, the space on the right is the kernel of the linear map

$$N_p \times T_uE \to M_{f(p)},$$

$$(v, w) \mapsto f_{*p}v - \pi_{*u}w.$$

This map is onto, since it is already so when restricted to $0 \times T_uE$. Thus, its kernel has dimension equal to $\dim N_p + \dim T_uE - \dim M_{f(p)} = \dim N + \operatorname{rank} E = \dim T_{(p,u)}f^*E$, which establishes the claim.     $\square$

**PROOF OF THEOREM 1.2.** Lemma 1.1 implies that $\mathcal{V}(f^*E) = 0 \times VE$, and that $f^*\mathcal{H} = \{(v, w) \in TN \times \mathcal{H} \mid f_*v = \pi_*w\}$. Since $w \in TE$ decomposes uniquely as $w = w^h + w^v \in \mathcal{H} \oplus VE$, any $(v, w) \in Tf^*E$ decomposes as $(v, w) = (v, w^h) + (0, w^v) \in f^*\mathcal{H} \oplus \mathcal{V}(f^*E)$. Thus, the first condition of Definition 1.1 is satisfied. The second condition holds because $\mathcal{H}$ is a connection, and $\mu_{a*}(v, w) = (v, \mu_{a*}w)$ for $(v, w) \in f^*\mathcal{H}$.     $\square$

**DEFINITION 1.2.** Let $\xi = \pi : E \to M$ be a vector bundle. A *section of $\xi$ along* $f : N \to M$ is a map $X : N \to E$ such that $\pi \circ X = f$. Given a connection $\mathcal{H}$ on $\xi$, the section $X$ is said to be *parallel along* $f$ if $X_*N_p \subset \mathcal{H}_{X(p)}$ for all $p \in N$. When $\xi = \tau M$, $X$ is also referred to as a *vector field along* $f$.

Notice that $X$ is a section of $\xi$ along $f$ iff $p \mapsto \tilde{X}(p) := (p, X(p))$ is a section of the pullback bundle $f^*\xi$. Since $\pi_2 \circ \tilde{X} = X$, $X$ is parallel along $f$ iff $\tilde{X}$ is parallel with respect to the induced connection $f^*\mathcal{H}$ on $f^*\xi$.

PROPOSITION 1.1. *Let $\mathcal{H}$ be a connection on a vector bundle $\xi = \pi : E \to M$, $c : [a,b] \to M$ a curve in $M$. For any $u \in E_{c(a)}$, there exists a unique parallel section $X_u$ of $\xi$ along $c$ such that $X_u(a) = u$. Furthermore, the map $P_c : E_{c(a)} \to E_{c(b)}$ that assigns to $u$ the vector $X_u(b)$ is an isomorphism, called parallel translation in $\xi$ along $c$.*

PROOF. Let $D$ denote the standard coordinate vector field on $[a,b]$, $\bar{D}$ its $c^*\mathcal{H}$-horizontal lift to $c^*E$. If $c_u$ denotes the integral curve of $\bar{D}$ with $c_u(a) = (a, u)$, then

$$\widetilde{\pi_1 \circ c_u} = \pi_{1*} \circ \dot{c}_u = \pi_{1*} \circ \bar{D} \circ c_u = D \circ \pi_1 \circ c_u,$$

and $\pi_1 \circ c_u$ is an integral curve of $D$; i.e., $c_u(t) = (t, \pi_2 \circ c_u(t))$. Thus, $X_u := \pi_2 \circ c_u$ is a section of $\xi$ along $c$, which is parallel because $\dot{c}_u$ is horizontal. To show uniqueness, suppose $\tilde{X}_u$ is any parallel section along $c$ with $\tilde{X}_u(a) = u$. Then $t \mapsto \tilde{c}_u(t) := (t, \tilde{X}_u(t))$ is a parallel section of $c^*\xi$, and $\dot{\tilde{c}}_u$ is horizontal. Furthermore, $\pi_{1*} \circ \dot{\tilde{c}}_u = D \circ \pi_1 \circ \tilde{c}_u$, so that $\tilde{c}_u$ is also an integral curve of $\bar{D}$. Since it coincides with $c_u$ at $a$, $\tilde{c}_u = c_u$, and thus, $\tilde{X}_u = X_u$.

It remains to show that $P_c : E_{c(a)} \to E_{c(b)}$ is an isomorphism. But $P_c$ is invertible with inverse $P_{c^{-1}}$, where $c^{-1}(t) = c(a + b - t)$, so we only need to establish linearity. Now, given $a \in \mathbb{R}$, the field $aX_u$ is parallel because

$$\widetilde{aX_u} = \mu_{a*}\dot{X}_u \text{ is horizontal. Thus, } P_c(au) = aP_c(u).$$ For brevity of notation, let us denote by the same letter $\mathcal{J}_0$ the canonical isomorphisms of $E_{c(a)}$ and $E_{c(b)}$ with their tangent spaces at 0. We claim that $P_c$ is the composition $\mathcal{J}_0^{-1}(P_c)_{*0}\mathcal{J}_0$ of linear transformations, and is therefore linear. To see this, consider $w$ in the tangent space of $E_{c(a)}$ at 0, so that $w = \dot{\phi}(0)$, where $\phi(t) = tv$,

$$v = \mathcal{J}_0^{-1}w. \text{ Then } (P_c)_{*0}w = \widetilde{P_c \circ \phi}(0); \text{ but } (P_c \circ \phi)(t) = P_c(tv) = tP_c(v), \text{ so}$$
$(P_c)_{*0}w = \mathcal{J}_0 P_c v = \mathcal{J}_0 P_c \mathcal{J}_0^{-1}w$, as claimed. $\qquad\square$

If $c_1 : [a,b] \to M$, $c_2 : [b,c] \to M$ are curves in $M$ with $c_1(b) = c_2(b)$, define parallel translation along the piecewise-smooth curve $c_1 * c_2$ by $P_{c_1 * c_2} = P_{c_2} \circ P_{c_1}$. Then the set $G(p)$ of isomorphisms $E_p \to E_p$ consisting of parallel translation along piecewise-smooth closed curves based at $p$ is a subgroup of $GL(E_p)$, called the *holonomy group* of $\mathcal{H}$ at $p$. If $\tilde{c}$ is a curve from $p$ to $q$, then $G(p)$ is isomorphic to $G(q)$ via $P_c \mapsto P_{\tilde{c}^{-1} * c * \tilde{c}}$.

A connection on $\xi$ is said to be a *trivial* connection if for any $u \in E(\xi)$, there exists a parallel section $X$ of $\xi$ with $X(\pi(u)) = u$. Clearly, $\xi$ admits a trivial connection iff it is a trivial bundle.

PROPOSITION 1.2. *If $\mathcal{H}$ is a connection on a bundle over a connected manifold, then $\mathcal{H}$ is trivial iff its holonomy group is trivial.*

PROOF. The only if part of the statement is clear. Conversely, if the holonomy group is trivial, then the parallel translate of any vector $u \in E_p$ to a point $q$ is independent of the chosen curve, and therefore defines a parallel section of the bundle. $\qquad\square$

DEFINITION 1.3. A connection on a vector bundle is said to be *flat* if it is integrable as a distribution, cf. Chapter 1, Definition 9.1.

If $\mathcal{H}$ is trivial, then it is flat: Given $u \in E$, the parallel section $X$ of $\xi$ with $X(\pi(u)) = u$ is an integral manifold of $\mathcal{H}$. Conversely:

PROPOSITION 1.3. *Let $\mathcal{H}$ be a connection on a vector bundle $\xi = \pi : E \to M$. The following statements are equivalent:*

(1) *$\mathcal{H}$ is flat.*
(2) *For any open, simply connected subset $U$ of $M$, the restriction of $\mathcal{H}$ to $\pi^{-1}(U)$ is a trivial connection on $\xi_{|U}$.*

PROOF. We have already remarked that a locally trivial connection is flat. Conversely, if $\mathcal{H}$ is flat, then given $u \in \pi^{-1}(U)$, there exists by Frobenius' theorem (Theorem 9.2 of Chapter 1), a maximal connected integral manifold $\tilde{U}$ of $\mathcal{H}_{|\pi^{-1}(U)}$ through $u$. By Proposition 1.1, any curve $c$ in $U$ can be uniquely lifted horizontally to $\tilde{U}$. This implies that $\pi : \tilde{U} \to U$ is a covering map (see, e.g., Spanier's "Algebraic Topology" 2.7.8 and 2.4.10), and therefore a diffeomorphism. Its inverse is then a parallel section of $\xi_{|U}$ through $u$.  $\square$

EXAMPLES AND REMARKS 1.1. (i) It follows from Proposition 1.3 that a flat connection on a bundle over a simply connected manifold is a trivial connection. In particular, both the bundle and the holonomy group of the connection are trivial.

(ii) More generally, one defines the *restricted holonomy group* at $p \in M$ of a connection to be the subgroup $G_0(p)$ of $G(p)$ obtained by considering only those closed curves that are homotopic to the identity. A homotopy $H$ between the closed curve $c$ and the constant curve $p$ induces a curve $t \mapsto P_{H \circ_t}$ joining $P_c$ to $1_{E_p}$. Thus, the restricted holonomy group is a path-connected subgroup of the Lie group $GL(E_p)$, and is then itself a Lie group. Notice that $G_0(p)$ is a normal subgroup of $G(p)$, and that the map

$$\pi_1(M, p) \to G(p)/G_0(p),$$
$$[c] \mapsto P_c \cdot G_0(p)$$

is a surjective homomorphism. Thus, $G(p)/G_0(p)$ is countable, and $G(p)$ is also a Lie group. If the connection is flat, then by Proposition 1.3, $G_0(p)$ is trivial, and there is an epimorphism $\pi_1(M, p) \to G(p)$.

(iii) The canonical connection on the tangent bundle of $S^n$: If $\imath : S^n \to \mathbb{R}^{n+1}$ denotes inclusion, then by Exercise 14 in Chapter 1, $\imath_* S_p^n = \{u \in \mathbb{R}_p^{n+1} \mid \langle p, \mathcal{J}_p^{-1} u \rangle = 0\}$ for $p \in S^n$. The equivalence $\mathbb{R}^{n+1} \times \mathbb{R}^{n+1} \cong T\mathbb{R}^{n+1}$ mapping $(u, v) \in \mathbb{R}^{n+1} \times \mathbb{R}^{n+1}$ to $\mathcal{J}_u v \in T\mathbb{R}^{n+1}$ allows us to identify $TS^n$ with the set of all $(p, u) \in S^n \times \mathbb{R}^{n+1}$ with $u$ orthogonal to $p$. Under this identification, the tangent field $\dot{c}$ to a curve $c$ in $S^n$ is $\dot{c} = (c, c')$, since $\mathcal{J}_{c(t)} c'(t) = \dot{c}(t)$.

In order to specify a connection on $\tau S^n$, we need a similar description of $T(TS^n)$. If $X = (c, x)$ is a section of $\tau S^n$ along $c$, then $\dot{X} = (\dot{c}, \dot{x}) = (c, x, c', x')$. Since $\langle c, x \rangle = 0$, $\langle c', x \rangle + \langle c, x' \rangle = \langle c, x \rangle' = 0$. Furthermore, $\langle c, c' \rangle = 0$ because $\langle c, c \rangle \equiv 1$. Thus, if $X(0) = (p, u)$, then $TS^n_{(p,u)}$ is contained in the space of all $(p, u, v, w)$, where $(v, w) \in \mathbb{R}^{2n+2}$ satisfies $\langle p, v \rangle = 0$ and $\langle u, v \rangle + \langle p, w \rangle = 0$. By dimension considerations,

(1.1)
$$(TS^n)_{(p,u)} = \{((p,u),(v,w)) \in TS^n \times \mathbb{R}^{2n+2} \mid \langle u,v \rangle + \langle p,w \rangle = \langle p,v \rangle = 0\}.$$

Now, $X$, as a section of $\tau\mathbb{R}^{n+1}$ along $c$, is parallel (with respect to the canonical trivial connection on $\tau\mathbb{R}^{n+1}$) if $x' = 0$. But if $x$ is constant, then in general $\langle c,x \rangle$ will not be zero, and $X$ does not remain tangent to the sphere. The most one can hope for is that the orthogonal projection of $x'$ onto $S^n$ be zero; i.e., $x'$ should be a multiple of $c$. In terms of Equation (1.1), this means that $((p,u),(v,w))$ should be horizontal if $w$ is a multiple of $p$. But then the condition $\langle u,v \rangle + \langle p,w \rangle = 0$ forces $w$ to equal $-\langle u,v \rangle p$. We therefore define the horizontal space at $(p,u)$ to be

(1.2)
$$\mathcal{H}_{(p,u)} = \{((p,u),(v,-\langle u,v \rangle p)) \mid v \in \mathbb{R}^{n+1}, \langle p,v \rangle = 0\}.$$

The verification that (1.2) does indeed define a connection on $\tau S^n$ is left as an exercise.

When $n = 2$, it is easy to describe the parallel fields along great circles $c$ in $S^2$, since they are determined by their length and the angle they make with $c$. Let $c(t) = (\cos t)p + (\sin t)q$, where $p,q \in S^2$, $\langle p,q \rangle = 0$, and consider a parallel field $X = (c,x)$ along $c$. Since $\langle c,x \rangle = 0$ and $x' = -\langle x,c' \rangle c$, it follows that $\langle x,x' \rangle = 0$, and $|x|$ is constant. In particular, the holonomy group is a subgroup of $O(2)$. Since $S^2$ is simply connected, the holonomy group is path connected, and is actually a subgroup of $SO(2)$. Furthermore, $\langle x,c' \rangle' = \langle x',c' \rangle + \langle x,c'' \rangle = 0 - \langle x,c \rangle = 0$, so that $\langle x,c' \rangle$ is constant; i.e., the parallel fields along $c$ have constant length and make a constant angle with $c$.

By parallel translating any vector in $(S^2)_p$ along $c$ from $p$ to $-p = c(\pi)$, and then parallel translating it back to $p$ along a different great circle, one easily sees that the holonomy group $G(p)$ acts transitively on the unit circle in the tangent space at $p$; i.e., the orbit of any point is the whole circle. Thus, $G(p) \cong SO(2)$.

EXERCISE 86. Let $\mathcal{H}$ be a connection on $\xi = \pi : E \to M$, and denote by $\overline{X}$ the horizontal lift of a vector field $X$ on $M$.

(a) Show that for $X,Y \in \mathfrak{X}M$, $\alpha,\beta \in \mathbb{R}$,
$$\overline{\alpha X + \beta Y} = \alpha\overline{X} + \beta\overline{Y}, \qquad \overline{[X,Y]} = [\overline{X},\overline{Y}]^h.$$

(b) Prove that if $[\overline{X},\overline{Y}]^v = 0$ for all $X, Y \in \mathfrak{X}M$, then $\mathcal{H}$ is flat.

EXERCISE 87. (a) Consider the bundle projection $\pi : TS^n \to S^n$. With notation as in Examples and Remarks 1.1 (iii), show that $\pi_* : TTS^n \to TS^n$ is given by
$$\pi_*((p,u),(v,w)) = (p,v).$$
Deduce that the vertical space at $(p,u)$ is
$$\{((p,u),(0,w)) \in TS^n \times 0 \times \mathbb{R}^{n+1} \mid \langle p,w \rangle = 0\}.$$

(b) Prove that (1.2) defines a connection on $\tau S^n$.

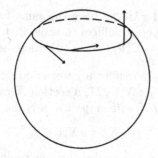

FIGURE 1. A parallel field along a circle of latitude.

EXERCISE 88. Describe the parallel fields along a circle of latitude
$$c(t) = (\cos t \sin \phi, \sin t \sin \phi, \cos \phi)$$
in $S^2$, $\phi \in (-\pi, \pi)$.

## 2. Covariant Derivatives

Let $\mathcal{H}$ be a connection on $\xi = \pi : E \to M$, so that $\tau E = \mathcal{V} \oplus \mathcal{H}$, where by abuse of notation $\mathcal{H}$ also denotes the bundle $\mathcal{H} \to E$. A vector $w \in TE$ decomposes as $w = w^v + w^h$. The vertical component measures the amount by which a vector fails to horizontal. It is more convenient to replace it by a vector in $E$ rather than in $TE$. This can be done as follows: by Exercise 59, $\mathcal{V}$ is equivalent to $\pi^*\xi$. Under this identification, the second factor projection $\pi_2 : E(\mathcal{V}) = \pi^*E \to E$ is a bundle map ($E(\mathcal{V})$ denotes the total space of the bundle $\mathcal{V}$).

DEFINITION 2.1. The *connection map* $\kappa : TE \to E$ of $\mathcal{H}$ is given by
$$\kappa(w) = \pi_2(w^v), \qquad w \in TE.$$

Alternatively, for $w \in T_u E$, let $p := \pi(u)$, and $\imath : E_p \to E$ be the inclusion. Then by Example 5.1 in Chapter 2,
$$E(\mathcal{V})_u = \imath_*(E_p)_u = \{\imath_* \mathcal{J}_u v \mid v \in E_p\}.$$
Thus,

(2.1)
$$\kappa(w) = (\imath_* \mathcal{J}_u)^{-1} w^v.$$

Notice that $\mathcal{H} = \ker \kappa$. This implies the following:

PROPOSITION 2.1. $(\pi_*, \kappa) : \tau E \to \tau M \oplus \xi$ *is a bundle map covering* $\pi : E \to M$.

$$
\begin{array}{ccc}
TE & \xrightarrow{(\pi_*,\kappa)} & E(\tau M \oplus \xi) \\
{\scriptstyle \pi_E}\downarrow & & \downarrow \\
E & \xrightarrow{\pi} & M
\end{array}
$$

PROOF. Both $\tau E$ and $\tau M \oplus \xi$ have the same fiber dimension, and since $(\pi_*, \kappa)$ is linear on each fiber, it suffices to show that it has trivial kernel. But $\ker(\pi_*, \kappa) = \ker \pi_* \cap \ker \kappa = E(\mathcal{V}) \cap \mathcal{H} = \{0\}$ fiberwise.                     □

DEFINITION 2.2. Let $\mathcal{H}$ denote a connection on $\xi = \pi : E \to M$ with connection map $\kappa$. Given $f : N \to M$, a section $X$ of $\xi$ along $f$, and $u \in TN$, the *covariant derivative of $X$ with respect to $u$* is the vector

$$\nabla_u X := \kappa X_* u \in E.$$

When $N = M$ and $f = 1_M$, $\nabla$ is called the *covariant derivative operator of $\mathcal{H}$*.

Notice that if $u \in N_p$, then $\nabla_u X \in E_{f(p)}$. Thus, for $U \in \mathfrak{X}N$, $\nabla_U X$ is a section of $\xi$ along $f$, where $\nabla_U X(p) := \nabla_{U(p)} X$. Furthermore, $X$ is parallel along $f$ iff $\nabla_U X = 0$ for all $U \in \mathfrak{X}N$, since $\ker \kappa = \mathcal{H}$.

We have already observed that the space $\Gamma_f$ of sections of $\xi$ along $f$ is canonically isomorphic to the space $\Gamma_{f^*\xi}$ of sections of $f^*\xi$ via

$$\Gamma_f \longrightarrow \Gamma_{f^*\xi},$$
$$X \longmapsto (1_N, X).$$

By the remark following Definition 1.2, the covariant derivative $f^*\nabla$ of the induced connection on $f^*\xi$ is given by

$$(f^*\nabla)_u(1_N, X) = (1_N, \nabla_u X).$$

THEOREM 2.1. *Suppose $\mathcal{H}$ is a connection on $\xi$ with covariant derivative operator $\nabla$. Given $f : N \to M$, $u, v \in N_p$, and sections $X, Y$ of $\xi$ along $f$,*

(1) $\nabla_u(X + Y) = \nabla_u X + \nabla_u Y$.
(2) $\nabla_{au+v} X = a\nabla_u X + \nabla_v X, \quad a \in \mathbb{R}$.
(3) $\nabla_u h X = u(h)X(p) + h(p)\nabla_u X, \quad h \in \mathcal{F}N$.
(4) *If $g : L \to N$, and $w \in TL$, then $\nabla_w(X \circ g) = \nabla_{g_* w} X$.*

The proof of the above theorem requires a couple of lemmas:

LEMMA 2.1. $\kappa \circ \mu_{a*} = \mu_a \circ \kappa, \quad a \in \mathbb{R}$.

PROOF OF LEMMA 2.1. It suffices to consider vertical vectors, since both sides vanish when applied to horizontal ones. So let $w \in (E_p)_u$, $v := \mathcal{J}_u^{-1} w \in E_p$. If $\imath : E_p \hookrightarrow E$ denotes inclusion, then $\mu_a \circ \kappa(\imath_* w) = \mu_a \circ \kappa(\imath_* \mathcal{J}_u v) = av$ by (2.1). On the other hand, $\kappa \circ \mu_{a*}(\imath_* w) = \kappa \circ \mu_{a*}(\imath_* \mathcal{J}_u v) = \kappa \circ \imath_* \circ \mu_{a*} \mathcal{J}_u v$, because $\imath \circ \mu_a = \mu_a \circ \imath$. Thus, $\kappa \circ \mu_{a*} \circ \imath_* w = \kappa \circ \imath_* \circ \mu_{a*} \mathcal{J}_u v = \kappa \circ \imath_* \mathcal{J}_{au} av = av$.      □

LEMMA 2.2. *Let $(\pi, \phi) : \pi^{-1}(U) \to U \times \mathbb{R}^k$ be a bundle chart of $\xi$, where $(U, x)$ is a chart of $M$. Set $\overline{x} = x \circ \pi$, so that $(\overline{x}, \phi) : \pi^{-1}(U) \to x(U) \times \mathbb{R}^k$ is a (manifold) coordinate map of $E$.*
*If $u, v \in E_p$, $x(p) = 0$, then*

$$\kappa\left(\frac{\partial}{\partial \overline{x}^i}(u+v)\right) = \kappa\left(\frac{\partial}{\partial \overline{x}^i}(u)\right) + \kappa\left(\frac{\partial}{\partial \overline{x}^i}(v)\right), \quad 1 \leq i \leq n.$$

PROOF OF LEMMA 2.2. Consider the map $L : E_p \to E_p$ given by $L(u) = \kappa(\partial/\partial \overline{x}^i(u))$. Notice that $\partial/\partial \overline{x}^i(tu) = \mu_{t*}\partial/\partial \overline{x}^i(u)$ for $t \in \mathbb{R}$: Indeed, $\partial/\partial \overline{x}^i(tu)$

may be written as $\dot{c}_{i,tu}(0)$, where $c_{i,tu}(s) = (\overline{x}, \phi)^{-1}(s\mathbf{e}_i, tu) = t(\overline{x}, \phi)^{-1}(s\mathbf{e}_i, u)$. Thus,

$$\frac{\partial}{\partial \overline{x}^i}(tu) = \dot{c}_{i,tu}(0) = \mu_{t*}\dot{c}_{i,u}(0) = \mu_{t*}\frac{\partial}{\partial \overline{x}^i}(u).$$

It follows that the derivative $DL$ of $L$ at 0 satisfies $DL(u) = c'(0)$, where $c(t) = \kappa(\partial/\partial \overline{x}^i(tu)) = \kappa(\mu_{t*}\partial/\partial \overline{x}^i(u)) = t\kappa(\partial/\partial \overline{x}^i(u))$ by Lemma 2.1; i.e., $DL(u) = L(u)$, so that $L$ is linear, and the lemma is proved.                    □

PROOF OF THEOREM 2.1. We will again use the coordinate map $(\overline{x}, \phi)$ : $\pi^{-1}(U) \to x(U) \times \mathbb{R}^k$ from Lemma 2.2, where $f(p) \in U$. Define sections $U_i$ of $\xi_{|U}$ by $\phi(U_i(q)) = \mathbf{e}_i$, $q \in U$, $1 \leq i \leq k$, and write $X_{|f^{-1}(U)} = \sum X^i U_i \circ f$. Then

$$X_*u = \sum_{i=1}^{n} X_*u(\overline{x}^i)\frac{\partial}{\partial \overline{x}^i}(X(p)) + \sum_{j=1}^{k} X_*u(\phi^j)\frac{\partial}{\partial \phi^j}(X(p))$$

$$(2.2) \qquad = \sum_{i} u(\overline{x}^i \circ X)\frac{\partial}{\partial \overline{x}^i}(X(p)) + \sum_{j} u(\phi^j \circ X)\frac{\partial}{\partial \phi^j}(X(p))$$

$$= \sum_{i} u(x^i \circ f)\frac{\partial}{\partial \overline{x}^i}(X(p)) + \sum_{j} u(X^j)\frac{\partial}{\partial \phi^j}(X(p)).$$

The vector $Y_*u$ can be expressed in a similar way. On the other hand,

$$(X+Y)_*u = \sum_{i} u(x^i \circ f)\frac{\partial}{\partial \overline{x}^i}(X(p)+Y(p)) + \sum_{j} u(X^j+Y^j)\frac{\partial}{\partial \phi^j}(X(p)+Y(p)).$$

Next, observe that $U_j \circ \pi = \kappa\partial/\partial \phi^j$: Indeed, recall that $U_j(p) = (\phi_{|E_p})^{-1}\mathbf{e}_j$. If $u \in E_p$ and $\imath : E_p \to E$ denotes inclusion, then

$$\frac{\partial}{\partial \phi^j}(u) = \imath_*(\phi_{|E_p})_*^{-1}D_j(\phi(u)) = \imath_*(\phi_{|E_p})_*^{-1}\mathcal{J}_{\phi(u)}\mathbf{e}_j = \imath_*\mathcal{J}_u(\phi_{|E_p})^{-1}\mathbf{e}_j$$

$$= \imath_*\mathcal{J}_u U_j(p),$$

where the third equality uses the fact that $\phi$ is linear. Applying (2.1) now yields the claim. Thus,

$$\kappa X_*u + \kappa Y_*u = \sum_{i} u(x^i \circ f)\left[\kappa\left(\frac{\partial}{\partial \overline{x}^i}(X(p))\right) + \kappa\left(\frac{\partial}{\partial \overline{x}^i}(Y(p))\right)\right]$$

$$+ \sum_{j}[u(X^j) + u(Y^j)]U_j(f(p))$$

$$= \sum_{i} u(x^i \circ f)\kappa\frac{\partial}{\partial \overline{x}^i}(X(p)+Y(p)) + \sum_{j} u(X^j+Y^j)U_j(f(p))$$

$$= \kappa(X+Y)_*u,$$

where the second equality makes use of Lemma 2.2. This proves (1). Statement (2) is an immediate consequence of the definition of $\kappa$. For (3), observe that

$\overline{x}^i \circ hX = \overline{x}^i \circ X = x^i \circ f$, so that (2.2) yields

$$(hX)_* u = \sum_i u(x^i \circ f) \frac{\partial}{\partial \overline{x}^i} (h(p)X(p))$$

$$+ \sum_j [u(h)X^j(p) + h(p)u(X^j)] \frac{\partial}{\partial \phi^j} (h(p)X(p))$$

$$= \left[ \sum_i u(x^i \circ f) \mu_{h(p)*} \frac{\partial}{\partial \overline{x}^i} (X(p)) + h(p) \sum_j u(X^j) \frac{\partial}{\partial \phi^j} (h(p)X(p)) \right]$$

$$+ u(h) \sum_j X^j(p) \frac{\partial}{\partial \phi^j} (h(p)X(p)).$$

Applying $\kappa$ to the expression inside brackets and using Lemma 2.1 then yields $h(p)\kappa X_* u$, while the last line becomes $u(h)X(p)$ under $\kappa$, thus establishing (3). Finally, (4) is an immediate consequence of the chain rule. $\qquad \square$

It is often useful to express an arbitrary cross-section along a curve in terms of parallel ones: Consider a map $f : N \to M$, and let $\xi$ be a rank $k$ vector bundle with connection over $M$. Given a basis $x_1, \ldots, x_k$ of the fiber of $\xi$ over some point $f(p)$, $p \in N$, and a curve $c : [a, b] \to N$ with $c(a) = p$, there exist by Proposition 1.1 parallel sections $X_1, \ldots, X_k$ of $\xi$ along $f \circ c$ with $X_i(a) = x_i$. Thus, any section $X$ along $f \circ c$ can be written as $X = \sum h^i X_i$ for some functions $h^i : [a, b] \to \mathbb{R}$. Let us use this fact to show that $\nabla_u X$ only depends on the values of $X$ along $any$ curve tangent to $u$:

PROPOSITION 2.2. $Let$ $\xi$ $be$ $a$ $vector$ $bundle$ $with$ $connection$ $over$ $M$, $f :$ $N \to M$, $u \in N_p$, $and$ $X$ $a$ $section$ $of$ $\xi$ $along$ $f$. $Given$ $a$ $curve$ $c : I \to N$ $with$ $\dot{c}(0) = u$, $denote$ $by$ $X_t$ $the$ $parallel$ $section$ $along$ $f \circ c$ $with$ $X_t(t) = X \circ c(t)$. $Then$

$$\nabla_u X = \lim_{t \to 0} \frac{X_t(0) - X \circ c(0)}{t}.$$

PROOF. Choose linearly independent parallel sections $X_1, \ldots, X_k$ of $\xi$ along $f \circ c$, so that $X \circ c = \sum h^i X_i$, and $X_t(s) = \sum h^i(t)X_i(s)$. Then

$$\lim_{t \to 0} \frac{X_t(0) - X \circ c(0)}{t} = \lim_{t \to 0} \sum \frac{h^i(t) - h^i(0)}{t} X_i(0) = \sum h^{i\prime}(0)X_i(0).$$

On the other hand,

$$\nabla_u X = \nabla_{\dot{c}(0)} X = \nabla_{D(0)}(X \circ c) = \nabla_{D(0)} \sum h^i X_i = \sum D(0)(h^i)X_i(0)$$

$$= \sum h^{i\prime}(0)X_i(0).$$

$\qquad \square$

EXAMPLES AND REMARKS 2.1. (i) We have seen how a connection $\mathcal{H}$ on $\xi$ determines a covariant derivative operator $\nabla$. The two concepts are in fact equivalent; i.e., if $\nabla$ is a covariant derivative operator on $\xi$ (in the sense that it satisfies (1)–(3) of Theorem 2.1 for the case $N = M$ and $f = 1_M$), then there exists a connection $\mathcal{H}$ on $\xi$ whose covariant derivative coincides with $\nabla$: given $w \in E(\xi)$, define

$$\mathcal{H}_w = \{\mathcal{J}_w \nabla_u X - X_* u \mid X \in \Gamma\xi, \ X(\pi(w)) = w, \ u \in M_{\pi(w)}\}.$$

It is straightforward to verify that $\mathcal{H}$ is a connection. In order to establish that its covariant derivative is $\nabla$, it suffices to check that $\nabla_u X = 0$ iff $X_* u \in \mathcal{H}$ since the vertical distribution is the same for all connections. But this is immediate from the definition of $\mathcal{H}$.

In view of this, we shall use the word connection to denote either $\mathcal{H}$ or $\nabla$.

(ii) If $\xi$ is a vector bundle over $M$, a *differential $k$-form on $M$ with values in* $\xi$ is a section of the bundle $\mathrm{Hom}(\Lambda_k(M), \xi)$, where by abuse of notation $\Lambda_k(M)$ refers both to the bundle and to its total space. The space of these sections is denoted $A_k(M, \xi)$. Notice that in the notation of Section 11 in Chapter 1, the space of $k$-forms $A_k(M) = A_k(M, \epsilon^1)$, where $\epsilon^1$ is the trivial line bundle over $M$.

If $\nabla$ and $\tilde{\nabla}$ are two connections on $\xi$, then $\omega := \nabla - \tilde{\nabla} \in A_1(M, \mathrm{End}\,\xi)$, where $\mathrm{End}\,\xi = \mathrm{Hom}(\xi, \xi)$: Given $u \in M_p$, $x \in E(\xi)_p$, $\omega(u)x = \nabla_u X - \tilde{\nabla}_u X \in E(\xi)_p$ for any section $X$ of $\xi$ with $X(p) = x$; $\omega$ is well-defined by Proposition 11.3 of Chapter 1, together with the fact that $\nabla_u(fX) - \tilde{\nabla}_u(fX) = f(\nabla_u X - \tilde{\nabla}_u X)$ for $f \in \mathcal{F}(M)$ by Theorem 2.1(3). $\omega(u) : E_p \to E_p$ is clearly linear. Conversely, if $\nabla$ is a connection and $\omega \in A_1(M, \mathrm{End}\,\xi)$, then $\tilde{\nabla} := \nabla + \omega$ is again a connection. In other words, the set of connections on $\xi$ is an affine space modeled on the vector space $A_1(M, \mathrm{End}\,\xi)$.

(iii) At the beginning of Section 1, we defined parallel translation in $\mathbb{R}^n$ by letting the standard coordinate vector fields $D_i$ be parallel. Thus, if $X \in \mathfrak{X}\mathbb{R}^n$, so that $X = \sum X^i D_i$, where $X^i = X(u^i)$, then $\nabla_U X = \sum U(X^i)D_i$. Alternatively, identify $T\mathbb{R}^n$ with $\mathbb{R}^n \times \mathbb{R}^n$ by mapping $(u, w) \in \mathbb{R}^n \times \mathbb{R}^n$ to $\mathcal{J}_u w \in \mathbb{R}^n_u$. Then a vector field $X$ along a curve $c : I \to \mathbb{R}^n$ becomes identified with $(c, x)$, where $x : I \to \mathbb{R}^n$, and $\nabla_D X = (c, x')$.

(iv) In Examples and remarks 1.1(iii), we defined the canonical connection on $\tau S^n$ first in terms of parallel sections: A section $X$ of $\tau S^n$ along $c$ can be viewed as a section of $\tau\mathbb{R}^{n+1}$ along $c$ via the derivative of the inclusion map $\imath : S^n \to \mathbb{R}^{n+1}$, and may thus be written as $X = (c, x)$. $X$ was then said to be parallel along $c$ if the component of $x'$ tangent to $S^n$ is zero. The connection $\nabla$ on $\tau S^n$ can therefore be expressed in terms of the Euclidean connection $\tilde{\nabla}$ as follows: Given $p \in S^n$, $v \in \mathbb{R}^{n+1}_p$, let $v^\perp$ denote the orthogonal projection (with respect to the usual Euclidean metric on $\tau\mathbb{R}^{n+1}$) of $v$ onto $\imath_* S^n_p$. Then for $u \in \tau S^n$, $X \in \mathfrak{X}S^n$,

$$(2.3) \qquad\qquad \imath_* \nabla_u X = (\tilde{\nabla}_u \imath_* X)^\perp.$$

More generally, (2.3) makes sense in the context of any submanifold $M$ of $\mathbb{R}^n$, and defines the so-called *canonical connection* on $\tau M$.

(v) A connection $\nabla$ on $\xi$ determines a connection (which will be also denoted $\nabla$) on the dual bundle $\xi^*$ by

$$(\nabla_u \omega)X := u(\omega(X)) - \omega(\nabla_u X), \qquad u \in TM, \quad \omega \in \Gamma\xi^*, \quad X \in \Gamma\xi.$$

If $c$ is a curve in $M$, then $\omega \circ c$ is parallel iff $\omega \circ c(X)$ is constant for any parallel section $X$ of $\xi$ along $c$.

Similarly, if $\nabla^i$ are connections on the bundles $\xi_i$ over $M$, $i = 1, 2$, define a corresponding connection $\nabla$ on the tensor product $\xi_1 \otimes \xi_2$ by

$$\nabla_u(X_1 \otimes X_2) = (\nabla^1_u X_1) \otimes X_2(p) + X_1(p) \otimes (\nabla^2_u X_2), \qquad u \in M_p, \quad X_i \in \Gamma\xi_i,$$

and one on the Whitney sum $\xi_1 \oplus \xi_2$ by

$$\nabla_u(X_1, X_2) = (\nabla^1_u X_1, \nabla^2_u X_2).$$

(vi) Let $g$ denote a Euclidean metric on $\xi$. Since $g$ is a section of $(\xi \otimes \xi)^*$, (v) implies that

$$(\nabla_u g)(X \otimes Y) = u(g(X \otimes Y)) - g(\nabla_u X \otimes Y(p)) - g(X(p) \otimes \nabla_u Y),$$

for $u \in M_p$, $X, Y \in \Gamma\xi$. $\nabla$ is said to be a *Riemannian connection* if $g$ is parallel. Writing $\langle X, Y \rangle$ instead of $g(X \otimes Y)$, the connection is Riemannian iff

$$u\langle X, Y \rangle = \langle \nabla_u X, Y(p) \rangle + \langle X(p), \nabla_u Y \rangle.$$

In this case, $\langle X, Y \rangle$ is constant for parallel sections $X$, $Y$ of $\xi$ along a curve $c$, so that the holonomy group is a subgroup of the orthogonal group.

EXERCISE 89. Prove that $\mathcal{H}$ as defined in (i) is indeed a connection.

EXERCISE 90. Show that if $\nabla$ is a connection on $\xi$ and $\omega \in A_1(M, \operatorname{End}\xi)$, then $\tilde{\nabla} = \nabla + \omega$ is again a connection.

EXERCISE 91. Let $p, q \in S^n$, $p \perp q$, and consider the great circle $c$ : $[0, 2\pi] \to S^n$ given by $c(t) = (\cos t)p + (\sin t)q$. Prove that $\nabla_D \dot{c} = 0$ for the canonical connection $\nabla$ on $\tau S^n$.

EXERCISE 92. Let $\xi_i$ be vector bundles over $M$ with connections $\nabla^i$. Show that the induced connection $\nabla$ on $\operatorname{Hom}(\xi_1, \xi_2)$ is given by

$$(\nabla_u L)X = \nabla^2_u(LX) - L(\nabla^1_u X), \qquad u \in TM, \quad L \in \Gamma(\operatorname{Hom}(\xi_1, \xi_2)), \quad X \in \Gamma\xi_1.$$

Deduce that $L \circ c$ is parallel ($c : I \to M$) iff $LX$ is a parallel section of $\xi_2$ along $c$ whenever $X$ is a parallel section of $\xi_1$ along $c$.

EXERCISE 93. Let $\xi_1$, $\xi_2$ be vector bundles over $M$ with connections $\nabla_1$, $\nabla_2$. Define a "natural" connection on the product $\xi_1 \times \xi_2$ so that the Whitney sum connection from Examples and Remarks 2.1 becomes its pullback via the diagonal imbedding of $M$ into $M \times M$.

## 3. The Curvature Tensor of a Connection

In the previous section, we saw that flat connections are those for which parallel translation is, at least locally, independent of the chosen curve. Deviation from flatness is measured by a tensor field called the curvature.

DEFINITION 3.1. The *curvature tensor* $R : \mathfrak{X}M \times \mathfrak{X}M \times \Gamma\xi \to \Gamma\xi$ of a connection $\nabla$ on $\xi = \pi : E \to M$ is given by

$$R(U, V)X = \nabla_U \nabla_V X - \nabla_V \nabla_U X - \nabla_{[U,V]}X, \qquad U, V \in \mathfrak{X}M, \quad X \in \Gamma\xi.$$

PROPOSITION 3.1. *The operator $R$ is a 2-form on $M$ with values in $\operatorname{End}\xi$; i.e., $R \in A_2(M, \operatorname{End}\xi)$, cf. Examples and Remarks 2.1(ii). In particular, $R$ is tensorial in all three arguments.*

PROOF. It is clear that $R(U_1 + U_2, V)X = R(U_1, V)X + R(U_2, V)X$. Since $R$ is skew-symmetric in the first two arguments, a similar identity holds for the second argument. We apply Proposition 11.3 of Chapter 1 to show tensoriality: Given $f \in \mathcal{F}M$,

$$
\begin{aligned}
R(fU, V)X &= \nabla_{fU}\nabla_V X - \nabla_V \nabla_{fU} X - \nabla_{[fU,V]} X \\
&= f\nabla_U \nabla_V X - \nabla_V (f\nabla_U X) - f\nabla_{[U,V]} X + \nabla_{V(f)U} X \\
&= f\nabla_U \nabla_V X - (Vf)\nabla_U X - f\nabla_V \nabla_U X - f\nabla_{[U,V]} X + V(f)\nabla_U X \\
&= fR(U, V)X.
\end{aligned}
$$

Similarly,

$$
\begin{aligned}
R(U, V)hX &= \nabla_U \nabla_V hX - \nabla_V \nabla_U hX - \nabla_{[U,V]} hX \\
&= \nabla_U (h\nabla_V X + V(h)X) - \nabla_V (h\nabla_U X + U(h)X) - h\nabla_{[U,V]} X \\
&\quad - ([U,V]h)X \\
&= h\nabla_U \nabla_V X + U(h)\nabla_V X + V(h)\nabla_U X + UV(h)X \\
&\quad - h\nabla_V \nabla_U X - V(h)\nabla_U X - U(h)\nabla_V X - VU(h)X \\
&\quad - h\nabla_{[U,V]} X - ([U,V]h)X \\
&= hR(U, V)X + (UV(h) - VU(h) - [U,V](h))X \\
&= hR(U, V)X.
\end{aligned}
$$

$\square$

A useful interpretation of $R$ can be given in terms of the so-called exterior covariant derivative operator of a connection:

DEFINITION 3.2. Given a connection $\nabla$ on $\xi$, its *exterior covariant derivative operator* is the collection of maps $d^\nabla : A_k(M, \xi) \to A_{k+1}(M, \xi)$ given by

$$
d^\nabla(X)U = \nabla_U X, \qquad X \in A_0(M, \xi) = \Gamma(\xi), \quad U \in \mathfrak{X}M,
$$

and by

$$
(d^\nabla \omega)(U_0, \ldots, U_k) = \sum_{i=0}^{k} (-1)^i \nabla_{U_i}(\omega(U_0, \ldots, \hat{U}_i, \ldots, U_k))
$$
$$
+ \sum_{i<j} (-1)^{i+j} \omega([U_i, U_j], U_0, \ldots, \hat{U}_i, \ldots, \hat{U}_j, \ldots, U_k)
$$

for $\omega \in A_k(M, \xi)$, $k \geq 1$, and $U_i \in \mathfrak{X}M$.

The operator $d^\nabla$ is the generalization to vector bundles of the exterior derivative operator $d$ on $A(M)$. In fact, if $\xi = \epsilon^1$ is the trivial line bundle over $M$ with the canonical connection, then a section $X$ of $\xi$ can be written as $X = (1_M, f)$ for some $f \in \mathcal{F}M$, and $d^\nabla(X)U = \nabla_U X = (1_M, U(f))$ corresponds to $df(U)$. Similarly, when $k \geq 1$, the expression for $d^\nabla \omega$ is identical to that for $d\omega$ in Theorem 11.3 from Chapter 1. Thus, $d^\nabla = d$ in this case.

The connection on $\epsilon^1$ is flat, and we will soon see that flat connections have vanishing curvature. This turns out to be the reason why $d^2 = 0$:

THEOREM 3.1. *Let $R$ denote the curvature of a connection $\nabla$ on $\xi$. Given $X \in \Gamma\xi$, define $R_X \in A_2(M, \xi)$ by $R_X(U, V) = R(U, V)X$. Then $R_X = d^\nabla \circ d^\nabla X$.*

PROOF.
$$d^{\nabla 2}(X)(U, V) = \nabla_U((d^\nabla X)V) - \nabla_V((d^\nabla X)U) - d^\nabla X([U, V])$$
$$= \nabla_U \nabla_V X - \nabla_V \nabla_U X - \nabla_{[U,V]} X$$
$$= R(U, V)X.$$

$\square$

In order to provide a geometric interpretation of curvature, we need the following:

LEMMA 3.1 (Cartan's Structure Equation). *Given $f : N \to M$, $U, V \in \mathfrak{X}N$, and a section $X$ of $\xi$ along $f$,*
$$R(f_* U, f_* V)X = \nabla_U \nabla_V X - \nabla_V \nabla_U X - \nabla_{[U,V]} X,$$
*where $\nabla$ denotes the covariant derivative along $f$.*

PROOF. Let $p \in N$, $x$ a coordinate map of $M$ around $f(p)$. Then locally,
$$f_* U = \sum_i (f_* U)(x^i) \frac{\partial}{\partial x^i} \circ f = \sum_i U(x^i \circ f) \frac{\partial}{\partial x^i} \circ f.$$

Similarly, if $E_1, \ldots, E_k$ are local sections of $\xi$ that are linearly independent, then (the restriction of) $X = \sum h^i E_i \circ f$. Thus, by tensoriality of $R$, we may assume that there exist $\tilde{U}, \tilde{V} \in \mathfrak{X}M$, $\tilde{X} \in \Gamma\xi$ such that $f_* U = \tilde{U} \circ f$, $f_* V = \tilde{V} \circ f$, and $X = \tilde{X} \circ f$. Now,
$$\nabla_U \nabla_V X = \nabla_U \nabla_V (\tilde{X} \circ f) = \nabla_U (\nabla_{f_* V} \tilde{X}) = \nabla_U (\nabla_{\tilde{V} \circ f} \tilde{X}) = \nabla_U (\nabla_{\tilde{V}} \tilde{X}) \circ f$$
$$= \nabla_{f_* U} (\nabla_V \tilde{X}) = (\nabla_{\tilde{U}} \nabla_{\tilde{V}} \tilde{X}) \circ f.$$

Similarly,
$$\nabla_{[U,V]} X = \nabla_{[U,V]} (\tilde{X} \circ f) = \nabla_{f_* [U,V]} \tilde{X} = \nabla_{[\tilde{U}, \tilde{V}] \circ f} \tilde{X} = (\nabla_{[\tilde{U}, \tilde{V}]} \tilde{X}) \circ f.$$

Thus,
$$R(f_* U, f_* V)X = (R(\tilde{U}, \tilde{V})\tilde{X}) \circ f = (\nabla_{\tilde{U}} \nabla_{\tilde{V}} \tilde{X} - \nabla_{\tilde{V}} \nabla_{\tilde{U}} \tilde{X} - \nabla_{[\tilde{U}, \tilde{V}]} \tilde{X}) \circ f$$
$$= \nabla_U \nabla_V X - \nabla_V \nabla_U X - \nabla_{[U,V]} X.$$

$\square$

The above lemma may be interpreted in terms of the induced connection on $f^* \xi$: Given $\omega \in A(M, \text{End }\xi)$, define $f^* \omega \in A(N, \text{End }f^* \xi)$ by
$$(f^* \omega)(U_1, \ldots, U_k)(1_N, X) := (1_N, \omega(f_* U_1, \ldots, f_* U_k)X), \qquad U_i \in \mathfrak{X}N, \quad X \in \Gamma_f.$$

If $R_f$ denotes the curvature tensor of $f^* \xi$, Cartan's equation states that $R_f = f^* R$.

The next proposition clarifies the remark at the beginning of the section to the effect that the curvature tensor of a connection measures deviation from flatness:

PROPOSITION 3.2. *Let $u, v \in M_p$, $x \in E_p$. Consider a map $f$ from a neighborhood of the origin in $\mathbb{R}^2$ into $M$ with $f(0) = p$, $f_*D_1(0) = u$, $f_*D_2(0) = v$. For $t$, $s$ small enough, let $x_{t,s}$ denote the vector in $E_p$ which is obtained after parallel translating $x$ along*

(1) $\tau \mapsto f(\tau, 0)$ *from $p$ to $f(t, 0)$, then along*
(2) $\sigma \mapsto f(t, \sigma)$ *from $f(t, 0)$ to $f(t, s)$, then along*
(3) $\tau \mapsto f(t - \tau, s)$ *from $f(t, s)$ to $f(0, s)$, and finally along*
(4) $\sigma \mapsto f(0, s - \sigma)$ *from $f(0, s)$ to $p$.*

*Then*

$$R(u, v)x = -\lim_{t,s \to 0} \frac{x_{t,s} - x}{ts}.$$

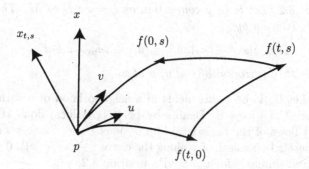

FIGURE 2. Curvature via parallel translation.

PROOF. Define a section $X$ along $f$ by letting $X(t, s)$ be the parallel translate of $x$ along the curves in (1) and (2) from $p$ to $f(t, s)$. By Lemma 3.1,

$$
\begin{aligned}
R(u, v)x &= R(f_*D_1(0), f_*D_2(0))x \\
&= \nabla_{D_1(0)} \nabla_{D_2} X - \nabla_{D_2(0)} \nabla_{D_1} X \quad \text{(since } [D_1, D_2] = 0) \\
&= -\nabla_{D_2(0)} \nabla_{D_1} X \quad \text{(since } X \text{ is parallel along the curves (2)).}
\end{aligned}
$$

If $P_s$ is parallel translation along $\sigma \mapsto f(0, \sigma)$ from $p$ to $f(0, s)$, then by Proposition 2.2,

$$\nabla_{D_2(0)} \nabla_{D_1} X = \lim_{s \to 0} \frac{P_s^{-1} \nabla_{D_1(0,s)} X - \nabla_{D_1(0)} X}{s} = \lim_{s \to 0} \frac{P_s^{-1} \nabla_{D_1(0,s)} X}{s}.$$

Now, if $P_{t,s}$ denotes parallel translation along $\tau \mapsto f(\tau, \sigma)$ from $f(0, s)$ to $f(t, s)$, then

$$\nabla_{D_1(0,s)} X = \lim_{t \to 0} \frac{P_{t,s}^{-1} X(t, s) - X(0, s)}{t}.$$

Thus,

$$R(u, v)x = -\lim_{t,s \to 0} \frac{P_s^{-1} P_{t,s}^{-1} X(t, s) - P_s^{-1} X(0, s)}{ts} = -\lim_{t,s \to 0} \frac{x_{t,s} - x}{ts}.$$

$\square$

The curvature tensor of a connection $\mathcal{H}$ measures the amount by which the distribution $\mathcal{H}$ fails to be integrable: Recall that $\mathcal{H}$ is integrable iff $[\bar{U}, \bar{V}]$ is horizontal for all $U, V \in \mathfrak{X}M$, where $\bar{U}, \bar{V}$ are the horizontal lifts of $U, V$ to $TE$. Notice that $A_{\bar{U}}\bar{V} := \frac{1}{2}[\bar{U}, \bar{V}]^v$ is tensorial: If $f \in \mathcal{F}(TE)$, then

$$A_{f\bar{U}}\bar{V} = \frac{1}{2}[f\bar{U}, \bar{V}]^v = \frac{1}{2}(f[\bar{U}, \bar{V}]^v - \bar{V}(f)\bar{U}^v) = \frac{1}{2}f[\bar{U}, \bar{V}]^v = fA_{\bar{U}}\bar{V}.$$

$A$ is then tensorial in the second argument by skew-symmetry. We may therefore define

$$(3.1) \qquad A_u v := \frac{1}{2}[U, V]^v(x), \qquad u, v \in \mathcal{H}_x, \quad x \in E,$$

where $U, V$ are horizontal vector fields on $E$ with $U_x = u$, $V_x = v$.

THEOREM 3.2. *Let $\mathcal{H}$ be a connection on $\xi = \pi : E \to M$. The curvature tensor $R$ of $\mathcal{H}$ is given by*

$$R(u, v)x = -2\kappa A_{\bar{u}}\bar{v}, \qquad u, v \in M_p, \quad x \in E_p,$$

*where $\bar{u}, \bar{v}$ are the horizontal lifts of $u, v$ at $x$.*

PROOF. Let $U, V$ be vector fields in a neighborhood of $p$ with $U(p) = u$, $V(p) = v$, and $[U, V](p) = 0$. Denote by $\{\phi_t\}$, $\{\psi_s\}$ local flows of $U, V$, and by $\{\bar{\phi}_t\}$, $\{\bar{\psi}_s\}$ flows of the horizontal lifts $\bar{U}, \bar{V}$ of $U, V$. Since $\pi \circ \bar{\phi}_t = \phi_t \circ \pi$, $\bar{\phi}_t(x)$ is the parallel translate of $x$ along the curve $\tau \mapsto \phi_\tau(\pi(x))$, $0 \le \tau \le t$, for any $x \in E_p$, and similarly for $\bar{\psi}_s$. By Proposition 3.2,

$$R(u, v)x = -\lim_{t,s \to 0} \frac{x_{t,s} - x}{ts} = -\lim_{t \to 0^+} \frac{x_{\sqrt{t},\sqrt{t}} - x}{t} = -\lim_{t \to 0^+} \frac{c(t) - c(0)}{t},$$

where $c(t) = \bar{\psi}_{-\sqrt{t}} \circ \bar{\phi}_{-\sqrt{t}} \circ \bar{\psi}_{\sqrt{t}} \circ \bar{\phi}_{\sqrt{t}}(x)$. Thus, given $f \in \mathcal{F}(TE)$,

$$\imath_* \mathcal{J}_x R(u, v)x(f) = -\lim_{t \to 0} \frac{(f \circ c)(t) - (f \circ c)(0)}{t} = -[\bar{U}, \bar{V}](x)(f)$$

by Theorem 8.3 in Chapter 1. Finally, $[U, V](p) = 0$, so $[\bar{U}, \bar{V}](x)$ is vertical, and equals $2A_{\bar{u}}\bar{v}$. The somewhat annoying factor $1/2$ in the definition of $A$ comes from the traditional terminology used for Riemannian submersions in the next chapter. $\square$

The following is an immediate consequence:

COROLLARY 3.1. *A connection is flat iff its curvature tensor vanishes.*

PROPOSITION 3.3. *Let $\nabla$ and $\tilde{\nabla}$ denote two connections on $\xi$, $\omega := \tilde{\nabla} - \nabla \in A_1(M, \operatorname{End}\xi)$. Then*

$$\tilde{R} = R + d^\nabla \omega + [\omega, \omega],$$

*where $[\omega, \omega] \in A_2(M, \operatorname{End}\xi)$ is given by $[\omega, \omega](U, V) = \omega(U)\omega(V) - \omega(V)\omega(U)$, $U, V \in \mathfrak{X}M$.*

PROOF. Given $U, V \in \mathfrak{X}M$,

$$\tilde{\nabla}_U \tilde{\nabla}_V X = \nabla_U(\nabla_V X + \omega(V)X) + \omega(U)(\nabla_V X + \omega(V)X)$$
$$= \nabla_U \nabla_V X + \nabla_U(\omega(V)X) + \omega(U)(\nabla_V X) + \omega(U)\omega(V)X.$$

Thus,

$$\tilde{R}(U,V)X = R(U,V)X + \nabla_U(\omega(V)X) - \omega(V)(\nabla_U X) - \nabla_V(\omega(U)X)$$
$$+ \omega(U)(\nabla_V X) - \omega[U,V]X + [\omega(U),\omega(V)]X$$
$$= R(U,V)X + (\nabla_U\omega(V) - \nabla_V\omega(U) - \omega[U,V])X + [\omega(U),\omega(V)]X$$
$$= R(U,V)X + d^\nabla\omega(U,V)X + [\omega(U),\omega(V)]X.$$

$\square$

EXAMPLES AND REMARKS 3.1. (i) Since the standard connection on $\tau\mathbb{R}^n$ is flat, its curvature tensor is zero.

(ii) (Curvature tensor of $\tau S^2$) Consider the circle of latitude given by $t \mapsto c(t) = (a\cos t, a\sin t, b)$ on $S^2$, where $0 < a \le 1$, $-1 < b < 1$, $a^2 + b^2 = 1$. Let $X = \frac{1}{a}\dot{c}$, and $t \mapsto Y(t) := -b\cos t D_1(c(t)) - b\sin t D_2(c(t)) + aD_3(c(t))$ the northward-pointing unit vector field along $c$ orthogonal to $X$. Then the parallel vector field $E$ along $c$ with $E(0) = \cos c_0 X(0) + \sin c_0 Y(0)$ is given by

$$(3.2) \qquad E(t) = \cos(c_0 - bt)X(t) + \sin(c_0 - bt)Y(t),$$

cf. Exercise 88. To see this, recall from Examples and Remarks 2.1 that the connection $\nabla$ on $\tau S^2$ satisfies

$$\imath_* \nabla_D E = (\tilde{\nabla}_D \imath_* E)^\perp,$$

where $\tilde{\nabla}$ is the connection on the tangent bundle of Euclidean space, and $\imath : S^2 \to \mathbb{R}^3$ is inclusion. A straightforward computation yields

$$\tilde{\nabla}_{D(t)}\imath_* E = -a\cos(c_0 - bt)P \circ c,$$

where $P$ is the position vector field $u \mapsto P(u) := \mathcal{J}_u u$ on $\mathbb{R}^3$. Since $P^\perp = 0$, $E$ is parallel, as claimed.

Let $p = (1,0,0)$, $u = D_2(p)$, and $v = D_3(p)$. We will use Proposition 3.2 to compute $R(u,v)u$: Define a map $f$ from a neighborhood of $0 \in \mathbb{R}^2$ into $S^2$ by $f(t,s) = (\cos s \cos t, \cos s \sin t, \sin s)$, so that $f(0,0) = p$. Then

$$R(u,v)u = -\lim_{t\to 0}\frac{u_{t,t} - u}{t^2},$$

where $u_{t,t}$ is the parallel translate of $u$ along

(1) $\tau \mapsto f(\tau,0)$ from $p$ to $f(t,0)$,
(2) $\sigma \mapsto f(t,\sigma)$ from $f(t,0)$ to $f(t,t)$,
(3) $\tau \mapsto f(t-\tau,t)$ from $f(t,t)$ to $f(0,t)$, and
(4) $\sigma \mapsto f(0,t-\sigma)$ from $f(0,t)$ to $p$.

Notice that the curves (1), (2), and (4) are great circles, so that by Examples and Remarks 1.1(iii), parallel fields have constant length and make a constant angle with the tangent field to the curve. The curve (3) is the circle of latitude described above with $b = \sin t$, traveled in the opposite direction. The parallel translate of $u$ along the first two curves is then $X(t)$. Parallel translating $X(t)$ along curve (3) yields $E(0)$, where $E$ is the parallel field along the circle of latitude with $E(t) = X(t)$. By (3.2), $E(0) = \cos bt X(0) + \sin bt Y(0)$. Finally, since curve (4) is a great circle, we obtain

$$u_{t,t} = (\cos bt)u + (\sin bt)v = \cos(t\sin t)u + \sin(t\sin t)v,$$

and

$$R(u,v)u = -\lim_{t \to 0} \frac{[\cos(t \sin t) - 1]u + \sin(t \sin t)v}{t^2} = -v.$$

Since $R$ is skew-symmetric in the first two arguments, $R(u,v)v = -R(v,u)v = u$, and thus,

$$(3.3) \qquad R(u,v)w = \langle v,w \rangle u - \langle u,w \rangle v, \qquad w \in S_p^2,$$

where $\langle,\rangle$ is the inner product on $S_p^2$ induced by the one on $\mathbb{R}_p^3$ and $\imath_*$; i.e., $\langle u,v \rangle = \langle \imath_* u, \imath_* v \rangle$. Using the fact that $u$, $v$ in (3.3) form a basis of the tangent space of $S^2$ at $p$, it is easy to check that (3.3) holds for arbitrary $u$, $v \in S_p^2$.

If $q$ is another point on $S^2$, choose some $A \in SO(3)$ with $A(p) = q$, and replace $f$ by $A \circ f$ in the above discussion. Since $A$ maps great circles to great circles and circles of latitude to circles of latitude (possibly around a different axis), we conclude that $R$ is given by (3.3) at any point of $S^2$.

(iii) Let $\nabla$ be a connection on a bundle $\xi$ over $M$, $U \subset M$ such that $\xi_{|U}$ is trivial. A section $X$ of $\xi_{|U}$ can then be written as $X = (1_U, x)$, where $x : U \to \mathbb{R}^k$. The standard flat connection $D$ on $\xi_{|U}$ is given by

$$D_u(1_U, x) = (\pi(u), (x \circ c)'(0)),$$

where $c$ is a curve in $U$ with $\dot{c}(0) = u$. It follows that $\nabla = D + A$, for some $A \in A_1(U, \operatorname{End} \xi_{|U})$. Since the flat connection has zero curvature, Proposition 3.3 implies that the curvature tensor $R$ of $\nabla$ can be locally written as

$$R_{|U} = DA + [A, A].$$

EXERCISE 94. Prove the Bianchi identity: If $R$ is the curvature tensor of a connection $\nabla$, then $d^\nabla R = 0$.

EXERCISE 95. Let $M$ denote an $n$-dimensional affine subspace of $\mathbb{R}^{n+1}$. Compute the curvature tensor of the canonical connection on $\tau M$ given by (2.3).

## 4. Connections on Manifolds

If $M$ is a manifold, a *connection on $M$* is a connection $\nabla$ on the tangent bundle of $M$. Those curves $c$ in $M$ whose tangent field $\dot{c}$ is parallel along $c$ play a key role in the study of $\nabla$.

DEFINITION 4.1. A curve $c$ in $M$ is said to be a *geodesic* if $\nabla_D \dot{c} = 0$.

We will see below that the tangent fields of geodesics are integral curves of a certain vector field on $TM$, and therefore enjoy the usual existence, uniqueness, and smooth dependence on initial conditions properties. These properties can also be shown to hold locally with the help of a chart $(U, x)$: Let $X_i = \partial/\partial x^i$, and define the *Christoffel symbols* to be the functions $\Gamma_{ij}^k \in \mathcal{F}U$ given by

$$\Gamma_{ij}^k = dx^k(\nabla_{X_i} X_j), \qquad 1 \le i, j, k \le n.$$

Since $\dot{c} = c_* D = \sum_i c_* D(x^i) X_i \circ c = \sum_i D(x^i \circ c) X_i \circ c = \sum_i (x^i \circ c)' X_i \circ c$,

$$\nabla_D \dot{c} = \sum_i D(x^i \circ c)' X_i \circ c + (x^i \circ c)' \nabla_D (X_i \circ c)$$

$$= \sum_i (x^i \circ c)'' X_i \circ c + (x^i \circ c)' \nabla_{\dot{c}} X_i$$

$$= \sum_i (x^i \circ c)'' X_i \circ c + (x^i \circ c)' \sum_j (x^j \circ c)' (\nabla_{X_j} X_i) \circ c$$

$$= \sum_i (x^i \circ c)'' X_i \circ c + (x^i \circ c)' \sum_{j,k} (x^j \circ c)' (\Gamma_{ji}^k \circ c) X_k \circ c.$$

Thus, $c$ is a geodesic iff

$$(x^k \circ c)'' + \sum_{i,j} (x^i \circ c)'(x^j \circ c)' \Gamma_{ij}^k \circ c = 0, \qquad 1 \le k \le n.$$

Existence and uniqueness of geodesics for initial conditions in $c$ and $\dot{c}$ are then guaranteed by classical theorems on differential equations. A connection is said to be *complete* if its geodesics are defined on all of $\mathbb{R}$.

EXAMPLES AND REMARKS 4.1. (i) If $M = \mathbb{R}^n$ with the standard flat connection, then $\Gamma_{ij}^k = 0$, and the geodesics are the straight lines $t \mapsto at + b$, $a$, $b \in M$. The connection is complete.

(ii) More generally, let $M$ be a parallelizable manifold, $X_1, \ldots, X_k \in \mathfrak{X}M$ a parallelization of $M$. Any $X \in \mathfrak{X}M$ can be written $X = \sum_i X^i X_i$, $X^i \in \mathcal{F}M$. The formula

$$\nabla_u X = \sum_i u(X^i) X_i(\pi(u)), \qquad u \in TM,$$

defines a connection on $M$ such that $X \in \mathfrak{X}M$ is a parallel section of $\tau M$ iff $X$ is a constant linear combination $\sum a_i X_i$, $a_i \in \mathbb{R}$. This can be seen by checking axioms (1)–(4) of Theorem 2.1. Alternatively, one can define a horizontal distribution $\mathcal{H}$ whose value at $u = \sum a_i X_i(\pi(u))$ is $\mathcal{H}_u := X_{*\pi(u)} M_{\pi(u)}$, where $X := \sum a_i X_i \in \mathfrak{X}M$. Since $\pi \circ X = 1_M$, $\pi_* \mathcal{H}_u = \pi_* X_* M_{\pi(u)} = M_{\pi(u)}$. Furthermore, $\mathcal{H}_{au} = (aX)_* M_{\pi(u)} = (\mu_a \circ X)_* M_{\pi(u)} = \mu_{a*} \mathcal{H}_u$ for $a \in \mathbb{R}$, so that $\mathcal{H}$ is a connection.

Given $p \in M$, $u = \sum a_i X_i(p) \in M_p$, the geodesic $c$ of $M$ with $\dot{c}(0) = u$ is the integral curve of the vector field $X := \sum a_i X_i$ passing through $p$ when $t = 0$: Notice that $\nabla_X X = 0$, so that if $\gamma$ is an integral curve of $X$, then

$$\nabla_D \dot{\gamma} = \nabla_D (X \circ \gamma) = \nabla_{\dot{\gamma}} X = (\nabla_X X) \circ \gamma = 0.$$

(iii) In case $M$ is a Lie group $G$, the connection from (ii) obtained by choosing as parallelization a basis of the Lie algebra $\mathfrak{g}$ is called the *left-invariant connection* of $G$. It is independent of the chosen basis, because for $u \in TG$, $\mathcal{H}_u = X_{*\pi(u)} G_{\pi(u)}$, where $X$ is the element of $\mathfrak{g}$ with $X_{\pi(u)} = u$.

The geodesics $c$ with $c(0) = e$ coincide with the Lie group homomorphisms $c : \mathbb{R} \to G$ (and in particular the connection is complete): If $c : I \to G$ is an integral curve of $X$ with $c(0) = e$, let $s$, $t$ be numbers such that $s$, $t$, $s + t \in I$. Then the curve $\gamma$ defined by $\gamma(t) = c(s)c(t)$ is an integral curve of $X$, since

$$\dot{\gamma}(t) = L_{c(s)*} X(c(t)) = X(L_{c(s)} c(t)) = X(\gamma(t)).$$

But $t \mapsto c(s + t)$ is also an integral curve of $X$ which coincides with $\gamma$ when $t = 0$, so that $c(s + t) = c(s)c(t)$. A similar argument shows that if $c : I \to G$ is a maximal integral curve of $X$, then $s + t \in I$ whenever $s, t \in I$. In particular, $I = \mathbb{R}$.

Conversely, suppose $c : \mathbb{R} \to G$ is a homomorphism, and let $X$ be the left-invariant vector field with $X(e) = \dot{c}(0)$. If $t_0 \in \mathbb{R}$, and $\gamma$ is the curve given by $\gamma(t) = c(t_0 + t) = c(t_0)c(t)$, then as above,

$$\dot{c}(t_0) = \dot{\gamma}(0) = L_{c(t_0)*}\dot{c}(0) = L_{c(t_0)*}X(e) = X(c(t_0)),$$

so that $c$ is an integral curve of $X$.

The curvature tensor of a left-invariant connection on a Lie group is zero, since the connection is flat.

(iv) (Geodesics on $S^n$) Let $p, q \in S^n$, $p \perp q$. Then the great circle $t \mapsto c(t) = (\cos t)p + (\sin t)q$ is a geodesic: In fact, $\dot{c}(t) = \mathcal{J}_{c(t)}c'(t) = -(\sin t)\mathcal{J}_{c(t)}p + (\cos t)\mathcal{J}_{c(t)}q$, and $t \mapsto \mathcal{J}_{c(t)}p$ is parallel along $c$ for the connection $\tilde{\nabla}$ on $\mathbb{R}^{n+1}$. Thus,

$$\tilde{\nabla}_{D(t)}\dot{c} = -(\cos t)\mathcal{J}_{c(t)}q - (\sin t)\mathcal{J}_{c(t)}q = -P \circ c(t),$$

where $P$ is the position vector field, and $\imath_* \nabla_D\dot{c} = (\tilde{\nabla}_D\dot{c})^\perp = 0$. Since $\imath_{*p}S_p^n = \mathcal{J}_p(p^\perp)$, this describes, up to reparametrization, all geodesics passing through $p$ at $t = 0$, cf. Exercise 96.

DEFINITION 4.2. A vector field $S$ on $TM$ is called a *spray* on $M$ if

(1) $\pi_* \circ S = 1_{TM}$, and
(2) $S \circ \mu_a = a\mu_{a*}S$, for $a \in \mathbb{R}$.

By Theorem 7.5 in Chapter 1, the maximal flow $\Phi : W \to TM$ of $S$ is defined on an open set $W \subset \mathbb{R} \times TM$ containing $0 \times TM$, and if $\Phi_v : I_v \to TM$ denotes the maximal integral curve of $S$ with $\Phi_v(0) = v$, then $\Phi_v(t) = \Phi(t, v)$.

Let $\widetilde{TM}$ denote the open subset of $TM$ consisting of all $v$ such that $1 \in I_v$, and define the *exponential map* $\exp : \widetilde{TM} \to M$ of the spray $S$ by

(4.1)                    $\exp(v) := \pi \circ \Phi(1, v).$

For $p \in M$, $\exp_p$ will denote the restriction of $\exp$ to $\widetilde{M}_p := \widetilde{TM} \cap M_p$.

THEOREM 4.1. *Let $S$ be a spray on $M^n$ with exponential map $\exp : \widetilde{TM} \to M$. Then for any $p \in M$,*

(1) $\widetilde{M}_p$ *is a star-shaped neighborhood of $0 \in M_p$: If $v \in \widetilde{M}_p$, then $tv \in \widetilde{M}_p$ for $0 \leq t \leq 1$, and $\exp(tv) = \pi \circ \Phi_v(t)$.*
(2) $\exp_p$ *has rank $n$ at $0 \in M_p$, and therefore maps a neighborhood of $0$ in $M_p$ diffeomorphically onto a neighborhood of $p$ in $M$.*
(3) $(\pi, \exp) : \widetilde{TM} \to M \times M$ *has rank $2n$ at $0 \in M_p$, and therefore maps a neighborhood of $0$ in $TM$ diffeomorphically onto a neighborhood of $(p, p)$ in $M \times M$. If $s : M \to TM$ denotes the zero section of $\tau M$, then there exists a neighborhood $U$ of $s(M)$ in $\widetilde{TM}$ such that $(\pi, \exp)$ maps $U$ diffeomorphically onto a neighborhood of the diagonal $\{(q, q) \mid q \in M\}$ in $M \times M$.*

PROOF. Notice first that if $v \in TM$, and $s \in \mathbb{R}$, then the curve $\psi :=$ $\mu_s \circ \Phi_v \circ \mu_s : \frac{1}{s} I_v \to TM$ is an integral curve of $S$. Indeed,

$$\dot{\psi}(t) = \mu_{s*} \Phi_{v*} \mu_{s*} D(t) = s\mu_{s*} \Phi_{v*} D(st) = s\mu_{s*} S(\Phi_v(st)) = S \circ \mu_s \circ \Phi_v \circ \mu_s(t)$$
$$= S \circ \psi(t).$$

But $\psi(0) = s\Phi_v(0) = sv = \Phi_{sv}(0)$, and by uniqueness of integral curves, $\Phi_{sv}(t) = \psi(t) = s\Phi_v(st)$ for $st \in I_v$. Now if $v \in \widetilde{M}_p$, then $s \cdot 1 \in I_v$ for any $s \in [0,1]$, and $\Phi_{sv}(1) = s\Phi_v(s)$. Thus,

$$\exp(sv) = \pi \circ \Phi_{sv}(1) = \pi \circ \mu_s \circ \Phi_v(s) = \pi \circ \Phi_v(s),$$

which establishes (1).

For (2), let $c(t) = \exp_p(tv)$. Then, by (1),

$$(4.2) \qquad \exp_{p*} \mathcal{J}_0 v = \dot{c}(0) = \pi_* \circ \dot{\Phi}_v(0) = \pi_* S(v) = v,$$

and $\exp_{p*}$ is an isomorphism at $0 \in M_p$.

In order to prove (3), consider a chart $(U, x)$ of $M$ around $p$, and the associated chart $(\pi^{-1}(U), \bar{x})$ of $TM$ from Proposition 4.2 in Chapter 1. If $y := x \times x$, then $(U \times U, y)$ is a chart of $M \times M$ around $(p, p)$, and

$$(\pi, \exp)_* \frac{\partial}{\partial \bar{x}^k}(0) = \sum_{i=1}^{2n} \frac{\partial}{\partial \bar{x}^k}(0)(y^i \circ (\pi, \exp)) \frac{\partial}{\partial y^i}(p, p).$$

But for $i \le n$, $y^i \circ (\pi, \exp) = x^i \circ \pi = \bar{x}^i$, and $y^{n+i} \circ (\pi, \exp) = x^i \circ \exp$. Thus, for $k \le n$,

$$(4.3) \qquad (\pi, \exp)_* \frac{\partial}{\partial \bar{x}^k}(0) = \frac{\partial}{\partial y^k}(p, p) + \sum_{i=1}^{n} \frac{\partial}{\partial \bar{x}^k}(0)(x^i \circ \exp) \frac{\partial}{\partial y^{n+i}}(p, p),$$

$$(4.4) \qquad (\pi, \exp)_* \frac{\partial}{\partial \bar{x}^{n+k}}(0) = \sum_{i=1}^{n} \frac{\partial}{\partial \bar{x}^{n+k}}(0)(x^i \circ \exp) \frac{\partial}{\partial y^{n+i}}(p, p).$$

The right side of (4.4) can be rewritten as follows: Notice that $\exp_* \partial/\partial \bar{x}^{n+k}(0) = \partial/\partial x^k(p)$; Indeed, if $\imath : M_p \hookrightarrow TM$ is inclusion, and $c_k$ is the curve $t \mapsto t\partial/\partial x^k(p)$ in $M_p$, then

$$(\imath \circ c_k)_* D(0) = \sum_{j=1}^{2n} D(0)(\bar{x}^j \circ \imath \circ c_k) \frac{\partial}{\partial \bar{x}^j}(0) = \frac{\partial}{\partial \bar{x}^{n+k}}(0),$$

since $\bar{x}^j \circ \imath \circ c_k = x^j(p)$, and $\bar{x}^{n+j} \circ \imath \circ c_k(t) = t\,dx^j(p)\partial/\partial x^k(p)$ for $j \le n$. But we also have $(\imath \circ c_k)_* D(0) = \imath_* \mathcal{J}_0 \partial/\partial x^k(p)$, so that

$$\exp_* \frac{\partial}{\partial \bar{x}^{n+k}}(0) = \exp_* \imath_* \mathcal{J}_0 \frac{\partial}{\partial x^k}(p) = \exp_{p*} \mathcal{J}_0 \frac{\partial}{\partial x^k}(0) = \frac{\partial}{\partial x^k}(p)$$

by (4.2), as claimed. On the other hand,

$$\exp_* \frac{\partial}{\partial \bar{x}^{n+k}}(0) = \sum_{i=1}^{n} \frac{\partial}{\partial \bar{x}^{n+k}}(x^i \circ \exp) \frac{\partial}{\partial x^i}(p),$$

and comparing with the previous expression, we deduce that $(\partial/\partial \bar{x}^{n+k})(0)(x^i \circ \exp) = \delta_{ik}$. Substituting in (4.4) yields

$$(4.5) \qquad (\pi, \exp)_* \frac{\partial}{\partial \bar{x}^{n+k}}(0) = \frac{\partial}{\partial y^{n+k}}(p, p).$$

By (4.3) and (4.5), the matrix of $(\pi, \exp)_{*0}$ with respect to the bases $\{\partial/\partial \bar{x}^i\}$ and $\{\partial/\partial y^j\}$ is

$$\begin{pmatrix} I_n & 0 \\ * & I_n \end{pmatrix},$$

where $I_n$ denotes the $n \times n$ identity matrix. This matrix has rank $2n$. The last statement of (3) then follows from Lemma 1.1 in Chapter 3.  □

THEOREM 4.2. *Let $\nabla$ be a connection on $M$ with connection map $\kappa$. Then there exists a unique horizontal spray $S$ on $M$; i.e., $\kappa \circ S = s \circ \pi$, where $s$ denotes the zero section of $\tau M$. A curve $c : I \to M$ is a geodesic iff there exists an integral curve $\tilde{c} : I \to TM$ of $S$ for which $c = \pi \circ \tilde{c}$. In this case, $\dot{c} = \tilde{c}$. $S$ is called the geodesic spray of $M$.*

PROOF. $S$ is a horizontal spray iff

(1) $\pi_* S(v) = v, \quad v \in TM$,
(2) $\kappa(S(v)) = 0 \in M_{\pi(v)}$, and
(3) $S(av) = a\mu_{a*} S(v), \quad a \in \mathbb{R}$.

The first two conditions determine $S$ uniquely, since by Proposition 2.1, $(\pi_*, \kappa) : \tau TM \to \tau M \oplus \tau M$ is a bundle map covering $\pi : \tau M \to M$. So let $S(v) = (\pi_*, \kappa)^{-1}(v, s \circ \pi(v))$, where $s$ denotes the zero section of $\tau M$. Then $S$ satisfies (1) and (2), and is a differentiable vector field on $TM$, being a composition $(\pi_*, \kappa)^{-1} \circ (1_{TM}, s \circ \pi)$ of differentiable maps. In order to establish (3), it suffices to show that

$$(4.6) \qquad \pi_* S(av) = \pi_*(a\mu_{a*} S(v)),$$

and

$$(4.7) \qquad \kappa S(av) = \kappa(a\mu_{a*} S(v)).$$

But $\pi \circ \mu_a = \pi$, so that

$$\pi_*(a\mu_{a*} S(v)) = a\pi_* \mu_{a*} S(v) = a\pi_* S(v) = av = \pi_* S(av),$$

which proves (4.6). For (4.7), observe that $\kappa S(av) = 0$ because $S$ is horizontal. On the other hand, Lemma 2.1 implies that

$$\kappa(a\mu_{a*} S(v)) = a(\kappa \circ \mu_{a*}) S(v) = a(\mu_a \circ \kappa) S(v) = 0.$$

□

EXAMPLES AND REMARKS 4.2. (i) It follows from Theorems 4.1 and 4.2 that for $v \in M_p$, the geodesic $c$ with $c(0) = p$ and $\dot{c}(0) = v$ is given by $c(t) = \exp_p(tv)$.

(ii) On $\mathbb{R}^n$, $\exp_p v = p + \mathcal{J}_p^{-1} v$, $v \in \mathbb{R}_p^n$.

(iii) On $S^n$, if $v \in S_p^n$ has norm 1, then $\exp_p(\pi v) = -p$ by Examples and Remarks 4.1(iv). Thus, $\exp$ is not, in general, one-to-one.

(iv) The terminology for the exponential map is derived from the classical exponential map on the space $M_{n,n}$ of $n \times n$ matrices. The Lie group $G =$

$GL(n)$ is an open subset of $M_{n,n}$, so that there is a canonical identification $\mathcal{J}_B : M_{n,n} \to G_B$ for $B \in G$. The left-invariant vector field $X$ with $X(e) = \mathcal{J}_e A$ is given by $X(B) = \mathcal{J}_B(BA)$; this follows from Examples and Remarks 4.1(iv) in Chapter 1, together with the fact that $L_B : M_{n,n} \to M_{n,n}$, where $L_B(C) = BC$, is a linear map. Define

$$e^A := \sum_{n=0}^{\infty} \frac{A^n}{n!}.$$

Then the curve $t \mapsto c(t) = e^{tA}$ has derivative $c'(t) = c(t)A$, and $\dot{c}(t) = \mathcal{J}_{c(t)}(c(t)A) = X \circ c(t)$. $c$ is therefore an integral curve of $X$, and by Examples and Remarks 4.1(iii), $c$ is a geodesic of the left-invariant connection on $G$. Thus, $\exp(t\mathcal{J}_e A) = e^{tA}$.

More generally, the *exponential map* $\exp : \mathfrak{g} \to G$ of a Lie group $G$ is given by $\exp(X) := c(1)$, where $c$ is the integral curve of $X$ with $c(0) = e$. It follows that $c(t) = \exp(tX)$ for all $t \in \mathbb{R}$, so that $\exp = \exp_e$: Given $a \in \mathbb{R}$, the curve $\phi : s \mapsto c(as)$ is a 1-parameter subgroup of $G$ with $\dot{\phi}(0) = aX(e)$, and $\phi$ is therefore the integral curve of $aX$ passing through $e$ when $t = 0$. Thus, $c(a) = \phi(1) = \exp(aX)$.

EXERCISE 96. Prove that if $c$ is a geodesic, then so is any affine reparametrization $t \mapsto c(at + b)$ for $a, b \in \mathbb{R}$.

EXERCISE 97. Let $M = \mathbb{R} \setminus \{0\} = GL(1)$, with its left-invariant connection. Determine $\nabla_D D$.

EXERCISE 98. Show that two connections $\nabla$ and $\tilde{\nabla}$ on $M$ have the same geodesics iff the connection difference 1-form $\omega = \nabla - \tilde{\nabla}$ is skew-symmetric: i.e., $\omega(u)u = 0$, $u \in TM$.

EXERCISE 99. Prove that if $M$ is $n$-dimensional with tangent bundle $\pi : TM \to M$, then $\pi_* : TTM \to TM$ admits a rank $2n$ vector bundle structure. Notice that the fibers of $\pi_*$ do not coincide with those of $\tau(TM)$, even though both bundles share the same total and base spaces.

## 5. Connections on Principal Bundles

Although the approach followed here has been to study connections on vector bundles, many authors prefer to do so on principal bundles. This is essentially a matter of taste, and in this section, we show how to go from one to the other and back.

Let $\xi = \pi : E \to M$ be a vector bundle over $M$, $Fr(\xi) = \pi_P : P \to M$ its frame bundle; i.e., the associated principal $GL(n)$-bundle over $M$. If $\mathcal{H}$ is a connection on $\xi$, there is a natural way of transferring it to the frame bundle: An element $b \in P$ is a basis $u_1, \ldots, u_n$ of the fiber $E_{\pi_P(b)}$ of $\xi$; equivalently $b : \mathbb{R}^n \to E_{\pi_P(b)}$ is an isomorphism, where $u_i = b(\mathbf{e}_i)$. Given a curve $c : [0, a] \to M$ with $c(0) = \pi_P(b)$, consider the parallel sections $U_i$ of $\xi$ along $c$ with $U_i(0) = u_i$. Then $\gamma : [0, a] \to P$, where $\gamma(t)$ is the basis $U_1(t), \ldots, U_n(t)$, is a section of $Fr(\xi)$ along $c$, and it seems reasonable to say that $\gamma$ is the parallel section along $c$ with $\gamma(0) = b$. Furthermore, $\gamma$ enables us to recover all parallel sections of $\xi$ along $c$: If $u = \sum_i a_i u_i \in E_{\pi_P(b)}$, then the parallel section $U$ along $c$ with $U(0) = u$ is just $U = \sum a_i U_i = \rho(\gamma, a)$, where $\rho : P \times \mathbb{R}^n \to E = P \times_{Gl(n)} \mathbb{R}^n$ is the projection, and $a = (a_1, \ldots, a_n) \in \mathbb{R}^n$. However, $X$ can also be written as $\rho(\gamma g, g^{-1} a)$ for any $g \in GL(n)$, so if this definition is to make sense, we must require that $\gamma g$ be parallel whenever $\gamma$ is. In other words, if $\dot{\gamma}(t)$ is horizontal, then so is $R_{*g} \dot{\gamma}(t)$ for any $g \in GL(n)$ (here, $R_g$ is the principal bundle equivalence given by $R_g(b) = bg$).

DEFINITION 5.1. A *connection* on a principal $G$-bundle $\pi : P \to M$ is a distribution $\mathcal{H}$ on $P$ such that:

(1) $TP = \ker \pi_* \oplus \mathcal{H}$.
(2) $R_{g*} \mathcal{H} = \mathcal{H} \circ R_g$ for all $g \in G$.

As in the vector bundle case, the splitting in (1) determines a decomposition $u = u^v + u^h \in \ker \pi_* \oplus \mathcal{H}$ of any $u \in TP$ as a sum of a vertical and a horizontal vector.

By the above definition, any connection $\mathcal{H}$ on a vector bundle $\xi$ determines a connection $\tilde{\mathcal{H}} = \{u \in TP \mid \rho_*(u, 0) \in \mathcal{H}\}$ on the principal $GL(n)$-bundle $Fr(\xi)$: If $\pi_E$ denotes the vector bundle projection, and $\pi_1 : P \times \mathbb{R}^n \to P$ the projection onto the first factor, then $\pi \circ \pi_1 = \pi_E \circ \rho$. Since $\pi_{E*|\mathcal{H}}$ is onto, $\pi_* \tilde{\mathcal{H}} = \pi_* \pi_{1*}(\tilde{\mathcal{H}} \times 0) = \pi_{E*} \rho_*(\tilde{\mathcal{H}} \times 0) = \pi_{E*}(\mathcal{H}) = TM$, so that $\tilde{\mathcal{H}}$ is complementary to $\ker \pi_*$. Furthermore, if $\gamma$ is a basis of parallel fields along a curve in $M$ and $g \in GL(n)$, then each element of $\gamma g$ is a constant linear combination of the fields in $\gamma$, and is therefore parallel. Thus, $\mathcal{H}$ is invariant under $R_g$.

Conversely, given a principal $GL(n)$-bundle $P \to M$ and a connection $\mathcal{H}$ on the bundle, we obtain a connection on the vector bundle $\xi : E = P \times_{GL(n)} \mathbb{R}^n \to M$ by requiring that $\rho(\gamma, u)$ be parallel along $c$ whenever $\gamma$ is parallel along $c$ and $u \in \mathbb{R}^n$; i.e., we claim that $\tilde{\mathcal{H}} := \rho_*(\mathcal{H} \times 0)$ is a connection on $\xi$: Clearly, $\tilde{\mathcal{H}} + \mathcal{V}\xi = TE$. To see that $\tilde{\mathcal{H}}$ is invariant under multiplication $\mu_a$ by $a \in \mathbb{R}$, recall that the map $\rho \circ (R_g \times 1_{\mathbb{R}^n})$ on $P \times \mathbb{R}^n$ equals $\rho \circ (1_P \times g)$. Thus,

$$\tilde{\mathcal{H}}_{a\rho(b,u)} = \tilde{\mathcal{H}}_{\rho(b,au)} = \tilde{\mathcal{H}}_{\rho(baI_n,u)} = \rho_*(\mathcal{H}_{baI_n} \times 0_u)$$
$$= \rho_* \circ (R_{aI_n} \times 1_{\mathbb{R}^n})_*(\mathcal{H}_b \times 0_u) = \rho_* \circ (1_P \times aI_n)_*(\mathcal{H}_b \times 0_u).$$

But $\rho \circ (1_P \times aI_n) = \mu_a \circ \rho$, so that

$$\tilde{\mathcal{H}}_{a\rho(b,u)} = \mu_{a*} \circ \rho_*(\mathcal{H}_b \times 0_u) = \mu_{a*} \tilde{\mathcal{H}}_{\rho(b,u)}$$

as claimed.

For $b \in P$, the map $l_b : G \to P$ given by $l_b(g) = R_g(b) = bg$ is an imbedding onto the fiber of $P$ through $b$ by Lemma 10.1 in Chapter 5. If $U \in \mathfrak{g}$, the *fundamental vector field* $\tilde{U} \in \mathfrak{X}P$ determined by $U$ is defined by

$$\tilde{U}(b) = l_{b*} U(e), \qquad b \in P.$$

In analogy with the vector bundle case, define the *horizontal lift* of $X \in \mathfrak{X}M$ to be the unique horizontal $\bar{X} \in \mathfrak{X}P$ that is $\pi$-related to $X$. Such an $\bar{X}$ is said to be *basic*.

PROPOSITION 5.1. *The map* $\psi : \mathfrak{g} \to \mathfrak{X}P$ *which assigns to* $U \in \mathfrak{g}$ *the fundamental vector field* $\tilde{U}$ *determined by* $U$ *is a Lie algebra homomorphism. Furthermore,* $[\tilde{U}, X]$ *is horizontal if* $X$ *is, and is zero if* $X$ *is basic.*

PROOF. $\psi$ is by definition linear. To see that it is a homomorphism, notice first of all that the flow of $\tilde{U}$ is $R_{\exp(tU)}$: In fact, if $\gamma$ is the curve $t \mapsto \exp(tU)$ in $G$ and $c(t) = R_{\exp(tU)}(b) = l_b \circ \gamma(t)$, then

$$\dot{c}(t) = l_{b*} \circ \dot{\gamma}(t) = l_{b*}U_{\exp(tU)} = l_{b*} \circ L_{(\exp tU)*}U(e) = l_{b\exp(tU)*}U(e) = \tilde{U} \circ c(t).$$

Thus, by definition of the Lie bracket,

$$[\tilde{U}, \tilde{V}](b) = \lim_{t \to 0} \frac{1}{t}(R_{\exp(-tU)*}l_{b\exp(tU)*}V(e) - l_{b*}V(e)).$$

If $\tau_a$ denotes conjugation by $a$ in $G$, then

$$R_{\exp(-tU)} \circ l_{b\exp(tU)}(g) = b\exp(tU)g\exp(-tU) = l_b \circ \tau_{\exp(tU)}(g).$$

By Example 8.1(iii) in Chapter 1,

$$[\tilde{U}, \tilde{V}](b) = l_{b*} \lim_{t \to 0} \frac{1}{t}(\mathrm{Ad}_{\exp(tU)} V(e) - V(e)),$$

and it remains to show that the latter limit is $[U, V](e)$. But if $R$ now denotes right translation in $G$, then

$$\lim_{t \to 0} \frac{1}{t}(\mathrm{Ad}_{\exp(tU)} V(e) - V(e)) = \lim_{t \to 0} \frac{1}{t}(R_{\exp(-tU)*} \circ L_{\exp(tU)*}V(e) - V(e))$$

$$= \lim_{t \to 0} \frac{1}{t}(R_{\exp(-tU)*} \circ V \circ R_{\exp(tU)}(e) - V(e))$$

$$= [U, V](e),$$

since $U$ has flow $R_{\exp(tU)}$ and $V$ is left-invariant. This shows that $\psi$ is a Lie algebra homomorphism. At this stage, it is worth noting that the above argument establishes the following:

OBSERVATION. *Denote by* $\mathrm{ad} : \mathfrak{g} \to \mathfrak{gl}(\mathfrak{g})$ *the derivative at the identity of* $\mathrm{Ad} : G \to GL(\mathfrak{g})$*; i.e., for* $U \in \mathfrak{g}$*,* $\mathrm{ad}_U = \mathrm{Ad}_{*e} U$*. Then* $\mathrm{ad}_U V = [U, V]$*.*

We now proceed to the second part of the proposition: If $X$ is horizontal, then as above,

$$[\tilde{U}, X](b) = \lim_{t \to 0} \frac{1}{t}(R_{\exp(-tU)*} \circ X \circ R_{\exp(tU)}(b) - X(b))$$

is horizontal, since $\mathcal{H}$ is invariant under $R_g$. Finally, if $\bar{X}$ is basic and $\pi$-related to $X \in \mathfrak{X}M$, then $\pi_*[\tilde{U}, \bar{X}] = [0, X] \circ \pi = 0$, since vertical fields are $\pi$-related to the zero field on $M$. Thus, the horizontal component of $[\tilde{U}, \bar{X}]$, and by the above, $[\tilde{U}, \bar{X}]$ itself, must vanish. $\qquad\square$

We next discuss an analogue for principal bundles of the connection map $\kappa : TE \to E$ for vector bundles: Recall that $\kappa$ essentially picked out the vertical component $u^v$ of $u \in TE$. Since $u^v \in \ker \pi_*$, it can be identified with an element $\kappa(v)$ of $E$. A similar property holds for principal bundles: If $u \in T_bP$ is vertical, that is, $u \in \ker \pi_*$, then it is tangent to the orbit of $b$ which is diffeomorphic to $G$, and hence parallelizable. In other words, there exists a unique $U \in \mathfrak{g}$ with $\tilde{U}(b) = u$, so we may define $\kappa(u) = U$. It is customary to use the letter $\omega$ instead:

DEFINITION 5.2. The *connection form* $\omega$ of a connection on a principal $G$-bundle $P \to M$ is the $\mathfrak{g}$-valued 1-form given by

$$\omega(u) = (l_{b*e})^{-1} u^v, \qquad u \in T_b P, \quad b \in P.$$

(Strictly speaking, $\omega \in A_1(P, \eta)$, where $\eta$ denotes the trivial bundle over $P$ with total space $P \times \mathfrak{g}$.)

PROPOSITION 5.2. *The connection form $\omega$ of a connection $\mathcal{H}$ satisfies*

(1) $\omega_{|\mathcal{H}} \equiv 0, \qquad l_{b*} \circ \omega_{|\ker \pi_*} = 1_{\ker \pi_*},$
(2) $R_g^* \omega = \mathrm{Ad}_{g^{-1}} \circ \omega, \qquad g \in G.$

*Conversely, if $\omega$ is a $\mathfrak{g}$-valued 1-form on $P$ satisfying the first part of (1) and part (2), then $\ker \omega$ is a connection on $P \to M$.*

PROOF. Part (1) is immediate from Definition 5.2. It suffices to verify (2) for a vertical vector $u \in \ker \pi_{*b}$, since both sides vanish when applied to a horizontal one. Now, by (1),

$$(R_g^* \omega)(b)(u) = (\omega \circ R_g)(b) R_{g*} u = l_{bg*}^{-1} R_{g*} u = l_{bg*}^{-1} R_{g*} l_{b*} \omega u.$$

Notice that $R_g \circ l_b = l_{bg} \circ L_{g^{-1}} \circ R_g$, where the $R_g$ on the right side is right translation by $g$ in $G$, so that $l_{bg}^{-1} \circ R_g \circ l_b = L_{g^{-1}} \circ R_g$ is conjugation by $g^{-1}$. The derivative of the latter at $e$ is $\mathrm{Ad}_{g^{-1}}$, which establishes (2).

For the converse, (1) implies that $TP = \ker \omega \oplus \ker \pi_*$, whereas (2) ensures that $R_{g*} \ker \omega \subset \ker \omega$ (and hence $R_{g*} \ker \omega = \ker \omega$) for all $g \in G$. Thus, $\ker \omega$ is a connection. $\square$

Just as in Section 3, the assignment $(X, Y) \mapsto [X, Y]^v$ is tensorial for horizontal vector fields $X, Y$ on $P$. The following definition should be compared with Theorem 3.2.

DEFINITION 5.3. The *curvature form* $\Omega$ of a connection $\mathcal{H}$ is the $\mathfrak{g}$-valued 2-form on $P$ defined by

$$\Omega(b)(x, y) = -\omega[X, Y]^v(b), \qquad b \in P, \quad x, y \in T_b P,$$

where $X, Y$ are horizontal vector fields on $P$ with $X(b) = x$, $Y(b) = y$.

Here again, $\Omega$ is actually a form on $P$ with values in the trivial bundle $\eta : P \times \mathfrak{g} \to P$, and we identify $\alpha \in A(P, \eta)$ with $\pi_2 \circ \alpha : \mathfrak{X} P \times \cdots \times \mathfrak{X} P \to \mathfrak{g}$, where $\pi_2 : P \times \mathfrak{g} \to \mathfrak{g}$ is projection. In what follows, we consider the trivial connection on $\eta$; for example, any $\alpha \in A_0(P, \eta)$ can be written as $\pi_2 \circ \alpha = \sum f^i X_i$ with $f^i \in \mathcal{F} P$ and $X_i \in \mathfrak{g}$. The exterior covariant derivative operator is then given by $d\alpha(x) = \sum x(f^i) X_i$.

THEOREM 5.1 (Cartan's Structure Equation). *If $\omega$ and $\Omega$ denote the connection and curvature forms respectively of a connection on a principal $G$-bundle $P \to M$, then*

$$\Omega = d\omega + [\omega, \omega].$$

PROOF. Consider $X, Y \in \mathfrak{X} M$ with basic lifts $\bar{X}, \bar{Y}$, and $U, V \in \mathfrak{g}$ with fundamental vector fields $\tilde{U}, \tilde{V}$. Since both sides of the above equation are tensorial, it suffices to check its validity for various combinations of the above fields, keeping in mind that (a) $\omega\tilde{U}$ and $\omega\tilde{V}$ are the constant functions $U$, $V \in A_0(P,\eta)$, and (b) $\omega\bar{X} = \omega\bar{Y} = 0$. Now,

$$d\omega(\tilde{U},\tilde{V}) + [\omega\tilde{U},\omega\tilde{V}] = \tilde{U}(\omega\tilde{V}) - \tilde{V}(\omega\tilde{U}) - \omega[\tilde{U},\tilde{V}] + [\omega\tilde{U},\omega\tilde{V}]$$

$$= -\omega\widetilde{[U,V]} + [U,V] = 0 = \Omega(\tilde{U},\tilde{V}),$$

whereas

$$d\omega(\tilde{U},\bar{X}) + [\omega\tilde{U},\omega\bar{X}] = -\bar{X}(\omega\tilde{U}) - \omega[\tilde{U},\bar{X}] = 0 = \Omega(\tilde{U},\bar{X})$$

since $[\tilde{U},\bar{X}] = 0$ by Proposition 5.1. Finally,

$$d\omega(\bar{X},\bar{Y}) + [\omega\bar{X},\omega\bar{Y}] = -\omega[\bar{X},\bar{Y}] = \Omega(\bar{X},\bar{Y}).$$

$\square$

THEOREM 5.2 (Bianchi's Identity). *If $\Omega$ denotes the curvature form of a connection $\mathcal{H}$ on a principal bundle $P \to M$, then $d\Omega_{|\mathcal{H}} = 0$.*

PROOF. Since the connection on $\eta$ is the trivial one, differentiating the structure equation yields $d\Omega = d[\omega,\omega]$. But $\omega$ vanishes on $\mathcal{H}$, and therefore so does $d[\omega,\omega]$. $\square$

Consider a rank $n$ vector bundle $\xi : E \to M$ with covariant derivative $\nabla$ and curvature tensor $R$, and its frame bundle $Fr(\xi) = \pi : P \to M$ together with the associated connection form $\omega$ and curvature form $\Omega$. We wish to describe the relationship between $\nabla$ and $\omega$, and between $R$ and $\Omega$.

Consider a curve $c : I \to M$, $0 \in I$, and a section $\gamma = (X_1,\ldots,X_n)$ of $Fr(\xi)$ along $c$. If $\rho : P \times \mathbb{R}^n \to E = P \times_{GL(n)} \mathbb{R}^n$ denotes the projection, then for any $a = (a_1,\ldots,a_n) \in \mathbb{R}^n$, $X := [\gamma,a] = \rho(\gamma,a)$ is a section of $\xi$ along $c$. Conversely, given a section $X$ along $c$, there exists a section $\gamma$ of $Fr(\xi)$ along $c$ and $a \in \mathbb{R}^n$ such that $X = [\gamma,a]$. Let $\epsilon = (E_1,\ldots,E_n)$ denote the parallel section of $Fr(\xi)$ along $c$ with $\epsilon(0) = \gamma(0)$, and $g$ the curve $g : I \to GL(n)$ satisfying $\gamma = \epsilon g$, so that $X = \sum_j a_j X_j = \sum_{i,j} a_j g_{ij} E_i$. We have

$$(5.1) \quad (\nabla_D X)(0) = \sum_{i,j} a_j g'_{ij}(0) E_i(0) = \rho(\epsilon(0),g'(0)a) = \rho(\gamma(0),\mathcal{J}_e^{-1}\dot{g}(0)a).$$

For simplicity of notation, identify the tangent space of $GL(n)$ at $g(0) = I_n$ with the space $M_{n,n}$ of $n \times n$ matrices via $\mathcal{J}_e^{-1}$, and write $\nabla_{D(0)}X = [\gamma(0),\dot{g}(0)a]$. Now, $\gamma = \epsilon g = \mu(\epsilon,g)$, where $\mu : P \times G \to P$ denotes the action of $G$ on $P$. Thus,

$$\dot{\gamma}(0) = \mu_*(\dot{\epsilon}(0),\dot{g}(0)) = \mu_*(\dot{\epsilon}(0),0) + \mu_*(0,\dot{g}(0)) = R_{g(0)*}\dot{\epsilon}(0) + l_{\epsilon(0)*}\dot{g}(0)$$

$$= \dot{\epsilon}(0) + l_{\epsilon(0)*}\dot{g}(0),$$

and applying $\omega$ to both sides, we obtain $\omega\dot{\gamma}(0) = \dot{g}(0)$. Substituting in (5.1) then yields $\nabla_D X(0) = \rho(\gamma(0),\omega(\dot{\gamma}(0))a)$. We therefore have the following:

PROPOSITION 5.3. *Let $\gamma$ be a section of $Fr(\xi)$ along a curve $c$. Then for $a \in \mathbb{R}^n$, $\nabla_D[\gamma,a] = [\gamma,(\omega\dot{\gamma})a]$.*

PROPOSITION 5.4. *Given $b \in P$, and $x$, $y \in T_b P$, the matrix of $R(\pi_* x,\pi_* y) \in \mathfrak{o}(E_{\pi(b)})$ with respect to the basis $b$ is $\Omega(b)(x,y)$.*

PROOF. It must be shown that $R(\pi_* x, \pi_* y)[b, u] = [b, \Omega(b)(x, y)u]$ for $u \in$ $\mathbb{R}^n$. Both sides of the equation vanish if $x$ or $y$ are vertical, so we may assume the vectors are horizontal. Extending $\pi_* x$ and $\pi_* y$ locally to vector fields on $M$, denote by $X$, $Y$ (respectively $\bar{X}$, $\bar{Y}$) their horizontal lifts to $E$ (respectively $P$). Let $z := \rho(b, u) = [b, u]$. The map $\rho_u : P \to E$, $\rho_u(p) = \rho(p, u)$, is a section of $\xi$ along $\pi : P \to M$. Since horizontal lifts are unique, $\bar{X}$, $\bar{Y}$ are $\rho_u$-related to $X$, $Y$. Now, by Theorem 3.2, $R(\pi_* x, \pi_* y)z = -\kappa[X, Y](z)$. Thus,

$$R(\pi_* x, \pi_* y)z = -\kappa[X, Y] \circ \rho_u(b) = -\kappa \rho_{u*}[\bar{X}, \bar{Y}](b) = -\nabla_{[\bar{X}, \bar{Y}](b)}\rho_u,$$

where $\nabla$ is the covariant derivative operator along $\pi$. If $c$ is a curve in $P$ with $\dot{c}(0) = [\bar{X}, \bar{Y}](b)$, then by Proposition 5.3,

$$\nabla_{[\bar{X}, \bar{Y}](b)}\rho_u = \nabla_{\dot{c}(0)}\rho_u = \nabla_{D(0)}(\rho_u \circ c) = [b, (\omega \dot{c}(0))u] = [b, (\omega[\bar{X}, \bar{Y}](b))u]$$
$$= -[b, \Omega(b)(x, y)u],$$

thereby completing the argument.                                         $\square$

EXERCISE 100. Use Theorem 5.2 to prove the Bianchi identity $d^\nabla R = 0$ for vector bundles.

EXERCISE 101. Show that a fundamental vector field $\tilde{U}$ on $P$ associated to $U \in \mathfrak{g}$ satisfies $R_{g*}\tilde{U} = \mathrm{Ad}_{g^{-1}} U \circ R_g$ for $g \in G$.

EXERCISE 102. Let $H$ be a subgroup of $G$, and $Q \to M$ a principal $H$-subbundle of a principal $G$-bundle $P \to M$. Suppose that $\mathfrak{g}$ admits a decomposition $\mathfrak{g} = \mathfrak{h} + \mathfrak{m}$, where $\mathfrak{h}$ is a subspace invariant under the adjoint action of $H$; i.e., $\mathrm{Ad}_h \mathfrak{m} \subset \mathfrak{m}$ for all $h \in H$. Denote by $p : \mathfrak{g} \to \mathfrak{h}$ the projection induced by this decomposition. Show that if $\mathcal{H}$ is a connection on $P \to M$ with connection form $\omega$, then $\ker(\imath^* p\omega)$ is a connection on $Q \to M$ (here $\imath : Q \hookrightarrow P$ denotes inclusion).

# CHAPTER 5

# Metric Structures

## 1. Euclidean Bundles and Riemannian Manifolds

A *Euclidean bundle* is a vector bundle together with a Euclidean metric $g$. Recall from Definition 4.2 in Chapter 2 that a Euclidean metric on the tangent bundle of a manifold is called a *Riemannian metric*. A *Riemannian manifold* is a differentiable manifold together with a Riemannian metric. We will often write $\langle u, v \rangle$ instead of $g(u, v)$, and $|u|$ for $\langle u, u \rangle^{1/2}$. Maps that preserve metric stuctures are of fundamental importance in Riemannian geometry:

DEFINITION 1.1. Let $(\xi_i, \langle, \rangle_i)$, $i = 1, 2$, be Euclidean bundles over $M_i$. A map $h : E(\xi_1) \to E(\xi_2)$ is said to be *isometric* if

(1) $h$ maps each fiber $\pi_1^{-1}(p_1)$ linearly into a fiber $\pi_2^{-1}(p_2)$, for $p_i \in M_i$; and

(2) $\langle hu, hv \rangle_2 = \langle u, v \rangle_1$ for $u, v \in \pi_1^{-1}(p)$, $p \in M_1$.

Given Riemannian manifolds $(M_i, g_i)$, a map $f : M_1 \to M_2$ is said to be *isometric* if $f_* : TM_1 \to TM_2$ is isometric. An isometric diffeomorphism is called an *isometry*.

EXAMPLES AND REMARKS 1.1. (i) A parallelization $X_1, \ldots, X_n$ of $M^n$ induces a Riemannian metric on $M$ by defining $\langle X_i, X_j \rangle = \delta_{ij}$. The *canonical metric* on $\mathbb{R}^n$ is the one induced by the parallelization $D_1, \ldots, D_n$.

(ii) A *left-invariant metric* on a Lie group $G$ is one induced by a parallelization consisting of left-invariant vector fields; alternatively, it is a metric for which each left translation $L_g : G \to G$ is an isometry. Such metrics are therefore in bijective correspondence with inner products on $G_e$. When in addition, each right translation $R_g : G \to G$ is an isometry, the metric is called *bi-invariant*. In general, bi-invariant metrics are in bijective correspondence with inner products on $G_e \cong \mathfrak{g}$ which are Ad-invariant: If $\langle, \rangle$ is a left-invariant metric on $G$, then for $X, Y \in \mathfrak{g}$,

$$\langle R_{g*}X, R_{g*}Y \rangle = \langle L_{g^{-1}*} \circ R_{g*}X, L_{g^{-1}*} \circ R_{g*}Y \rangle = \langle \operatorname{Ad}_{g^{-1}} X, \operatorname{Ad}_{g^{-1}} Y \rangle.$$

Thus, a left-invariant metric on $G$ is right-invariant iff the induced inner product on $G_e$ is Ad-invariant.

It follows for example that any compact Lie group admits a bi-invariant metric: Fix an inner product $\langle, \rangle_0$ on $\mathfrak{g}$, and define for $X, Y \in \mathfrak{g}$,

$$\langle X, Y \rangle := \int_G f, \qquad f(g) := \langle \operatorname{Ad}_g X, \operatorname{Ad}_g Y \rangle_0.$$

$\langle,\rangle$ is clearly an inner product, and for $a \in G$,

$$\langle \text{Ad}_a\, X, \text{Ad}_a\, Y \rangle = \int_G f \circ R_a = \int_G f = \langle X, Y \rangle.$$

(iii) A Riemannian metric on a homogeneous space $M = G/H$ is said to be *G-invariant* if

$$\mathbb{L}_g : M \to M,$$
$$aH \mapsto gaH$$

is an isometry for every $g \in G$. Notice that if $\pi : G \to M$ is the projection, then $\mathbb{L}_g \circ \pi = \pi \circ L_g$. If $g = h \in H$, then $\mathbb{L}_h \circ \pi = \pi \circ L_h \circ R_{h^{-1}}$, so that

$$(1.1) \qquad\qquad \mathbb{L}_{h*} \circ \pi_* = \pi_* \circ \text{Ad}_h.$$

This implies that the $G$-invariant metrics on $M$ are in bijective correspondence with the inner products on $\mathfrak{g}/\mathfrak{h}$ which are $\text{Ad}_H$-invariant (and in particular, any bi-invariant metric on $G$ induces a $G$-invariant metric on $M$): In fact, $\pi_{*e} : \mathfrak{g}/\mathfrak{h} \to M_p$ is an isomorphism (here $p = \pi(e)$), and for each $h \in H$, $\text{Ad}_h$ induces a map $\text{Ad}_h : \mathfrak{g}/\mathfrak{h} \to \mathfrak{g}/\mathfrak{h}$, since $\text{Ad}_h(\mathfrak{h}) \subset \mathfrak{h}$. Thus, by (1.1), a $G$-invariant metric on $M$ induces via $\pi_{*e}$ an $\text{Ad}_H$-invariant inner product on $\mathfrak{g}/\mathfrak{h}$.

Conversely, any such inner product defines one on $M_p$ by requiring $\pi_*$ to be a linear isometry. By (1.1), the latter is invariant under each $\mathbb{L}_{h*p}$. It may then be extended to all of $M$ by setting $\langle \mathbb{L}_{g*}u, \mathbb{L}_{g*}v \rangle = \langle u, v \rangle$.

(iv) Although the group of diffeomorphisms of a manifold is not, in general, a Lie group, Myers and Steenrod have shown that the isometry group of a Riemannian manifold with the compact-open topology admits a Lie group structure.

(v) Let $c : [a, b] \to M$ be a differentiable curve on a Riemannian manifold $M$. Since the function $|\dot{c}| : [a, b] \to \mathbb{R}$ is continuous, we may define the *length* of $c$ to be $L(c) := \int_a^b |\dot{c}|$. If $f : M \to N$ is an isometry, then $L(f \circ c) = L(c)$.

(vi) Suppose $\xi_i = \pi_i : (E_i, \langle,\rangle_i) \to M_i$ are Euclidean vector bundles, $i = 1, 2$. The *product metric* on $\xi_1 \times \xi_2$ is defined by

$$\langle (u_1, v_1), (u_2, v_2) \rangle := \langle u_1, v_1 \rangle_1 + \langle u_2, v_2 \rangle_2.$$

When $\xi_i$ is the tangent bundle $\tau M_i$ of $M_i$, it is called the *Riemannian product metric* on $M_1 \times M_2$ (after identifying the tangent space of $M_1 \times M_2$ at $(m_1, m_2)$ with $(M_1)_{m_1} \times (M_2)_{m_2}$ via $(p_{1*}, p_{2*})$, where $p_i : M_1 \times M_2 \to M_i$ is the projection). Similarly, the *tensor product metric* on $\xi_1 \otimes \xi_2$ is given by

$$\langle u_1 \otimes u_2, v_1 \otimes v_2 \rangle := \langle u_1, v_1 \rangle_1 \cdot \langle u_2, v_2 \rangle_2,$$

on decomposable elements.

If $M = M_1 = M_2$, the *Whitney sum metric* on $\xi_1 \oplus \xi_2$ is the Euclidean metric for which $\pi_2 : E(\xi_1 \oplus \xi_2) \to E_1 \times E_2$ becomes isometric.

(vii) Since a Euclidean metric is a nonsingular pairing of $E = E(\xi)$ with itself (cf. Section 10 in Chapter 1), there are induced equivalences

$$\flat : E \to E^*, \qquad \sharp : E^* \to E,$$

where $u^b(v) = \langle u, v \rangle$, and $\alpha^\sharp$ is the unique element of $E$ satisfying $\langle \alpha^\sharp, v \rangle = \alpha(v)$ for all $v \in E$. The *Euclidean metric on the dual* $\xi^*$ is that metric for which the above musical equivalences become isometric.

If $\xi_i$ are Euclidean vector bundles over $M$, the *Euclidean metric on the bundle* $\mathrm{Hom}(\xi_1, \xi_2)$ is the metric for which the equivalence $\xi_1^* \otimes \xi_2 \cong \mathrm{Hom}(\xi_1, \xi_2)$ becomes isometric.

(viii) The *Euclidean metric on* $\Lambda_k(\xi)$ is the one given on decomposable elements by $\langle u_1 \wedge \cdots \wedge u_k, v_1 \wedge \ldots v_k \rangle = \det(\langle u_i, v_j \rangle)$.

EXERCISE 103. Show that $f : \mathbb{R}^n \to \mathbb{R}^n$ is an isometry (with respect to the canonical metric) iff there exist some $A \in O(n)$ and $b \in \mathbb{R}^n$ such that $f(a) = Aa + b$ for all $a \in \mathbb{R}^n$.

EXERCISE 104. The *length function* of a curve $c : J = [a, b] \to M$ in a Riemannian manifold $M$ is given by $l_c(t) = L(c_{|[a,t]}), a \leq t \leq b$. If $\phi : I \to J$ is a differentiable monotone function onto $J$, the curve $c \circ \phi : I \to M$ is called a *reparametrization* of $c$.

(a) Show that $l_{c \circ \phi} = l_c \circ \phi$ if $\phi' \geq 0$, and $l_{c \circ \phi} = L(c) - l_c \circ \phi$ if $\phi' \leq 0$. In particular, the length of a curve is invariant under reparametrization.

(b) Suppose that $c$ is a regular curve; i.e., $\dot{c}(t) \neq 0$ for all $t$. Prove that $c$ may be reparametrized by *arc-length*, meaning there exists a reparametrization $\tilde{c}$ of $c$ with $l_{\tilde{c}}(t) = t - a$.

EXERCISE 105. Let $\xi_i$ be Euclidean vector bundles over $M$, $i = 1, 2$, and suppose $L : E(\xi_1)_p \to E(\xi_2)_p \in \mathrm{Hom}(\xi_1, \xi_2)$. Show that $|L|^2 = \sum_i |Lv_i|^2$, where $\{v_i\}$ denotes an orthonormal basis of $E(\xi_1)_p$.

## 2. Riemannian Connections

Recall from Examples and Remarks 2.1(vi) in Chapter 4 that a connection on a Euclidean vector bundle $(\xi, \langle , \rangle)$ is called *Riemannian* if the metric $\langle , \rangle$ is parallel; i.e., if

$$(2.1) \qquad u \langle X, Y \rangle = \langle \nabla_u X, Y(\pi(u)) \rangle + \langle X(\pi(u)), \nabla_u Y \rangle,$$

for all $u \in TM$, and $X, Y \in \Gamma \xi$. In this section, we discuss further properties of Riemannian connections, and the extent to which these are preserved under isometric maps.

LEMMA 2.1. *Let $\nabla$ denote a Riemannian connection on $\xi$, $f : N \to M$, and $X$, $Y$ sections of $\xi$ along $f$. Then for $u \in N_p$, $u \langle X, Y \rangle = \langle \nabla_u X, Y(p) \rangle + \langle X(p), \nabla_u Y \rangle$.*

PROOF. Let $U_i$ be linearly independent sections of $\xi$ on a neighborhood of $f(p)$. Then locally, $X = \sum X^i U_i \circ f$ and $Y = \sum Y^i U_i \circ f$ for functions $X^i$, $Y^i$ defined on a neighborhood of $p$. Thus,

$$u \langle X, Y \rangle = u \left\langle \sum_i X^i U_i \circ f, \sum_j Y^j U_j \circ f \right\rangle = u \sum_{i,j} X^i Y^j \langle U_i, U_j \rangle \circ f$$

$$= \sum_{i,j} u(X^i Y^j) \langle U_i, U_j \rangle \circ f(p) + \sum_{i,j} (X^i Y^j)(p) u(\langle U_i, U_j \rangle \circ f).$$

The second summation may be rewritten as

$$\sum_{i,j}(X^iY^j)(p)u(\langle U_i, U_j\rangle \circ f) = \sum_{i,j}(X^iY^j)(p)f_*u\langle U_i, U_j\rangle$$

$$= \sum_{i,j}(X^iY^j)(p)(\langle \nabla_{f_*u}U_i, U_j \circ f(p)\rangle$$
$$+ \langle U_i \circ f(p), \nabla_{f_*u}U_j\rangle)$$
$$= \sum_{i,j}(X^iY^j)(p)(\langle \nabla_u(U_i \circ f), U_j \circ f(p)\rangle$$
$$+ \langle U_i \circ f(p), \nabla_u(U_j \circ f)\rangle).$$

On the other hand,

$$\langle \nabla_u X, Y(p)\rangle + \langle X(p), \nabla_u Y\rangle = \left\langle \nabla_u\left(\sum X^iU_i \circ f\right), \sum Y^j(p)U_j \circ f(p)\right\rangle$$
$$+ \left\langle \sum X^i(p)U_i \circ f(p), \nabla_u\left(\sum Y^jU_j \circ f\right)\right\rangle$$
$$= \sum_{i,j}(uX^i)Y^j(p)\langle U_i, U_j\rangle \circ f(p)$$
$$+ (X^iY^j)(p)\langle \nabla_u(U_i \circ f), U_j \circ f(p)\rangle$$
$$+ X^i(p)(uY^j)\langle U_i, U_j\rangle \circ f(p)$$
$$+ X^iY^j(p)\langle U_i \circ f(p), \nabla_u(U_j \circ f)\rangle,$$

and therefore equals the expression for $u\langle X, Y\rangle$.  $\qquad\square$

THEOREM 2.1. *A connection $\nabla$ on a Euclidean vector bundle $\xi : E \to M$ is Riemannian iff for any curve $c$ in $M$ and parallel sections $X$, $Y$ of $\xi$ along $c$, the function $\langle X, Y\rangle$ is constant.*

PROOF. If the connection is Riemannian and $X$, $Y$ are parallel along $c$, then by Lemma 2.1,

$$D\langle X, Y\rangle = \langle \nabla_D X, Y\rangle + \langle X, \nabla_D Y\rangle = 0.$$

Conversely, suppose $u \in TM$, $X$, $Y \in \Gamma\xi$. Consider a curve $c : I \to M$ with $c(0) = p$, $\dot{c}(0) = u$, and let $U_1, \ldots, U_k$ denote parallel sections of $\xi$ along $c$ such that $U_1(t), \ldots, U_k(t)$ is an orthonormal basis of $E_{c(t)}$ for $t \in I$. If $X^i := \langle X \circ c, U_i\rangle$, $Y^j := \langle Y \circ c, U_j\rangle$, then

$$\langle \nabla_u X, Y(p)\rangle = \langle \nabla_{\dot{c}(0)}X, Y(p)\rangle = \langle \nabla_{D(0)}(X \circ c), Y(p)\rangle$$
$$= \left\langle \sum_i X^{i\prime}(0)U_i(0), \sum_j Y^j(0)U_j(0)\right\rangle = \sum_i X^{i\prime}(0)Y^i(0).$$

Similarly, $\langle X(p), \nabla_u Y\rangle = \sum_i X^i(0)Y^{i\prime}(0)$. Thus,

$$u\langle X, Y\rangle = (\langle X, Y\rangle \circ c)'(0) = \sum(X^iY^i)'(0) = \langle \nabla_u X, Y(p)\rangle + \langle X(p), \nabla_u Y\rangle,$$

and the connection is Riemannian.  $\qquad\square$

In particular, the holonomy group of a Riemannian connection is a subgroup of the orthogonal group.

PROPOSITION 2.1. *Any Euclidean vector bundle admits a Riemannian connection.*

PROOF. Any trivial Euclidean bundle $\xi$ admits a Riemannian connection: If $X_1, \ldots, X_k$ is an orthonormal parallelization of $\xi$ and $u \in E(\xi)$, define a section $X^u := \sum \langle u, X_i(\pi(u)) \rangle X_i$ of $\xi$, and set $\mathcal{H}_u := X^u_* M_{\pi(u)}$. Then a section of $\xi$ will be parallel iff it is a constant linear combination of $X_1, \ldots, X_k$, so that $\mathcal{H}$ is Riemannian.

In the general case, recall that in the proof of Theorem 1.1 in Chapter 4, we constructed a connection on an arbitrary bundle $\xi$ by piecing together connections $\mathcal{H}^\alpha$ on $\xi_{|U_\alpha}$, where $\{U_\alpha\}$ is a locally finite cover of the base such that $\xi_{|U_\alpha}$ is trivial. We claim that when each $\mathcal{H}^\alpha$ is Riemannian, then the resulting connection $\mathcal{H}$ on $\xi$ also has that property: In fact, if $\nabla$ is the covariant derivative operator of $\mathcal{H}$, then by Examples and Remarks 2.1 in Chapter 4, $\nabla = \sum_\alpha \phi_\alpha \nabla^\alpha$, where $\{\phi_\alpha\}$ is a partition of unity subordinate to the cover, and $\nabla^\alpha$ is the covariant derivative of $\mathcal{H}^\alpha$ extended to be zero outside $U_\alpha$. To see this, write $u^\alpha$ for the $\mathcal{H}^\alpha$-component of $u \in TE$ and $u^h$ for its $\mathcal{H}$-component. Then

$$X_* u = \mathcal{J}_x \nabla^\alpha_u X + (X_* u)^\alpha$$

for any $\alpha$ with $\pi(x) \in U_\alpha$. Thus,

$$X_* u = \sum_\alpha \phi_\alpha(\pi(x)) X_* u = \sum_\alpha \phi_\alpha(\pi(x))(\mathcal{J}_x \nabla^\alpha_u X + (X_* u)^\alpha)$$

$$= \mathcal{J}_x \left( \sum_\alpha \phi_\alpha(\pi(x)) \nabla^\alpha_u X \right) + (X_* u)^h.$$

But $X_* u$ also equals $\mathcal{J}_x \nabla_u X + (X_* u)^h$, and the claim follows. Since each $\nabla^\alpha$ is Riemannian, so is $\nabla$. $\qquad\square$

PROPOSITION 2.2. *If $R$ denotes the curvature tensor of a Riemannian connection $\nabla$ on $\xi$, then*

$$\langle R(U,V)X, Y \rangle = -\langle R(U,V)Y, X \rangle, \qquad U, V \in \mathfrak{X}M, \quad X, Y \in \Gamma\xi.$$

PROOF. Given $u, v \in M_p$, $R(u, v)$ belongs to the Lie algebra of the holonomy group at $p$ by Proposition 3.2 in Chapter 4. Since the connection is Riemannian, the holonomy group is a subgroup of the orthogonal group $O(E_p)$, so that $R(u, v)$ is a skew-adjoint transformation of $E_p$. $\qquad\square$

The *torsion tensor field* $T$ of a connection $\nabla$ on a manifold $M$ (i.e., on the tangent bundle of $M$) is defined by

$$T(U,V) = \nabla_U V - \nabla_V U - [U, V], \qquad U, V \in \mathfrak{X}M.$$

It is straightforward to check that $T$ is indeed a tensor field on $M$; when it vanishes identically, the connection is said to be *torsion-free*.

THEOREM 2.2 (The Fundamental Theorem of Riemannian Geometry). *A Riemannian manifold $(M, \langle, \rangle)$ admits a unique Riemannian connection that is torsion-free.*

This connection is called the *Levi-Civita connection* of $M$.

PROOF. We first establish uniqueness. Given $X$, $Y$, $Z \in \mathfrak{X}M$, the Riemannian and torsion-free properties of $\nabla$ imply that

$$
\begin{aligned}
\langle \nabla_X Y, Z \rangle &= X \langle Y, Z \rangle - \langle Y, \nabla_X Z \rangle = X \langle Y, Z \rangle - \langle Y, \nabla_Z X + [X, Z] \rangle \\
&= X \langle Y, Z \rangle - Z \langle X, Y \rangle + \langle \nabla_Z Y, X \rangle + \langle Y, [Z, X] \rangle \\
&= X \langle Y, Z \rangle - Z \langle X, Y \rangle + \langle \nabla_Y Z + [Z, Y], X \rangle + \langle Y, [Z, X] \rangle \\
&= X \langle Y, Z \rangle - Z \langle X, Y \rangle + Y \langle Z, X \rangle - \langle Z, \nabla_Y X \rangle - \langle X, [Y, Z] \rangle \\
&\quad + \langle [Z, X], Y \rangle \\
&= X \langle Y, Z \rangle - Z \langle X, Y \rangle + Y \langle Z, X \rangle - \langle Z, \nabla_X Y - [X, Y] \rangle - \langle X, [Y, Z] \rangle \\
&\quad + \langle [Z, X], Y \rangle.
\end{aligned}
$$

Grouping the $\nabla$-terms, we obtain

$$
(2.2) \qquad
\begin{aligned}
\langle \nabla_X Y, Z \rangle = \frac{1}{2} \{ & X \langle Y, Z \rangle + Y \langle Z, X \rangle - Z \langle X, Y \rangle \\
& + \langle Z, [X, Y] \rangle + \langle Y, [Z, X] \rangle - \langle X, [Y, Z] \rangle \},
\end{aligned}
$$

which establishes uniqueness.

In order to show existence, define for fixed $X$, $Y \in \mathfrak{X}M$ a map $\alpha : \mathfrak{X}M \to \mathcal{F}(M)$, where $\alpha(Z)$ equals the right side of (2.2). Clearly, $\alpha(Z_1 + Z_2) = \alpha(Z_1) + \alpha(Z_2)$. Furthermore, given $f \in \mathcal{F}(M)$,

$$
\begin{aligned}
\alpha(fZ) &= \frac{1}{2} \{ X \langle Y, fZ \rangle + Y \langle fZ, X \rangle - fZ \langle X, Y \rangle \\
&\qquad + \langle fZ, [X, Y] \rangle + \langle Y, [fZ, X] \rangle - \langle X, [Y, fZ] \rangle \} \\
&= f\alpha(Z) + \frac{1}{2} \{ (Xf) \langle Y, Z \rangle + (Yf) \langle Z, X \rangle - (Xf) \langle Y, Z \rangle - (Yf) \langle X, Z \rangle \} \\
&= f\alpha(Z).
\end{aligned}
$$

Thus, $\alpha$ is a 1-form on $M$, and we may define $\nabla_X Y := \alpha^\sharp$; i.e., $\nabla_X Y$ is the unique vector field on $M$ such that $\langle \nabla_X Y, Z \rangle = \alpha(Z)$ for $Z \in \mathfrak{X}M$. The operator

$$
\nabla : \mathfrak{X}M \times \mathfrak{X}M \to \mathfrak{X}M
$$

satisfies the axioms (1) through (3) for a covariant derivative operator (Chapter 4, Theorem 2.1): Axioms (1) and (2) are immediate. For (3),

$$
\begin{aligned}
2 \langle \nabla_X fY, Z \rangle &= X \langle fY, Z \rangle + fY \langle Z, X \rangle - Z \langle X, fY \rangle + \langle Z, [X, fY] \rangle \\
&\quad + \langle fY, [Z, X] \rangle - \langle X, [fY, Z] \rangle \\
&= 2 \langle f \nabla_X Y, Z \rangle + (Xf) \langle Y, Z \rangle - (Zf) \langle X, Y \rangle + (Xf) \langle Z, Y \rangle \\
&\quad + (Zf) \langle X, Y \rangle \\
&= 2 \langle f \nabla_X Y + (Xf) Y, Z \rangle,
\end{aligned}
$$

so that $\nabla_X fY = f \nabla_X Y + (Xf) Y$. Furthermore, $\nabla$ is Riemannian, as can be seen by writing out (2.2) for $\langle \nabla_X Y, Z \rangle$ and adding this to the corresponding expression for $\langle \nabla_X Z, Y \rangle$. The torsion-free property is verified in the same way. $\qquad \square$

PROPOSITION 2.3. *Let $M$ be a Riemannian manifold with Levi-Civita connection $\nabla$. If $f : N \to M$ and $U$, $V$, $W \in \mathfrak{X}N$, then*

$$\langle \nabla_U f_* V, f_* W \rangle = \frac{1}{2} \{ U\langle f_* V, f_* W \rangle + V\langle f_* W, f_* U \rangle - W\langle f_* U, f_* V \rangle$$
$$+ \langle f_* W, f_*[U, V] \rangle + \langle f_* V, f_*[W, U] \rangle - \langle f_* U, f_*[V, W] \rangle \}.$$

PROOF. By Lemma 2.1,

$$U\langle f_* V, f_* W \rangle = \langle \nabla_U f_* V, f_* W \rangle + \langle f_* V, \nabla_U f_* W \rangle.$$

The result then follows from the proof of the uniqueness part in Theorem 2.2, once we establish that

$$\nabla_U f_* V - \nabla_V f_* U = f_*[U, V].$$

Equivalently, $T(f_* U, f_* V) = \nabla_U f_* V - \nabla_V f_* U - f_*[U, V]$. Now, in a chart $x$ of $M$, $f_* U$ may be locally written as $\sum U(x^i \circ f)\frac{\partial}{\partial x^i} \circ f$. Since $T$ is tensorial, we may assume that there exist vector fields $\tilde{U}$, $\tilde{V}$ on $M$ such that $f_* U = \tilde{U} \circ f$ and $f_* V = \tilde{V} \circ f$. Then

$$\nabla_U f_* V - \nabla_V f_* U = \nabla_U(\tilde{V} \circ f) - \nabla_V(\tilde{U} \circ f) = \nabla_{f_* U}\tilde{V} - \nabla_{f_* V}\tilde{U}$$
$$= (\nabla_{\tilde{U}}\tilde{V} - \nabla_{\tilde{V}}\tilde{U}) \circ f = [\tilde{U}, \tilde{V}] \circ f = f_*[U, V].$$

$\square$

Let $(M, g)$ be a Riemannian manifold, $\imath : N \to M$ an immersion. Then $\imath^* g$ is the Riemannian metric on $N$ for which $\imath$ becomes isometric. If $X$ is a vector field along $\imath$, define a 1-form $\alpha_X$ on $N$ by

$$\alpha_X(U) := \langle X, \imath_* U \rangle, \qquad U \in \mathfrak{X}N.$$

In other words, $\alpha_X = \imath^* X^\flat$. The *tangential component* of $X$ with respect to $\imath$ is the vector field $X^\top$ along $\imath$ given by

$$X^\top := \imath_* \alpha_X^\sharp,$$

where $^\sharp$ denotes the musical isomorphism with respect to $\imath^* g$. Thus, $\langle X^\top, \imath_* U \rangle = \langle X, \imath_* U \rangle$ for all $U \in \mathfrak{X}N$. The *orthogonal component* of $X$ with respect to $\imath$ is $X^\perp := X - X^\top$. It is easy to see that $^\top, ^\perp$ are tensorial. In fact, the restriction $^\top : M_{\imath(p)} \to \imath_* N_p$ is just the orthogonal projection onto $\imath_* N_p$.

A key property of Riemannian connections is that they are preserved under isometric maps:

PROPOSITION 2.4. *Let $\imath : N \to M$ be an isometric immersion between Riemannian manifolds. If $\kappa_N$, $\kappa_M$ denote the connection maps of $\tau N$, $\tau M$, and $\nabla^N$, $\nabla^M$ are the respective Levi-Civita connections, then*

(1) $\imath_* \kappa_N w = (\kappa_M \imath_{**} w)^\top$ *for $w \in TTN$. If in addition $\dim N = \dim M$, then $\kappa_M \circ \imath_{**} = \imath_* \circ \kappa_N$; i.e., the diagram*

$$
\begin{array}{ccc}
TTN & \xrightarrow{\imath_{**}} & TTM \\
{\scriptstyle \kappa_N}\downarrow & & \downarrow{\scriptstyle \kappa_M} \\
TN & \xrightarrow{\imath_*} & TM
\end{array}
$$

*commutes.*

(2) *If $f : P \to N$, then for $U \in \mathfrak{X}P$ and any vector field $X$ along $f$,*

$$\imath_* \nabla^N_U X = (\nabla^M_U \imath_* X)^\top,$$

*and $\imath_* \nabla^N_U X = \nabla^M_U \imath_* X$ when the dimensions of $M$ and $N$ coincide.*

PROOF. Let $X, Y, Z \in \mathfrak{X}N$. By Theorem 2.2 and Proposition 2.3,

$$\langle \nabla^M_X \imath_* Y, \imath_* Z \rangle = \langle \nabla^N_X Y, Z \rangle$$

because $\imath$ is isometric. Furthermore, $\langle \nabla^N_X Y, Z \rangle = \langle \imath_* \nabla^N_X Y, \imath_* Z \rangle$, and therefore $(\nabla^M_X \imath_* Y)^\top = \imath_* \nabla^N_X Y$. This proves (2) for the case $P = N$, $f = 1_N$. It also implies that $(\kappa_M \imath_{**} Y_* X(p))^\top = \imath_* \kappa_N Y_* X(p)$ for $p \in N$. Since the set $\{Y_* v \mid Y \in \mathfrak{X}N, v \in M_p\}$ spans $(TN)_{Y(p)}$, (1) holds. (2) then follows from the definition of the connection map $\kappa$. $\qquad\square$

When $M$ and $N$ have the same dimension, Proposition 2.4 implies that an isometric map preserves parallel fields (and therefore geodesics) as well as curvature:

THEOREM 2.3. *Let $f : N \to M$ be an isometric map between Riemannian manifolds of the same dimension.*

(1) *If $X$ is a parallel vector field along a curve $c : I \to N$, then $f_* X$ is parallel along $f \circ c : I \to M$.*

(2) $\exp_M \circ f_* = f \circ \exp_N$.

(3) $f_* R^N(x, y)z = R^M(f_* x, f_* y)f_* z, \qquad x, y, z \in N_p, \quad p \in N$.

PROOF. (1) By Proposition 2.4(2), $\nabla^M_D f_* X = f_* \nabla^N_D X = 0$.

(2) Taking $X = \dot{c}$ in (1), we see that $f$ maps geodesics of $N$ to geodesics of $M$. If $v \in \widetilde{TN}$ and $c(t) = \exp_N(tv)$, then $f \circ c$ is the geodesic of $M$ with initial tangent vector $f_* v$. Thus, $f(\exp_N(tv)) = \exp_M(tf_* v)$. (3) follows from Proposition 2.4(2) and Cartan's Structure Equation 3.1 in Chapter 4. $\qquad\square$

EXAMPLES AND REMARKS 2.1. (i) The Levi-Civita connection of the canonical metric on $\mathbb{R}^n$ is the standard flat connection by (2.2).

(ii) Let $(M, g)$ be a Riemannian manifold. A submanifold $N$ of $M$ is said to be a *Riemannian submanifold* of $M$ if it is endowed with the metric $\imath^* g$, where $\imath : N \hookrightarrow M$ denotes inclusion. The Levi-Civita connection of $S^n$, as a Riemannian submanifold of Euclidean space, is the canonical connection from Examples and remarks 1.1(iii) in Chapter 4.

(iii) Let $G$ be a Lie group. The Levi-Civita connection of a left-invariant metric on $G$ is given by

$$(2.3) \quad \langle \nabla_X Y, Z \rangle = \frac{1}{2}\{\langle [X, Y], Z \rangle - \langle [Y, Z], X \rangle + \langle [Z, X], Y \rangle\}, \qquad X, Y, Z \in \mathfrak{g}.$$

This follows immediately from (2.2). Suppose that the metric is actually bi-invariant. We claim that the flow $\Phi_t$ of any left-invariant $X \in \mathfrak{g}$ consists of isometries of $M$: If $c$ is the integral curve of $X$ with $c(0) = e$, then the integral curve passing through $a \in G$ at $t = 0$ is $t \mapsto ac(t)$ because $X$ is left-invariant. Thus, $\Phi_t(a) = ac(t) = R_{c(t)}a$, and each $\Phi_t$ is an isometry. By Exercise 111, the assignment $Y \mapsto \nabla_Y X$ is then skew-adjoint for $Y \in \mathfrak{g}$. Consequently,

$$\langle [X, Y], Y \rangle = \langle \nabla_X Y, Y \rangle - \langle \nabla_Y X, Y \rangle = \frac{1}{2} X \langle Y, Y \rangle = 0,$$

and $\langle[\cdot,\cdot],\cdot\rangle$ is skew-adjoint in all three arguments. The last two terms on the right side of (2.3) then cancel, and

$$\nabla_X Y = \frac{1}{2}[X,Y], \qquad X,Y \in \mathfrak{g}.$$

The curvature tensor is therefore given by

$$(2.4) \qquad R(X,Y)Z = -\frac{1}{4}[[X,Y],Z], \qquad X,Y,Z \in \mathfrak{g}.$$

Notice also that $\nabla_X X = 0$ for any $X \in \mathfrak{g}$; i.e., the integral curves of left-invariant vector fields are the geodesics of $G$, just as in the case of the left-invariant connection from Examples and Remarks 4.1(iii) in Chapter 4.

(iv) If $M^n$ is an oriented Riemannian manifold, the *volume form* of $M$ is the $n$-form $\omega$ such that $\omega(v_1,\ldots,v_n) = 1$ for any positively oriented orthonormal basis $v_1,\ldots,v_n$ of $M_p$, $p \in M$, cf. Proposition 15.1 in Chapter 1. When $M$ is compact, it is customary to define the *integral of a function* $f$ on $M$ by

$$\int_M f := \int_M f\omega.$$

Suppose $X$ is a vector field on $M$. The Lie derivative $L_X\omega$ of $\omega$ in direction $X$ (see Exercise 33) is then again an $n$-form, and may therefore be expressed as $f\omega$ for some function $f$ on $M$. This function is called the *divergence* div $X$ of $X$. In other words, the divergence of $X$ is determined by the equation

$$(2.5) \qquad (\operatorname{div} X)\omega = L_X\omega.$$

Thus, the divergence is an infinitesimal measure of the amount by which the flow of a vector field fails to preserve volume. It is locally given by

$$(2.6) \qquad (\operatorname{div} X)_p = \sum_{i=1}^{n} \langle \nabla_{v_i} X, v_i \rangle,$$

where $\{v_i\}$ is an orthonormal basis of $M_p$, $p \in M$: To see this, extend the set $\{v_i\}$ to a local orthonormal basis $\{V_i\}$ of vector fields around $p$. Since $\omega(V_1,\ldots,V_n)$ is constant,

$$(L_X\omega)(p)(v_1,\ldots,v_n) = X_p\omega(V_1,\ldots,V_n) - \sum_{i=1}^{n}\omega(v_1,\ldots,[X,V_i]_p,\ldots,v_n)$$

$$= -\sum_i \omega(v_1,\ldots,\langle\nabla_{X_p}V_i - \nabla_{v_i}X, v_i\rangle v_i,\ldots,v_n)$$

$$= \sum_i \langle\nabla_{v_i}X, v_i\rangle\omega(p)(v_1,\ldots,v_n)$$

$$= \sum_i \langle\nabla_{v_i}X, v_i\rangle,$$

which establishes (2.6). Notice that by Exercise 34, the $n$-form

$$(2.7) \qquad L_X\omega = i(X)\circ d\omega + d\circ i(X)\omega = d\circ i(X)\omega$$

is exact, since $\omega$ is closed. Suppose next that $M$ is compact with boundary $\partial M$, so that the volume form $\omega'$ of the latter is given by $\omega' = i(N)\omega$, where

$N$ is the outward-pointing unit normal field on $\partial M$. Then $i(X)\omega = \langle X, N \rangle \omega'$, and Stokes' theorem together with (2.7) yields

$$(2.8) \qquad \int_M \operatorname{div} X = \int_{\partial M} \langle X, N \rangle.$$

(2.8) is known as the *divergence theorem*.

EXERCISE 106. Given a Riemannian connection on $\xi$, there is an induced connection on $\xi^*$ (by Examples and Remarks 2.1(v) in Chapter 4). Show that the latter is Riemannian with respect to the Euclidean metric on $\xi^*$ defined in Examples and Remarks 1.1(vii).

EXERCISE 107. Let $(M_i, g_i)$ be Riemannian manifolds, $i = 1, 2$, and consider $M := M_1 \times M_2$ with the Riemannian product metric from Examples and Remarks 1.1(vi). Show that the curvature tensor $R$ of $M$ is related to the curvatures $R_i$ of $M_i$ by the formula

$$R(X, Y)Z = (R_1(\pi_{1*}X, \pi_{1*}Y)\pi_{1*}Z, R_2(\pi_{2*}X, \pi_{2*}Y)\pi_{2*}Z),$$

where $\pi_i : M \to M_i$ denotes projection. Here, we identify $M_{(p_1, p_2)}$ with $M_{p_1} \times M_{p_2}$ via $(\pi_{1*}, \pi_{2*})$.

EXERCISE 108. Let $M^n$ be a Riemannian manifold. The *gradient* $\nabla f$ of $f \in \mathcal{F}M$ is the vector field $df^\sharp$; i.e., $\langle \nabla f, X \rangle = X(f)$ for $X \in \mathfrak{X}M$.

(a) Let $a \in \mathbb{R}$ be a regular value of $f$, so that $N := f^{-1}(a)$ is an $(n-1)$-dimensional submanifold of $M$. Show that for any $p \in N$, $\imath_* N_p = \nabla f(p)^\perp$, where $\imath : N \to M$ denotes inclusion. Thus, $(\nabla f)_{|N}$ is a nowhere-zero section spanning the normal bundle of $\imath$.

(b) Given $p \in M$ with $\nabla f(p) \neq 0$, show that $(\nabla f/|\nabla f|)(p)$ and $-(\nabla f/|\nabla f|)(p)$ represent the directions of "maximal increase" and "maximal decrease" of $f$ at $p$: For any $v \in M_p$ of unit length,

$$-\frac{\nabla f}{|\nabla f|}(p)(f) \leq vf \leq \frac{\nabla f}{|\nabla f|}(p)(f).$$

EXERCISE 109. The *Hessian tensor* of a function $f : M \to \mathbb{R}$ on a Riemannian manifold $M$ is the tensor field of type $(1, 1)$ given by

$$H_f(X) = \nabla_X \nabla f, \qquad X \in \mathfrak{X}M.$$

The associated quadratic form

$$h_f(X, Y) = \langle \nabla_X \nabla f, Y \rangle, \qquad X, Y \in \mathfrak{X}M,$$

is called the *Hessian form* of $f$.

(a) Prove that $H_f$ is a self-adjoint operator, so that the Hessian form is symmetric.

(b) Show that, at a critical point $p$ of $f$, $h_f(X, Y)(p) = X(p)Yf = Y(p)Xf$.

(c) Suppose that the Hessian form is positive definite at a critical point $p$ of $f$. Prove that $f$ has a local minimum at $p$.

EXERCISE 110. The *Laplacian* of a function $f$ on a Riemannian manifold $M$ is the function

$$\Delta f = \operatorname{div} \nabla f.$$

Prove that if $M$ is compact, oriented, without boundary, then

$$\int_M \Delta f = 0.$$

EXERCISE 111. If $T$ is a tensor field of type $(0,r)$ on $M$ and $X \in \mathfrak{X}M$, the *Lie derivative* of $T$ in direction $X$ is the tensor field $L_X T$ of the same type given by

$$L_X T(p) = \lim_{t \to 0} \frac{1}{t}[(\Phi_t^* X)(p) - X(p)], \qquad p \in M,$$

where $\Phi_t$ denotes the flow of $X$. Just as in the case of an $r$-form on $M$, it is not difficult to show that

$$L_X T(X_1, \ldots, X_r) = X(T(X_1, \ldots, X_r)) - \sum_{i=1}^{r} T(X_1, \ldots, L_X X_i, \ldots, X_r),$$

cf. Exercise 33 in Chapter 1. A vector field $X$ on a Riemannian manifold $(M, g)$ is said to be a *Killing field* if its flow consists of isometries of $M$.

(a) Show that $X$ is Killing iff $L_X g = 0$.

(b) Show that $X$ is Killing iff $\nabla X$ is skew-symmetric; i.e., $\langle \nabla_U X, U \rangle = 0$ for $U \in \mathfrak{X}M$.

(c) Prove that a Killing field is divergence-free. Give an example that shows the converse is not true.

## 3. Curvature Quantifiers

We have seen that the Levi-Civita connection of a Riemannian manifold is the unique torsion-free connection for which the metric is parallel. This translates into additional properties for its curvature tensor, properties which allow us to introduce other types of curvature commonly used in Riemannian geometry.

PROPOSITION 3.1. *Let $R$ denote the curvature tensor of a Riemannian manifold $M$. The following identities hold for any vector fields $X, Y, Z, U$ on $M$:*

(1) $R(X,Y)Z = -R(Y,X)Z.$

(2) $\langle R(X,Y)Z, U \rangle = -\langle R(X,Y)U, Z \rangle.$

(3) $R(X,Y)Z + R(Y,Z)X + R(Z,X)Y = 0.$

(4) $\langle R(X,Y)Z, U \rangle = \langle R(Z,U)X, Y \rangle.$

PROOF. Statement (1) is true for any connection and follows from the definition of $R$, whereas (2) holds for Riemannian connections and is the content of Proposition 2.2. Statement (3) is a consequence of the fact that the Levi-connection is torsion-free: We may assume that the vector fields involved have vanishing Lie brackets, since it is enough to show the property for, say, coordinate vector fields. Then

$$R(X,Y)Z = \nabla_X \nabla_Y Z - \nabla_Y \nabla_X Z,$$
$$R(Y,Z)X = \nabla_Y \nabla_Z X - \nabla_Z \nabla_Y X = \nabla_Y \nabla_X Z - \nabla_Z \nabla_Y X,$$
$$R(Z,X)Y = \nabla_Z \nabla_X Y - \nabla_X \nabla_Z Y = \nabla_Z \nabla_Y X - \nabla_X \nabla_Y Z,$$

and adding all three identities yields (3). Finally, (4) is an algebraic consequence of (1)–(3):

$$
\begin{aligned}
2\langle R(X,Y)Z,U \rangle &= \langle R(X,Y)Z,U \rangle + \langle R(X,Y)Z,U \rangle \\
&= -\langle R(Y,X)Z,U \rangle - \langle R(X,Y)U,Z \rangle \quad \text{by (1) and (2)} \\
&= \langle R(X,Z)Y,U \rangle + \langle R(Z,Y)X,U \rangle \\
&\quad + \langle R(Y,U)X,Z \rangle + \langle R(U,X)Y,Z \rangle \quad \text{by (3)} \\
&= \langle R(Z,X)U,Y \rangle + \langle R(Y,Z)U,X \rangle \\
&\quad + \langle R(U,Y)Z,X \rangle + \langle R(X,U)Z,Y \rangle \quad \text{by (1) and (2)} \\
&= \langle R(Z,X)U,Y \rangle + \langle R(X,U)Z,Y \rangle \\
&\quad + \langle R(U,Y)Z,X \rangle + \langle R(Y,Z)U,X \rangle \quad \text{rearranging terms} \\
&= -\langle R(U,Z)X,Y \rangle - \langle R(Z,U)Y,X \rangle \quad \text{by (3)} \\
&= \langle R(Z,U)X,Y \rangle + \langle R(Z,U)X,Y \rangle \quad \text{by (1) and (2)} \\
&= 2\langle R(Z,U)X,Y \rangle.
\end{aligned}
$$

$\square$

Notice that for $x,y \in M_p$, $R(x,y)$ is a linear transformation $R(x,y) : M_p \to M_p$. Thus, for vector fields $X$, $Y$ on $M$, $R(X,Y)$ is a section of the bundle $\mathrm{End}(M) = \mathrm{Hom}(\tau M, \tau M)$. By (2), it is actually a section of the bundle $\mathfrak{o}(M) = \{L \in \mathrm{End}(M) \mid L + L^t = 0\}$ of skew-adjoint endomorphisms. The latter bundle is in turn equivalent to $\Lambda_2(M)$ via $L : \Lambda_2(M) \to \mathfrak{o}(M)$, where

$$(3.1) \qquad\qquad L(x \wedge y)z := \langle y,z \rangle x - \langle x,z \rangle y;$$

skew-symmetry of $L(x \wedge y)$ follows from

$$\langle L(x \wedge y)z, u \rangle = \langle y,z \rangle \langle x,u \rangle - \langle x,z \rangle \langle y,u \rangle = \langle x \wedge y, u \wedge z \rangle,$$

cf. Examples and Remarks 1.1(viii). Since $L$ is one-to-one on each fiber, it is an equivalence by dimension considerations. But $R$ is also bilinear and skew-symmetric in its first two arguments, so that for each $p \in M$, $R$ may be viewed as a linear map $\rho : \Lambda_2(M_p) \to \Lambda_2(M_p)$. By (4), $\rho$ is symmetric.

DEFINITION 3.1. The *curvature operator* $\rho$ of a Riemannian manifold $M$ is the self-adjoint section of $\mathrm{End}(\Lambda_2 M)$ given by

$$\langle \rho(x \wedge y), z \wedge u \rangle = \langle R(x,y)u, z \rangle$$

on decomposable elements.

Let $k$ denote the quadratic form

$$k(x \wedge y) = \langle \rho(x \wedge y), x \wedge y \rangle$$

associated to $\rho$. If $P$ is a 2-plane in some tangent space, then $\Lambda_2 P$ is one-dimensional, and contains exactly 2 unit vectors: they can be written as $\pm x \wedge y/|x \wedge y|$, where $x$ and $y$ are linearly independent in $P$. Since $k(\alpha) = k(-\alpha)$ for $\alpha \in \Lambda_2 P$, we may associate to each 2-plane $P \subset M_p$ a unique number $k(\alpha)$, where $\alpha$ is a unit bivector in $\Lambda_2 P$.

DEFINITION 3.2. The *sectional curvature* of the plane spanned by $x, y \in M_p$ is

$$K_{x,y} := k\left(\frac{x \wedge y}{|x \wedge y|}\right) = \frac{\langle R(x,y)y, x\rangle}{|x|^2|y|^2 - \langle x,y\rangle^2}.$$

There are two additional types of curvature commonly used in Riemannian geometry: The *Ricci tensor* is the tensor field of type (0,2) given by

$$\operatorname{Ric}(x,y) := \operatorname{tr}(u \mapsto R(u,x)y).$$

It is symmetric, since in an orthonormal basis $\{u_i\}$ of $M_p$, (1), (2), and (4) imply

$$\operatorname{Ric}(x,y) = \sum_i \langle R(u_i, x)y, u_i\rangle = \sum_i \langle R(y, u_i)u_i, x\rangle = \sum_i \langle R(u_i, y)x, u_i\rangle$$
$$= \operatorname{Ric}(y, x).$$

If $x \neq 0$, the *Ricci curvature in direction $x$* is

$$\operatorname{Ric}(x) := \operatorname{Ric}\left(\frac{x}{|x|}, \frac{x}{|x|}\right) = \frac{1}{|x|^2}\sum_i K_{x, u_i}.$$

The *scalar curvature* $s(p)$ at $p \in M$ is the trace of the Ricci tensor:

$$s(p) := \sum_i \operatorname{Ric}(u_i) = 2\sum_{i<j} K_{u_i, u_j}.$$

Both Ricci and scalar curvatures are averages, and the curvature tensor $R$ cannot be reconstructed from them. $R$ can, however, be recovered from the sectional curvature:

$$\begin{aligned}
\langle R(x,y)z, u\rangle = \frac{1}{6}\{&k(x+u, y+z) - k(y+u, x+z) - k(x+u, y) - k(x+u, z)\\
&- k(x, y+z) - k(u, y+z) + k(y+u, x) + k(y+u, z)\\
&+ k(y, x+z) + k(u, x+z) + k(x, z) - k(y, z) + k(u, y)\\
&- k(u, x)\},
\end{aligned}$$

where we have replaced $k(x \wedge y)$ by $k(x, y)$, etc. This formula is readily verified by expanding the right side and using (1) through (4). It implies in particular that $R = 0$ whenever $k = 0$.

$M$ is said to be a *space of constant curvature* $\kappa \in \mathbb{R}$ if $K \equiv \kappa$. More generally, suppose there exists a function $K : M \to \mathbb{R}$ such that $K_P = K(p)$ for every plane $P \subset M_p$, $p \in M$. Then $k(X, Y) = K|X \wedge Y|^2 = \langle \rho(X \wedge Y), X \wedge Y\rangle$; i.e., $\rho = K1_{\Lambda_2 M}$. This in turn implies

$$\langle R(X,Y)Z, U\rangle = \langle \rho(X \wedge Y), U \wedge Z\rangle = K\langle X \wedge Y, U \wedge Z\rangle = K\langle L(X \wedge Y)Z, U\rangle,$$

so that

(3.2) $$R(X,Y)Z = K \cdot L(X \wedge Y)Z = K(\langle Y, Z\rangle X - \langle X, Z\rangle Y).$$

$\mathbb{R}^2$ with the standard metric has constant curvature 0, and by Examples and Remarks 3.1(ii) in Chapter 4, $S^2$ has constant curvature 1. In order to describe a 2-dimensional space of constant curvature $-1$, we will need the following concept: Two Riemannian metrics $g$ and $\tilde{g}$ on $M$ are said to be *conformally equivalent* if there exists a positive function $f$ such that $\tilde{g} = fg$. If $h = \log f$, then by (2.2), the respective Levi-Civita connections are related by

$$\tilde{\nabla}_X Y = \nabla_X Y + \frac{1}{2}(X(h)Y + Y(h)X - \langle X, Y \rangle \nabla h),$$

where $\nabla h$ is the gradient of $h$ in the metric $g$. A straightforward computation shows that the curvature tensors are related by

$$\tilde{R}(X,Y)Z = R(X,Y)Z + \frac{1}{2}\{\langle \nabla_X \nabla h, Z \rangle Y - \langle \nabla_Y \nabla h, Z \rangle X + \langle X, Z \rangle \nabla_Y \nabla h$$
$$- \langle Y, Z \rangle \nabla_X \nabla h\} + \frac{1}{4}\{((Yh)(Zh) - \langle Y, Z \rangle |\nabla h|^2)X - ((Xh)(Zh)$$
$$- \langle X, Z \rangle |\nabla h|^2)Y + ((Xh)\langle Y, Z \rangle - (Yh)\langle X, Z \rangle)\nabla h\}.$$

Thus, if $x, y$ form an orthonormal basis (in the metric $g$) of a plane $P$, then the sectional curvatures of $P$ satisfy

$$f\tilde{K}_P = K_P - \frac{1}{2}(\langle \nabla_x \nabla h, x \rangle + \langle \nabla_y \nabla h, y \rangle) - \frac{1}{4}(|\nabla h|^2 - (xh)^2 - (yh)^2),$$

and

$$(3.3) \qquad f\tilde{K}_P = K_P - \frac{1}{2}(\langle \nabla_x \nabla h, x \rangle + \langle \nabla_y \nabla h, y \rangle) = K_P - \frac{1}{2}\Delta h$$

when $\dim M = 2$.

Now, let $M = \{p \in \mathbb{R}^2 \mid u^2(p) > 0\}$ denote the upper half-plane, $g$ the standard Euclidean metric, and $\tilde{g} = (1/(u^2)^2)g$. By (3.3), $(M, \tilde{g})$ has constant sectional curvature $-1$.

EXERCISE 112. Use (3.3) to construct a 2-dimensional space of constant curvature $\kappa$, where $\kappa$ is an arbitrary real number.

EXERCISE 113. Let $G$ be a Lie group with bi-invariant metric. Prove that for $X, Y \in \mathfrak{g}$,

$$\langle R(X,Y)Y, X \rangle = \frac{1}{4}\|[X,Y]\|^2.$$

Thus, a Lie group with bi-invariant metric has nonnegative sectional curvature.

EXERCISE 114. Let $M_1$, $M_2$ be Riemannian manifolds, and $M = M_1 \times M_2$ together with the product metric. Show that if $M_i$ has nonnegative (resp. nonpositive) sectional curvature, $i = 1, 2$, then so does $M$. If $M_i$ has strictly positive or strictly negative curvature, is the same true for $M$?

EXERCISE 115. Let $F : \mathbb{R}^n \to S^n \subset \mathbb{R}^{n+1}$ denote the inverse of the stereographic projection from the north pole,

$$F(p) = \frac{1}{1 + |p|^2}(2p, |p|^2 - 1).$$

(a) Compute $F_*D_i$, and show that $\langle F_*D_i, F_*D_j \rangle = f\delta_{ij}$, where $f(p) = 4/(1 + |p|^2)^2$. Thus, if we endow $\mathbb{R}^n$ with the conformal metric $f\langle,\rangle$ (where $\langle,\rangle$ is the standard one), then $F : \mathbb{R}^n \to S^n$ is isometric.

(b) Show that $S^n$ has constant curvature 1.

## 4. Isometric Immersions

A large class of Riemannian manifolds consists of the submanifolds of Euclidean space together with the metric induced from the Euclidean one. In this section, we will investigate the sectional curvature of these spaces. Somewhat more generally, suppose $\imath : M \to \tilde{M}$ is an isometric immersion between Riemannian manifolds $M$ and $\tilde{M}$. If $\Gamma_\imath$ denotes the space of vector fields along $\imath$ and $\Gamma_\imath^\perp$ the subspace consisting of those $N \in \Gamma_\imath$ such that $N^T = 0$, then $\Gamma_\imath^\perp$ is naturally identified with the space $\Gamma\nu$ of sections of the normal bundle $\nu$ of $\imath$.

Recall from Section 2 that for each $X \in \Gamma_\imath$, we have an associated 1-form $\alpha_X := \imath^* X^\flat$. Define a vector field $\imath^* X$ on $M$ by

$$\imath^* X := \alpha_X^\sharp.$$

Thus, $\imath_*(\imath^* X) = X^T$, and for $Y \in \mathfrak{X}M$,

$$\langle \imath^* X, Y \rangle = \langle X, \imath_* Y \rangle.$$

Given $X \in \mathfrak{X}M$, $N \in \Gamma_\imath^\perp$, we define the *second fundamental tensor of $\imath$* with respect to $N$ to be the map $S_N : \mathfrak{X}M \to \mathfrak{X}M$ given by

$$(4.1) \qquad S_N X := -\imath^*(\tilde{\nabla}_X N), \qquad X \in \mathfrak{X}M,$$

where $\tilde{\nabla}$ denotes the Levi-Civita connection of $\tilde{M}$. Notice that, as the name suggests, $S : \Gamma_\imath^\perp \times \mathfrak{X}M \to \mathfrak{X}M$ is tensorial: $S$ is clearly bilinear over $\mathbb{R}$, and over $\mathcal{F}M$ in $X$; given $f \in \mathcal{F}M$,

$$S_{fN} X = -\imath^*(\tilde{\nabla}_X fN) = -\imath^*((Xf)N + f\tilde{\nabla}_X N) = -f\imath^*\tilde{\nabla}_X N = fS_N X.$$

We therefore obtain a map $S : E(\nu) \times TM \to TM$. Furthermore, given a unit vector $n \in E(\nu)_p$, the linear transformation $S_n : M_p \to M_p$ is self-adjoint: If $N$ is a local extension of $n$, then

$$\begin{aligned}
\langle S_N X, Y \rangle &= \langle -\imath^*(\tilde{\nabla}_X N), Y \rangle = -\langle \tilde{\nabla}_X N, \imath_* Y \rangle = \langle N, \tilde{\nabla}_X(\imath_* Y) \rangle \\
&= \langle N, \tilde{\nabla}_Y(\imath_* X) + \imath_*[X, Y] \rangle = \langle N, \tilde{\nabla}_Y(\imath_* X) \rangle = -\langle \tilde{\nabla}_Y N, \imath_* X \rangle \\
&= \langle S_N Y, X \rangle.
\end{aligned}$$

The eigenvalues of $S_n$ are called the *principal curvatures at $p$ in direction $n$*, and the corresponding unit eigenvectors the *principal curvature directions*. The *Gauss curvature $G_n$* of $M$ in direction $n$ is

$$G_n = \det S_n.$$

The second fundamental tensors of an isometric immersion $\imath : M \to \tilde{M}$ measure the difference between the curvatures of $M$ and $\tilde{M}$:

PROPOSITION 4.1 (The Gauss Equations). *Consider an isometric immersion $\imath : M \to \tilde{M}$, where $\dim \tilde{M} \geq 2$, and set $m = \dim \tilde{M} - \dim M$. Given $p \in M$, let $x, y, z \in M_p$, and $n_1, \ldots, n_m$ be an orthonormal basis of $\tilde{M}_{\imath(p)}^\perp$. Denote by $S_j : M_p \to M_p$ the second fundamental tensor of $\imath$ with respect to $n_j$, by $R$ and $\tilde{R}$ the curvature tensors of $M$, $\tilde{M}$, and by $k$, $\tilde{k}$ the associated quadratic* *forms from Section 3.* Then

$$R(x,y)z - \imath^*(\tilde{R}(\imath_*x, \imath_*y)\imath_*z) = \sum_{j=1}^{m} \langle S_j y, z \rangle S_j x - \langle S_j x, z \rangle S_j y,$$

$$k(x,y) - \tilde{k}(\imath_*x, \imath_*y) = \sum_{j=1}^{m} \det \begin{bmatrix} \langle S_j x, x \rangle & \langle S_j x, y \rangle \\ \langle S_j y, x \rangle & \langle S_j y, y \rangle \end{bmatrix}.$$

PROOF. We only prove the first equation, since the second one is an immediate consequence of it. Extend $x, y, z, n_j$ locally to $X, Y, Z, N_j$. Then

$$\tilde{\nabla}_Y \imath_* Z = (\tilde{\nabla}_Y \imath_* Z)^T + (\tilde{\nabla}_Y \imath_* Z)^{\perp} = \imath_* \nabla_Y Z + \sum_j \langle \tilde{\nabla}_Y \imath_* Z, N_j \rangle N_j$$

by Proposition 2.4. Taking the covariant derivative in direction $X$ yields

$$\tilde{\nabla}_X \tilde{\nabla}_Y \imath_* Z = \tilde{\nabla}_X \imath_* \nabla_Y Z + \sum_j (X \langle \tilde{\nabla}_Y \imath_* Z, N_j \rangle) N_j + \sum_j \langle \tilde{\nabla}_Y \imath_* Z, N_j \rangle \tilde{\nabla}_X N_j.$$

Thus,

$$(\tilde{\nabla}_X \tilde{\nabla}_Y \imath_* Z)^T = \imath_* \nabla_X \nabla_Y Z + \sum_j \langle S_j Y, Z \rangle (\tilde{\nabla}_X N_j)^T,$$

and

$$\imath^* \tilde{\nabla}_X \tilde{\nabla}_Y \imath_* Z = \nabla_X \nabla_Y Z - \sum_j \langle S_j Y, Z \rangle S_j X.$$

A similar identity holds when interchanging $X$ and $Y$. Finally,

$$\imath^* \tilde{\nabla}_{[X,Y]} \imath_* Z = \nabla_{[X,Y]} Z,$$

and substituting these expressions in the left side of the Gauss equation yields the right side.  □

EXAMPLES AND REMARKS 4.1. (i) The Gauss equation may be expressed in terms of the curvature operators $\rho, \tilde{\rho}$ of $M, \tilde{M}$, cf. Definition 3.1: Extend $S_j$ to a linear map $S_j : \Lambda_2 M_p \to \Lambda_2 M_p$, where $S_j(x \wedge y) := S_j x \wedge S_j y$ for a decomposable element $x \wedge y$. If we define $\imath^* \tilde{\rho} : \Lambda_2 M_p \to \Lambda_2 M_p$ by

$$\langle (\imath^* \tilde{\rho}) x \wedge y, w \wedge z \rangle := \langle \tilde{\rho}(\imath_* x \wedge \imath_* y), \imath_* w \wedge \imath_* z \rangle,$$

then

(4.2)      $\langle R(x,y)z, w \rangle - \langle \tilde{R}(\imath_*x, \imath_*y)\imath_*z, \imath_*w \rangle = \langle (\rho - \imath^* \tilde{\rho}) x \wedge y, w \wedge z \rangle.$

But by the Gauss equation, the left side of (4.2) equals

$$\sum_{j=1}^{m} \langle S_j y, z \rangle \langle S_j x, w \rangle - \langle S_j x, z \rangle \langle S_j y, w \rangle = \sum_j \langle S_j x \wedge S_j y, w \wedge z \rangle$$

$$= \left\langle \left( \sum_j S_j \right) x \wedge y, w \wedge z \right\rangle.$$

Thus,

(4.3)                        $\rho - \imath^* \tilde{\rho} = \sum_{j=1}^{m} S_j.$

(ii) When $M$ is 2-dimensional and $\tilde{M} = \mathbb{R}^3$, sectional and Gauss curvature coincide. More generally, when $\dim \tilde{M} - \dim M = 1$, there are two choices $\pm n \in \tilde{M}^{\perp}_{\imath(p)}$ for a unit normal vector at $\imath(p)$. Although $S_{-n} = -S_n$ as endomorphisms of $M_p$, $S_{-n} = S_n$ as endomorphisms of $\Lambda_2 M_p$, so that $S_{\pm n}$ extends to a unique linear map $S : \Lambda_2 M_p \to \Lambda_2 M_p$, and (4.3) reads

$$\rho - \imath^* \tilde{\rho} = S.$$

If $N$ is a normal field of unit length extending $n$, then $\langle \tilde{\nabla}_X N, N \rangle = \frac{1}{2} X \langle N, N \rangle = 0$, so that $\tilde{\nabla}_X N$ is tangential, and $\imath_* S_N X = -\tilde{\nabla}_X N$ measures the amount by which $N$ fails to be parallel in direction $X$.

(iii) (The sectional curvature of the round sphere of radius $r$). Let $S^n_r = \{p \in \mathbb{R}^{n+1} \mid |p| = r\}$. If $P$ is the position vector field on $\mathbb{R}^{n+1}$ given by $P(p) = \mathcal{J}_p p$, then $N := \frac{1}{r} P$ is a unit normal vector field when restricted to $S^n_r$. Since

$$\nabla_x P = \nabla_x \left( \sum u^i D_i \right) = \sum x(u^i) D_i = x,$$

the second fundamental tensor with respect to $N$ is given by $S_N x = -x/r$. The Gauss equation then yields

$$R(x,y)z = \frac{1}{r^2}(\langle y, z \rangle x - \langle x, z \rangle y),$$

and $S^n_r$ is a space of constant sectional curvature $1/r^2$.

EXERCISE 116. A Riemannian submanifold $M$ of $\tilde{M}$ is said to be *totally geodesic* if the geodesics of $M$ are also geodesics of $\tilde{M}$. Prove that $M$ is totally geodesic iff all the second fundamental tensors vanish.

EXERCISE 117. Show that a connected, complete submanifold of $\mathbb{R}^n$ is totally geodesic iff it is an affine subspace.

EXERCISE 118. Use the Gauss equations to show that the paraboloid of revolution $z = x^2 + y^2$ in $\mathbb{R}^3$ has curvature $K = 4/(4z+1)^2$.

EXERCISE 119. An $n$-dimensional submanifold of $\mathbb{R}^{n+1}$ is called a *hypersurface*. A hypersurface $M^n$ is said to be *(strictly) convex* if at any point of $M$ the second fundamental tensor with respect to a unit normal is definite; i.e., if all its eigenvalues have the same sign. This makes sense because there are exactly two such tensors, one being the negative of the other. Prove that a a strictly convex hypersurface of Euclidean space has positive definite curvature operator, and in particular positive sectional curvature.

## 5. Riemannian Submersions

Recall that a map $\pi : M^n \to B^k$ between manifolds $M$ and $B$ is a *submersion* if $\pi$ and $\pi_{*p}$ are onto for all $p \in M$. Submersions are topologically dual to immersions in the sense that both have derivatives of maximal rank, and thus generalize diffeomorphisms.

Before studying the interplay between submersions and metrics, we investigate the problem of lifting curves of $B$ to $M$. Let $c : I = [a, b] \to B$ be a regular curve; i.e., $\dot{c}(t) \neq 0$ for $t \in [a, b]$. Define

$$c^* M := \{(t, p) \in I \times M \mid \pi(p) = c(t)\},$$

and endow $c^* M$ with the subspace topology.

LEMMA 5.1. $c^* M$ is a differentiable manifold of dimension $n - k + 1$.

PROOF. We construct a chart $h$ in a neighborhood of $(t_0, p) \in c^* M$. Denote by $p_1 : c^* M \to I$, $p_2 : c^* M \to M$ the respective projections, and similarly for $\pi_1 : \mathbb{R}^n = \mathbb{R}^k \times \mathbb{R}^{n-k} \to \mathbb{R}^k$ and $\pi_2 : \mathbb{R}^n \to \mathbb{R}^{n-k}$. Since $\pi$ has maximal rank at $p$, there exist, by Theorem 6.1 in Chapter 1, charts $x : U \to x(U) = W \subset \mathbb{R}^n$ around $p$, and $y : V \to y(V) \subset \mathbb{R}^k$ around $\pi(p)$ such that $y \circ \pi \circ x^{-1} = \pi_{1|W}$. Define $h : (x \circ p_2)^{-1}(W) \to I \times \mathbb{R}^{n-k}$ by $h := (p_1, \pi_2 \circ x \circ p_2)$. We claim that $h$ is a homeomorphism onto its image: To see this, let $z$ be the map defined on the image $\operatorname{im} h$ of $h$ by

$$z(t, a_1, \ldots, a_{n-k}) = (t, x^{-1}((y \circ c)(t), a_1, \ldots, a_{n-k})).$$

Notice that $z(\operatorname{im} h) \subset c^* M$; i.e.,

(5.1)                                    $\pi \circ p_2 \circ z = c \circ p_1 \circ z.$

In fact,

$$y \circ \pi \circ p_2 \circ z(t, a_1, \ldots, a_{n-k}) = (y \circ \pi) \circ x^{-1}((y \circ c)(t), a_1, \ldots, a_{n-k})$$
$$= \pi_1((y \circ c)(t), a_1, \ldots, a_{n-k})$$
$$= (y \circ c)(t).$$

On the other hand, $y \circ c \circ p_1 \circ z(t, a_1, \ldots, a_{n-k}) = (y \circ c)(t)$. Since $y$ is a homeomorphism, $\pi \circ p_2 \circ z = c \circ p_1 \circ z$ as claimed, and $z(\operatorname{im} h) \subset c^* M$. Now,

$$(h \circ z)(t, a_1, \ldots, a_{n-k}) = h(t, x^{-1}((y \circ c)(t), a_1, \ldots, a_{n-k})) = (t, a_1, \ldots, a_{n-k}),$$

and

$$(z \circ h)(t, q) = z(t, (\pi_2 \circ x)(q)) = (t, x^{-1}((y \circ c)(t), (\pi_2 \circ x)(q)))$$
$$= (t, x^{-1}((y \circ \pi)(q), (\pi_2 \circ x)(q))) = (t, x^{-1}((\pi_1 \circ x)(q), (\pi_2 \circ x)(q)))$$
$$= (t, q).$$

Thus, $h$ is a homeomorphism with inverse $z$. We then obtain in the usual way an atlas with differentiable transition functions, and the projections $p_1$, $p_2$ are smooth for the induced differentiable structure.  $\square$

Suppose next that we are given a distribution $\mathcal{H}$ on $M$ that is complementary to the kernel $\mathcal{V}$ of $\pi_*$, so that $TM = \mathcal{V} \oplus \mathcal{H}$, and $\pi_* : \mathcal{H}_p \to B_{\pi(p)}$ is an isomorphism for each $p \in M$. Given a curve $c$ in $B$ as before, there exists for each $(t, p) \in c^* M$ a unique vector $Y(t, p) \in \mathcal{H}_p \subset M_p$ such that $\pi_* Y(t, p) = \dot{c}(t)$. We claim that there exists a unique vector field $X$ on $c^* M$ such that $p_{2*} X = Y$. To see this, consider a chart $h$ with inverse $z$ as above. By (5.1), $p_{2*} z_* D_1, \ldots, p_{2*} z_* D_{n-k}$ span the kernel of $\pi_*$, whereas $\pi_* p_{2*} z_* D = \dot{c}$. Thus, $Y - p_{2*} z_* D \in \mathcal{V} = \ker \pi_*$; i.e., for each $(t, p)$, $Y(t, p) \in p_{2*}(c^* M)_{(t,p)}$, and there exists some $X(t, p) \in (c^* M)_{(t,p)}$ with $p_{2*} X(t, p) = Y(t, p)$. To verify uniqueness, observe that any such vector field $X$ is $p_1$-related to the coordinate vector field $D$ on $I$: In fact, since $\pi \circ p_2 = c \circ p_1$, we have

$$c_* D \circ p_1(t, P) = c_* D(t) = \dot{c}(t) = \pi_* p_{2*} X(t, p) = c_* p_{1*} X(t, p),$$

and the claim follows from regularity of $c$. Since $\mathcal{H}$ is differentiable, so is $X$.

PROPOSITION 5.1. *Let* $\pi : M \to B$ *be a submersion,* $\mathcal{H}$ *a distribution complementary to* $\ker \pi_*$, *and* $c : [a, b] \to B$ *a regular curve. If* $M$ *is compact, then for any* $p \in \pi^{-1}(c(a))$, *there exists a unique curve* $\tilde{c} : [a, b] \to M$ *with* $\tilde{c}(a) = p$, $\pi \circ \tilde{c} = c$, *and* $\dot{\tilde{c}} \in \mathcal{H}$. *Such a curve* $\tilde{c}$ *is called the* horizontal lift *of* $c$ *at* $p$.

PROOF. Let $X$ be the vector field defined above, and consider the maximal integral curve $\gamma$ of $X$ with $\gamma(a) = (a, p)$. Now, $c^* M$ is compact because $M$ is, and by Exercise 18 in Chapter 1, $\gamma$ is defined on all of $[a, b]$. Furthermore, $X$ is $p_1$-related to $D$, so that

$$(p_1 \circ \gamma)_* D = p_{1*} \dot{\gamma} = p_{1*} \circ X \circ \gamma = D \circ (p_1 \circ \gamma),$$

and $p_1 \circ \gamma$ is an integral curve of $D$. Recalling that $(p_1 \circ \gamma)(a) = a$, we conclude that $(p_1 \circ \gamma)(t) = t$ for all $t \in [a, b]$. Define $\tilde{c} := p_2 \circ \gamma$. Then $\tilde{c}(a) = p$, $\pi \circ \tilde{c} = \pi \circ p_2 \circ \gamma = c \circ p_1 \circ \gamma = c$, and $\tilde{c}_* D = p_{2*} \dot{\gamma} = p_{2*} X \circ \gamma \in \mathcal{H}$. This establishes existence.

If $\bar{c}$ is any curve satisfying the conclusion of Proposition 5.1, then $t \mapsto (t, \bar{c}(t))$ is an integral curve of $X$, and $\bar{c} = \tilde{c}$ by uniqueness of integral curves. $\square$

Notice that when $M$ is not compact, the result is no longer necessarily true: Consider $M = \mathbb{R}^2 \setminus \{p\}$, $p \in \mathbb{R}^2$. The projection $u^1 : M \to \mathbb{R} \times 0 = \mathbb{R}$ is a submersion, but the identity curve $t \mapsto c(t) = t$ on $[u^1(p) - 1, u^1(p) + 1] \subset \mathbb{R}$ admits only a partial $\mathcal{H}$-lift at the point $p - (1, 0)$, if one takes $\mathcal{H}$ to be the span of $D_1$.

Suppose that $M$ is a Riemannian manifold, $\pi : M \to B$ a submersion. The *vertical distribution* is $\mathcal{V} := \ker \pi_*$, and the *horizontal distribution* is defined by $\mathcal{H} := \mathcal{V}^\perp$. Thus, $TM = \mathcal{H} \oplus \mathcal{V}$, and we write $e = e^h + e^v$ for the corresponding decomposition of $e \in TM$.

DEFINITION 5.1. A submersion $\pi : M^n \to B^k$ is said to be *Riemannian* if $|\pi_* e| = |e^h|$ for all $e \in TM$.

In the same way that an isometric immersion may be viewed as generalizing an isometry for $n < k$, a Riemannian submersion is the corresponding generalization to $n > k$.

Horizontal vectors will be denoted $x$, $y$, $z$, vertical ones $u$, $v$, $w$. A horizontal vector field $X$ on $M$ is said to be *basic* if it is $\pi$-related to a vector field on $B$. Thus, any vector field $X \in \mathfrak{X}B$ yields a unique basic field $\tilde{X} \in \mathfrak{X}M$ with $\pi_* \tilde{X} = X \circ \pi$.

LEMMA 5.2. *If* $X$ *is basic and* $U \in \mathfrak{X}M$ *is vertical, then* $[X, U]$ *is vertical.*

PROOF. Let $Y$ be the vector field on $B$ that is $\pi$-related to $X$. Since $U$ is $\pi$-related to 0, $\pi_*[X, U] = [Y, 0] \circ \pi = 0$. $\square$

The curvature tensors of $M$ and $B$ will be denoted $R_M$ and $R_B$; their Levi-Civita connections by $\nabla^M$ and $\nabla^B$.

LEMMA 5.3. *Suppose* $\tilde{X}, \tilde{Y} \in \mathfrak{X}M$ *are basic. Then* $(\nabla^M_{\tilde{X}} \tilde{Y})^h$ *is basic. In fact, if* $X, Y \in \mathfrak{X}B$ *are* $\pi$-*related to* $\tilde{X}, \tilde{Y}$, *then* $\nabla^B_X Y$ *is* $\pi$-*related to* $\nabla^M_{\tilde{X}} \tilde{Y}$; *i.e.,*

$$\pi_* \nabla^M_{\tilde{X}} \tilde{Y} = \nabla^B_X Y \circ \pi.$$

PROOF. This is an immediate consequence of (2.2) together with the fact that $\pi_*[\tilde{X}, \tilde{Y}] = [X, Y] \circ \pi$. □

DEFINITION 5.2. The *A-tensor* is the $(1,1)$ skew-adjoint tensor field $A :$ $\mathcal{H} \times \mathcal{H} \to \mathcal{V}$ given by

$$A_X Y = (\nabla^M_X Y)^v.$$

$A$ is clearly tensorial in the first argument; for the second one, observe that if $f \in \mathcal{F}M$, then

$$(\nabla^M_X fY)^v = f(\nabla^M_X Y)^v + (Xf)Y^v = f(\nabla^M_X Y)^v,$$

since $Y$ is horizontal. To check skew-symmetry, we may, by tensoriality of $A$, assume that $X$ is basic, so that by Lemma 5.2, $(\nabla^M_X U)^h = (\nabla^M_U X)^h$ for vertical $U$. Then

$$\langle A_X X, U \rangle = \langle \nabla^M_X X, U \rangle = -\langle X, \nabla^M_X U \rangle = -\langle X, \nabla^M_U X \rangle = -\frac{1}{2}U\langle X, X \rangle = 0,$$

and the claim follows.

The $A$-tensor represents the obstruction to the horizontal distribution being integrable: If we think of the horizontal distribution $\mathcal{H}$ as generalizing the notion of connection, then the $A$-tensor represents a multiple of the curvature, since

$$2A_X Y = (\nabla^M_X Y)^v + (\nabla^M_X Y)^v = (\nabla^M_X Y)^v - (\nabla^M_Y X)^v = [X, Y]^v.$$

Given $x \in TM$ or $TB$, denote by $\gamma_x$ the geodesic $\gamma_x(t) = \exp(tx)$. The next proposition says that geodesics which start out horizontally remain so for all time. Thus, even though $\mathcal{H}$ is not, in general, integrable, the geodesic spray of $M$ restricts to a vector field tangent to $\mathcal{H}$.

PROPOSITION 5.2. *If* $x \in \mathcal{H}$, *then* $\dot{\gamma}_x(t) \in \mathcal{H}$ *for all* $t$, *and* $\pi \circ \gamma_x = \gamma_{\pi_* x}$.

PROOF. Let $x \in \mathcal{H}_p$. It clearly suffices to prove the statement in a neighborhood of $p$; choosing this neighborhood to be compact, Proposition 5.1 guarantees the existence of a horizontal lift $c$ of $\gamma_{\pi_* x}$ at $p$. Extend $\dot{\gamma}_{\pi_* x}$ locally to a vector field $X$: For example, choose a neighborhood $U$ of $0 \in B_{\pi(p)}$ on which $\exp_{\pi(p)} : U \to V := \exp(U)$ is a diffeomorphism. Given $b = \exp(u) \in V$, define $X(b)$ to be the parallel translate of $x$ along the geodesic $t \mapsto \exp(tu)$, $0 \le t \le 1$. If $\tilde{X}$ is the basic lift of $X$, then by Lemma 5.3,

$$\pi_*(\nabla^M_D \dot{c}) = \pi_*(\nabla^M_D (\tilde{X} \circ c)) = \pi_*(\nabla^M_{\tilde{X}} \tilde{X} \circ c) = \nabla^B_X X \circ \gamma_{\pi_* x} = \nabla^B_D \dot{\gamma}_{\pi_* x} = 0,$$

so that $(\nabla^M_D \dot{c})^h = 0$. Similarly,

$$(\nabla^M_D \dot{c})^v = (\nabla^M_D (\tilde{X} \circ c))^v = (\nabla^M_{\tilde{X}} \tilde{X})^v \circ c = A_{\tilde{X}} \tilde{X} \circ c = 0.$$

Thus, the horizontal lift $c$ of $\gamma_{\pi_* x}$ is a geodesic, which implies that $c = \gamma_x$, thereby concluding the argument. □

In order to describe the relation between the curvatures of $M$ and $B$, it is convenient to introduce the pointwise adjoint $A_x^* : \mathcal{V} \to \mathcal{H}$ of $A_x : \mathcal{H} \to \mathcal{V}$. Notice that for basic $X$ and vertical $U$,

$$(5.2) \qquad (\nabla_X^M U)^h = (\nabla_U^M X)^h = -A_X^* U.$$

Indeed, if $\{X_i\}$ denotes a local orthonormal basis of horizontal fields, then for *any* horizontal $X$,

$$(\nabla_X^M U)^h = \sum_i \langle \nabla_X^M U, X_i \rangle X_i = -\sum_i \langle U, \nabla_X^M X_i \rangle X_i = -\sum_i \langle A_X X_i, U \rangle X_i$$

$$= -\sum_i \langle A_X^* U, X_i \rangle X_i = -A_X^* U.$$

When furthermore $X$ is *basic*, then $(\nabla_X^M U)^h = (\nabla_U^M X)^h$ by Lemma 5.2.

PROPOSITION 5.3. *Let $\tilde{X}, \tilde{Y}, \tilde{Z} \in \mathfrak{X}M$ be basic, and denote by $X, Y, Z \in \mathfrak{X}B$ the corresponding $\pi$-related vector fields on $B$. Then*

(1) $\pi_* R_M(\tilde{X}, \tilde{Y})\tilde{Z} = R_B(X, Y)Z \circ \pi + \pi_*(-A_{\tilde{X}}^* A_{\tilde{Y}} \tilde{Z} - A_{\tilde{Y}}^* A_{\tilde{Z}} \tilde{X}$
$\qquad + 2A_{\tilde{Z}}^* A_{\tilde{X}} \tilde{Y})$;

(2) $\pi_* R_M(\tilde{X}, \tilde{Y})\tilde{Y} = R_B(X, Y)Y \circ \pi + 3\pi_* A_{\tilde{Y}}^* A_{\tilde{X}} \tilde{Y}$; *and*

(3) *if $x, y \in \mathcal{H}_p$ are orthonormal, then $K_{\pi_* x, \pi_* y}^B = K_{x,y}^M + 3|A_x y|^2$.*

PROOF. Statements (2) and (3) are direct consequences of (1). For (1), we have that

$$R_M(\tilde{X}, \tilde{Y})\tilde{Z} = \nabla_{\tilde{X}}^M \nabla_{\tilde{Y}}^M \tilde{Z} - \nabla_{\tilde{Y}}^M \nabla_{\tilde{X}}^M \tilde{Z} - \nabla_{[\tilde{X},\tilde{Y}]}^M \tilde{Z}$$

$$= \nabla_{\tilde{X}}^M (\nabla_{\tilde{Y}}^M \tilde{Z})^h - \nabla_{\tilde{Y}}^M (\nabla_{\tilde{X}}^M \tilde{Z})^h - (\nabla_{\widetilde{[X,Y]}}^M \tilde{Z})^h$$

$$\quad + \nabla_{\tilde{X}}^M (\nabla_{\tilde{Y}}^M \tilde{Z})^v - \nabla_{\tilde{Y}}^M (\nabla_{\tilde{X}}^M \tilde{Z})^v - 2(\nabla_{A_X Y}^M \tilde{Z})^h$$

$$\quad - (\nabla_{\widetilde{[X,Y]}}^M \tilde{Z})^v - 2(\nabla_{A_X Y}^M \tilde{Z})^v,$$

where $\widetilde{[X,Y]}$ denotes the basic lift of $[X, Y]$. Take horizontal components on both sides. The first line in the second equality is then basic by Lemma 5.3 and $\pi$-related to $R_B(X, Y)Z$, whereas the third line vanishes. Applying $\pi_*$ to both sides and using (5.2) on the second line yields the result. $\qquad \square$

EXAMPLES AND REMARKS 5.1. (i) A Riemannian manifold $M$ is said to be *complete* if the geodesic spray of its Levi-Civita connection is complete; i.e., if $\exp_p$ is defined on all $M_p$ for each $p \in M$. We will also implicitly require that a complete manifold be connected. Let $\pi : M \to B$ be a submersion with $M$ complete, so that $B$ is also complete by Exercise 118 below. Then all the "fibers" $\pi^{-1}(b)$, $b \in B$, are diffeomorphic: In fact, if $c : [0, a] \to B$ is a geodesic, and $F_0$, $F_a$ denote the fibers over the endpoints of $c$, then the map $h_c$ from $F_0$ to $F_a$ which assigns to $p \in F_0$ the point $\tilde{c}_p(a)$, where $\tilde{c}_p$ is the horizontal lift of $c$ with $\tilde{c}_p(0) = p$, is a diffeomorphism. The claim then follows from the Hopf-Rinow theorem in Section 7, which guarantees that any two points of $B$ can be joined by a geodesic.

Given a Riemannian metric on $B$, $b \in B$, consider an open neighborhood $U$ of $b$ which is the diffeomorphic image under $\exp_b$ of some neighborhood of

$0_b$ in $B_b$. If $\gamma_m : [0,1] \to U$ denotes the geodesic from $b$ to $m$ in $U$, we obtain a diffeomorphism

$$h : \pi^{-1}(b) \times U \to \pi^{-1}(U),$$
$$(p,m) \mapsto h_{\gamma_m}(p).$$

In particular, the restriction of $\pi$ to $\pi^{-1}(U)$ is a fibration. By a standard result in topology, the submersion $\pi : M \to B$ itself is then a fibration.

(ii) Proposition 5.3 implies that if $\pi : M \to B$ is a Riemannian submersion and the sectional curvature $K_M$ of $M$ is nonnegative, then $B$ has nonnegative curvature as well. Apart from convex hypersurfaces in Euclidean space (those for which the second fundamental tensor $S_N$ with respect to a global outward-pointing normal field $N$ is nonnegative definite) and Lie groups with bi-invariant metrics, virtually all known examples of nonnegatively curved manifolds are constructed by means of submersions. One of the largest classes of such examples consists of the base spaces of *homogeneous submersions*, which we now describe: Let $M$ be a Riemannian manifold with $K_M \geq 0$. Suppose $G$ is a subgroup of the isometry group of $M$ that acts freely and properly on $M$, so that the orbit space $B := M/G$ admits a differentiable structure for which the projection $\pi : M \to B$ becomes a submersion according to Theorem 14.2 in Chapter 1. We claim that there exists a unique Riemannian metric on $B$ for which $\pi$ becomes Riemannian: Given $b \in B$, choose any $p \in \pi^{-1}(b)$, and define an inner product on $B_b$ by requiring that $\pi_{*p} : (\ker \pi_{*p})^\perp = \mathcal{H}_p \to B_b$ be a linear isometry. To see that this inner product is well-defined, consider some other point $q \in \pi^{-1}(b)$. Then there exists some $g \in G$ with $g(p) = q$, and since $g$ is an isometry which leaves $\pi^{-1}(b)$ invariant, $g_{*p}$ maps $\mathcal{H}_p$ isometrically onto $\mathcal{H}_q$. Now consider $x \in B_b$. If $y \in \mathcal{H}_p$ is the vector that gets mapped to $x$ by $\pi_*$, then $g_* y$ is the vector in $\mathcal{H}_q$ that is mapped to $x$, since $\pi \circ g = \pi$. But $|y| = |g_* y|$, so that the inner product is well-defined. Uniqueness, on the other hand, is immediate.

(iii) (Normal homogeneous spaces). Let $G$ be a Lie group, $H$ a closed subgroup of $G$, so that $\pi : G \to G/H$ is a principal $H$-bundle over $M := G/H$. Suppose that $G_e = \mathfrak{g}$ admits an inner product that is $\mathrm{Ad}_h$-invariant for all $h \in H$. By Examples and Remarks 1.1(ii), this inner product induces a left-invariant metric on $G$ that is right-invariant under $H$. Thus, $G/H$ is the orbit space of the free isometric action of $\{R_h \mid h \in H\}$ on $G$, and by (ii), there exists a unique Riemannian metric on $G/H$ such that $\pi$ becomes a Riemannian submersion. By Examples and Remarks 1.1(iii), the metric on $M$ is $G$-invariant.

We now consider the special case when the metric on $G$ is bi-invariant. The Riemannian manifold $M$ is then called a *normal homogeneous space*. $K_G \geq 0$ by Exercise 113, so that $M$ has nonnegative sectional curvature. Notice that any $X \in \mathfrak{h}^\perp$ is basic: In fact, the vector field $\bar{X}$ on $M$, defined by

$$\bar{X}(\pi(g)) = \mathbb{L}_{g*}\pi_* X(e)$$

is $\pi$-related to $X$ by Examples and Remarks 1.1(iii). Furthermore, $A_X Y = \frac{1}{2}[X,Y]^v$ is the projection $\frac{1}{2}[X,Y]^{\mathfrak{h}}$ of $\frac{1}{2}[X,Y]$ onto $\mathfrak{h}$ for $X, Y \in \mathfrak{h}^\perp$.

Consider orthonormal $X, Y \in \mathfrak{h}^{\perp}$. By (2.4),

$$\langle R(X,Y)Y, X \rangle = -\frac{1}{4}\langle [[X,Y],Y], X \rangle = \frac{1}{4}|[X,Y]|^2,$$

and the sectional curvature of the plane in $TM$ spanned by $\pi_* X$, $\pi_* Y$ is given by

$$K_{\pi_* X, \pi_* Y} = \frac{1}{4}|[X,Y]|^2 + \frac{3}{4}|[X,Y]^{\mathfrak{h}}|^2, \qquad X, Y \in \mathfrak{h}^{\perp}.$$

(iv) (Curvature of complex projective space). The restriction $N$ of the position vector field $P$ of $\mathbb{R}^{n+2}$, $P(p) = \mathcal{J}_p p$, to the unit sphere $S^{2n+1}$ is a unit normal field to the sphere. Identify $\mathbb{R}^{2n+2}$ with $\mathbb{C}^{n+1}$ via

$$(x_1, y_1, \ldots, x_{n+1}, y_{n+1}) \mapsto (x_1 + iy_1, \ldots, x_{n+1} + iy_{n+1}),$$

and consider the canonical complex structure $I$ on $\tau \mathbb{R}^{n+2}$ given by

$$I(\mathcal{J}_p v) = \mathcal{J}_p(iv), \qquad p, v \in \mathbb{C}^{n+1}.$$

Notice that $IN$ is a unit vector field on the sphere that spans the fibers of the Hopf fibration $\pi : S^{2n+1} \to \mathbb{C}P^n$. Moreover, $I$ is a parallel section of $\mathrm{End}(\tau \mathbb{R}^{n+2})$, so that

(5.3) $$\nabla_x IN = I\nabla_x N = Ix, \qquad x \in TS^{2n+1}.$$

The covariant derivatives in (5.3) are the Levi-Civita connection of Euclidean space, but since $Ix$ and $IN$ are both tangent to the sphere, the first covariant derivative also represents the Levi-Civita connection of the sphere.

The Hopf action of $S^1$ on $S^{2n+1}$ is by isometries, so that by (ii), there exists a unique metric on complex projective space for which the Hopf fibration becomes a Riemannian submersion. Since $IN$ is a unit field spanning the vertical distribution,

$$|A_x y|^2 = \langle A_x y, IN \rangle^2 = \langle y, A_x^* IN \rangle^2 = \langle y, (\nabla_x IN)^{\mathfrak{h}} \rangle^2 = \langle y, Ix \rangle^2$$

for horizontal $x$ and $y$. Here, we used the fact that if $x$ is horizontal, then so is $Ix$: In fact, given any $x \in TS^{2n+1}$, $\langle Ix, IN \rangle = -\langle x, I^2 N \rangle = \langle x, N \rangle = 0$. By Proposition 5.3(3),

$$K_{\pi_* x, \pi_* y} = 1 + 3\langle y, Ix \rangle^2$$

for orthonormal $x, y \in \mathcal{H}$. Thus, the sectional curvature $K$ of $\mathbb{C}P^n$ satisfies $1 \leq K \leq 4$. For any horizontal $x$, the plane spanned by $x$ and $Ix$ projects down to a plane of curvature 4 (such a plane is sometimes called a *holomorphic plane*), whereas the plane spanned by $x$ and any vector orthogonal to both $x$ and $Ix$ projects to a plane of curvature 1.

(v) Cheeger and Gromoll [11] have shown that every complete, noncompact Riemannian manifold $M$ with sectional curvature $K \geq 0$ contains a compact, totally geodesic submanifold $S$, called a *soul* of $M$. Since $S$ is totally geodesic, it too has nonnegative curvature by the Gauss equations. Furthermore, $M$ is diffeomorphic to the total space of the normal bundle $\nu(S)$ of $S$ in $M$. In particular, $M$ is homotopy equivalent to a compact manifold of nonnegative curvature. Let $\pi_\nu : E(\nu(S)) \to S$ denote the bundle projection, and $\exp_\nu :$

$E(\nu(S)) \to M$ the restriction of the exponential map of $\tau(M)$ to $\nu(S)$. Perelman [30] showed that there is a well-defined map $\pi : M \to S$ such that the diagram

$$
\begin{array}{ccc}
E(\nu(S)) & \xrightarrow{\exp_\nu} & M \\
\pi_\nu \downarrow & & \downarrow \pi \\
S & = & S
\end{array}
$$

commutes, and that $\pi$ is a Riemannian submersion. Notice that $\pi$ maps $p \in M$ to the unique point $\pi(p) \in S$ that is connected to $p$ by a geodesic orthogonal to $S$; in particular, the fibers of the submersion are totally geodesic at the soul. In Section 7, we will see that $\pi(p)$ is the point of $S$ that is closest to $p$, which justifies calling $\pi$ the *metric projection* onto the soul.

(vi) If $\imath : M \to N$ is an immersion of $M$ into a Riemannian manifold $N$, there exists a unique metric on $M$ for which $\imath$ becomes isometric. The dual problem for submersions may be phrased as follows: Let $\pi : M \to B$ be a submersion. If $M$ is a Riemannian manifold, does there exist a metric on $B$ for which $\pi$ becomes Riemannian? Clearly, such a metric is unique if it exists.

Notice that the vertical and horizontal distributions are still defined, as is the $A$-tensor. A necessary condition for $\pi$ to be Riemannian is that $A$ be skew-symmetric. We will show that this condition is in fact sufficient. Observe that the discussion in (ii) is a special case: If $B$ is the orbit space of an isometric action on $M$, then for any vertical Killing field $U$,

$$
\langle A_X Y, U \rangle = \langle \nabla_X Y, U \rangle = -\langle \nabla_X U, Y \rangle = \langle \nabla_Y U, X \rangle = -\langle \nabla_Y X, U \rangle
$$
$$
= \langle -A_Y X, U \rangle
$$

by skew-symmetry of $X \mapsto \nabla_X U$. $A$ is then also skew-symmetric because the vertical distribution is spanned by Killing fields.

To establish the above claim, consider the fiber $N$ of $\pi$ over some point of $B$. The *Bott connection* on the normal bundle $\nu$ of $N$ in $M$ is defined by

$$
\nabla_u^b X = [U, \bar{X}]^h, \qquad u \in TN, \quad X \in \Gamma\nu,
$$

where $U$ and $\bar{X}$ are vertical and horizontal fields extending $u$ and $X$ respectively. It is straightforward to verify that $\nabla^b$ is indeed a well-defined connection on $\nu$. For $Y \in \Gamma\nu$,

$$
\langle \nabla_u^b X, Y \rangle = \langle [U, \bar{X}], Y \rangle = \langle \nabla_u \bar{X}, Y \rangle - \langle \nabla_{\bar{X}} U, Y \rangle = \langle \nabla_u X, Y \rangle + \langle u, A_X Y \rangle,
$$

so that

$$
\langle \nabla_u^b X, Y \rangle + \langle X, \nabla_u^b Y \rangle = \langle \nabla_u X, Y \rangle + \langle X, \nabla_u Y \rangle + \langle u, A_X Y + A_Y X \rangle
$$
$$
= \langle \nabla_u X, Y \rangle + \langle X, \nabla_u Y \rangle
$$
$$
= u \langle X, Y \rangle,
$$

and $\nabla^b$ is Riemannian. The curvature tensor $R^b$ of the Bott connection is given by

$$\begin{aligned}
R^b(U,V)X &= \nabla^b_U \nabla^b_V X - \nabla^b_V \nabla^b_U X - \nabla^b_{[U,V]} X \\
&= [U,[V,\bar{X}]^h]^h - [V,[U,\bar{X}]^h]^h - [[U,V],\bar{X}]^h \\
&= [U,[V,\bar{X}]]^h + [V,[\bar{X},U]]^h + [\bar{X},[U,V]]^h \\
&= 0
\end{aligned}$$

by the Jacobi identity and the fact that vertical fields have vertical brackets. Thus, $\nu$ is a flat bundle, and in fact a trivial one: For any $x \in B_{\pi(N)}$, the assignment

$$p \mapsto X(p) := (\pi_{*|\mathcal{H}_p})^{-1} x, \qquad p \in N,$$

defines a global section $X$ of $\nu$ that is $\pi$-related to $x$. This section is then Bott-parallel, since any $U \in \mathfrak{X}N$ is $\pi$-related to 0, so that $\nabla^b_U X = [U,\bar{X}]^h = 0$. In particular, $X$ has constant norm along $N$. If we now define $|x| := |X|$, then $\pi$ becomes a Riemannian submersion.

EXERCISE 120. Let $M \to B$ be a Riemannian submersion. Prove that $B$ is complete whenever $M$ is.

EXERCISE 121. Show that the Riemannian submersions in Examples (iii) and (iv) have totally geodesic fibers.

EXERCISE 122. View $\mathbb{R}^3$ as $\mathbb{C} \times \mathbb{R}$, and consider the free isometric action of $\mathbb{R}$ on $\mathbb{R}^3$ given by $t(z,t_0) = (e^{it}z, t_0 + t)$, $t \in \mathbb{R}$, $(z,t_0) \in \mathbb{C} \times \mathbb{R}$. Compute the sectional curvature of the space $M = \mathbb{R}^3/\mathbb{R}$ of orbits, if $M$ is endowed with the metric for which the projection $\pi : \mathbb{R}^3 \to M$ becomes a Riemannian submersion. This example also shows that homogeneous submersions do not, in general, have totally geodesic fibers, see Figure 1.

EXERCISE 123. Use Example (ii) to construct a metric of nonnegative sectional curvature on the total space $TS^n = SO(n+1) \times_{SO(n)} \mathbb{R}^n$ of the tangent bundle of $S^n$.

EXERCISE 124. Prove that the Bott connection from Examples and Remarks 5.1(vi) is a well-defined connection. Explain why it is not even necessary to consider an extension $\bar{X}$ of $X$ in the definition.

EXERCISE 125. Let $M$ be a Riemannian manifold, $f : M \to \mathbb{R}$ a function that has maximal rank everywhere. Show that if $|\nabla f| \cong 1$, then $f$ is a Riemannian submersion with respect to the usual metric on $\mathbb{R}$.

## 6. The Gauss Lemma

One of the fundamental properties of geodesics is that they are, at least locally, length-minimizing, as we shall see in the next section. Since a geodesic $c$ is the image via the exponential map of a ray $t \mapsto tv$ in the tangent bundle, it is to be expected that such extremal properties follow from the behavior of the derivative of exp. This derivative can be conveniently expressed in terms of certain vector fields along $c$. If $Y$ is a vector field along a curve $c$, we will often abbreviate the covariant derivative $\nabla_D Y$ by $Y'$.

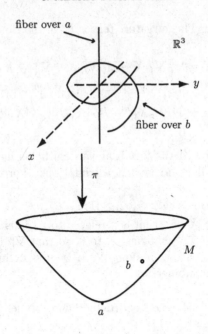

FIGURE 1

DEFINITION 6.1. Let $c$ be a geodesic in $M$. A vector field $Y$ along $c$ is called a *Jacobi field* along $c$ if

$$Y'' + R(Y, \dot{c})\dot{c} = 0.$$

Notice that the collection $\mathcal{J}_c$ of Jacobi fields along $c$ is a vector space that contains $\dot{c}$. The space of Jacobi fields orthogonal to $\dot{c}$ is the one of interest to us: If $X$ and $Y$ are Jacobi, then

$$\langle Y'', X \rangle = -\langle R(Y, \dot{c})\dot{c}, X \rangle = -\langle R(X, \dot{c})\dot{c}, Y \rangle = \langle X'', Y \rangle$$

by Proposition 3.1. Thus, $\langle X'', Y \rangle - \langle Y'', X \rangle = 0$, and $\langle X', Y \rangle - \langle Y', X \rangle$ must be constant. In particular, $\langle Y, \dot{c} \rangle' = \langle Y', \dot{c} \rangle = \langle Y', \dot{c} \rangle - \langle Y, \dot{c}' \rangle$ is constant, so that for a normal geodesic, the tangential component $Y^T$ of $Y$ is given by

$$Y^T = \langle Y, \dot{c} \rangle \dot{c} = (a + bt)\dot{c}, \qquad a = \langle Y, \dot{c} \rangle(0), \quad b = \langle Y, \dot{c} \rangle'(0),$$

and satisfies the Jacobi equation. It follows that the component $Y^\perp = Y - Y^T$ of $Y$ orthogonal to $\dot{c}$ is also a Jacobi field.

PROPOSITION 6.1. *Let* $c : I \to M$ *be a geodesic*, $t_0 \in I$. *For any* $v, w \in M_{c(t_0)}$ *there exists a unique Jacobi field* $Y$ *along* $c$ *with* $Y(t_0) = v$ *and* $Y'(t_0) = w$.

PROOF. Let $X_1, \ldots, X_n$ be parallel fields along $c$ such that $X_1(t_0), \ldots, X_{n-1}(t_0)$ form an orthonormal basis of $\dot{c}(t_0)^\perp$, and $X_n = \dot{c}$. Any vector field $Y$ along $c$ can then be expressed as

$$Y = \sum_i f^i X_i, \qquad f^i = \begin{cases} \langle Y, X_i \rangle, & \text{for } i \leq n-1, \\ \langle Y, \frac{X_n}{|X_n|^2} \rangle, & \text{for } i = n. \end{cases}$$

Since $X_i$ is parallel, $Y'' = \sum f^{i''} X_i$. Furthermore, $R(X_i, \dot{c})\dot{c} = \sum_{j=1}^{n-1} h_i^j X_j$, where $h_i^j = \langle R(X_i, \dot{c})\dot{c}, X_j \rangle$, so that $R(Y, \dot{c})\dot{c} = \sum_{i,j=1}^{n-1} f^i h_i^j X_j$. The Jacobi equation then reads

$$\sum_{j=1}^{n-1} \left( f^{j''} + \sum_{i=1}^{n-1} f^i h_i^j \right) X_j = 0, \qquad f^{n''} = 0,$$

or equivalently,

$$f^{j''} + \sum_{i=1}^{n-1} f^i h_i^j = 0, \qquad j = 1, \ldots, n,$$

if we set $h_i^n = \langle R(X_i, \dot{c})\dot{c}, \dot{c} \rangle = 0$. This is a homogeneous system of n linear second-order equations, which has a unique solution for given initial values $f^j(t_0) = \langle v, X_j(t_0) \rangle$, $f^{j'}(t_0) = \langle w, X_j(t_0) \rangle$ $(j < n)$, $f^n(t_0) = \langle v, (\dot{c}/|\dot{c}|^2)(t_0) \rangle$, and $f^{n'}(t_0) = \langle w, (\dot{c}/|\dot{c}|^2)(t_0) \rangle$. $\qquad\square$

Proposition 6.1 implies that the space $\mathcal{J}_c$ of Jacobi fields along $c$ is $2n$-dimensional, since the map

$$\mathcal{J}_c \to M_{c(t_0)} \times M_{c(t_0)},$$
$$Y \mapsto (Y(t_0), Y'(t_0))$$

is an isomorphism.

EXAMPLE 6.1. Let $M^n$ be a space of constant curvature $\kappa$, and let $c_\kappa$, $s_\kappa$ denote the solutions of the differential equation

$$f'' + \kappa f = 0$$

with $c_\kappa(0) = 1$, $c_\kappa'(0) = 0$, $s_\kappa(0) = 0$, $s_\kappa'(0) = 1$. For example, $c_1 = \cos$, and $s_1 = \sin$. Consider a normal geodesic $c : [0, b] \to M$. Given $v, w \in M_{c(0)}$ orthogonal to $\dot{c}(0)$, the Jacobi field $Y$ along $c$ with $Y(0) = v$ and $Y'(0) = w$ is given by

$$Y = c_\kappa E + s_\kappa F,$$

where $E$ and $F$ are the parallel fields along $c$ with $E(0) = v$ and $F(0) = w$: Indeed, $Y'' = c_\kappa'' E + s_\kappa'' F = -\kappa Y = -R(Y, \dot{c})\dot{c}$, so that $Y$ is a Jacobi field, and clearly satisfies the initial conditions at 0.

Jacobi fields essentially arise out of variations of geodesics: If $c : [a, b] \to M$ is a curve, and $I$ is an interval containing 0, a *variation* of $c$ is a smooth homotopy $V : [a, b] \times I \to M$ with $V(t, 0) = c(t)$ for $t \in [a, b]$. Notice that $V_* D_1(t, 0) = \dot{c}(t)$; the *variational vector field* $Y$ along $c$ is defined by $Y(t) = V_* D_2(t, 0)$.

PROPOSITION 6.2. *Let $c : [0, b] \to M$ be a geodesic. If $V$ is a variation of $c$ through geodesics—meaning that $t \mapsto V(t, s)$ is a geodesic for each $s$, then the variational vector field $t \mapsto V_* D_2(t, 0)$ is Jacobi along $c$. Conversely, let $Y$ be a Jacobi field along $c$. Then there exists a variation $V$ of $c$ through geodesics whose variational vector field equals $Y$.*

PROOF. Given a variation $V$ of $c$ through geodesics, define vector fields $\tilde{X}$ and $\tilde{Y}$ along $V$ by $\tilde{X} = V_* D_1$, $\tilde{Y} = V_* D_2$. By assumption, $\nabla_{D_1} \tilde{X} = 0$, so that

$$R(\tilde{Y}, \tilde{X})\tilde{X} = \nabla_{D_2}\nabla_{D_1}\tilde{X} - \nabla_{D_1}\nabla_{D_2}\tilde{X} = -\nabla_{D_1}\nabla_{D_2}\tilde{X}$$
$$= -\nabla_{D_1}\nabla_{D_2}V_*D_1 = -\nabla_{D_1}\nabla_{D_1}V_*D_2$$
$$= -\nabla_{D_1}\nabla_{D_1}\tilde{Y}.$$

When $s = 0$, the above expression becomes $R(Y, \dot{c})\dot{c} = -Y''$, and $Y$ is Jacobi.

Conversely, suppose $Y$ is a Jacobi field along $c$, and $v := Y(0)$, $w := Y'(0)$. Let $\gamma$ be a curve with $\dot{\gamma}(0) = v$, and $X$, $W$ parallel fields along $\gamma$ with $X(0) = \dot{c}(0)$, $W(0) = w$. Choose $\epsilon > 0$ small enough so that $t(X(s) + sW(s))$ belongs to the domain of $\exp_{\gamma(s)}$ for $(t, s) \in [0, b] \times (-\epsilon, \epsilon)$, and consider the variation

$$V : [0, b] \times (-\epsilon, \epsilon) \to M,$$
$$(t, s) \mapsto \exp_{\gamma(s)} t(X(s) + sW(s))$$

of $c$. Since the curves $t \mapsto V(t, s)$ are geodesics, the variational vector field $Z$ is Jacobi along $c$. Moreover, $V(0, s) = \gamma(s)$, so that $Z(0) = \dot{\gamma}(0) = v$. Finally,

$$Z'(0) = \nabla_{D_1(0,0)}V_*D_2 = \nabla_{D_2(0,0)}V_*D_1 = W(0) = w,$$

because $V_*D_1(0, s) = X(s) + sW(s)$, and $X$, $W$ are parallel along $\gamma$. By Proposition 6.1, $Z = Y$.                                                    □

In the special case when $Y(0) = 0$, the variation from Proposition 6.2 becomes $V(t, s) = \exp_{c(0)} t(\dot{c}(0) + sw)$; the Jacobi field $Y$ with initial conditions $Y(0) = 0$, $Y'(0) = w$ is given by

(6.1)                     $$Y(t) = \exp_{c(0)*}(t\mathcal{J}_{t\dot{c}(0)}w).$$

One can interpret this as follows: Let $p = c(0)$, $v = \dot{c}(0)$, and consider the manifold $M_p$ with the canonical Riemannian metric induced by the inner product on $M_p$. Then $t \mapsto t\mathcal{J}_{tv}w$ is the Jacobi field $F$ along the geodesic $t \mapsto tv$ in $M_p$ with $F(0) = 0$, $F'(0) = \mathcal{J}_v w$, and we have $Y = \exp_* F$.

We now use the above observation to show that the exponential map at a point is "radially" isometric:

LEMMA 6.1 (The Gauss Lemma). *Consider the canonical metric on* $M_p$, $p \in M$. *If* $v \in M_p$ *belongs to the domain of* $\exp_p$, *then*

$$\langle \exp_{p*} \mathcal{J}_v v, \exp_{p*} \mathcal{J}_v w \rangle = \langle \mathcal{J}_v v, \mathcal{J}_v w \rangle, \qquad w \in M_p.$$

PROOF. The right side of the above identity equals $\langle v, w \rangle$ by definition of the canonical metric on $M_p$, whereas the left side is $\langle \dot{c}(1), Y(1) \rangle$, where $c(t) = \exp(tv)$, and $Y$ is the Jacobi field along $c$ with initial conditions $Y(0) = 0$, $Y'(0) = w$. Since $Y$ is Jacobi, $\langle Y, \dot{c} \rangle(t) = \langle Y, \dot{c} \rangle(0) + t\langle Y', \dot{c} \rangle(0) = t\langle v, w \rangle$. Evaluating this expression at $t = 1$ yields the claim.                        □

EXAMPLE 6.2. Consider a Riemannian submersion $\pi : M \to B$. As noted in Examples and Remarks 5.1(i), if $c : [0, a] \to B$ is a geodesic, and $F_0$, $F_a$ denote the fibers over the endpoints of $c$, then the map $h : F_0 \to F_a$ which assigns to $p \in F_0$ the point $\tilde{c}_p(a)$, where $\tilde{c}_p$ is the horizontal lift of $c$ with $\tilde{c}_p(0) = p$, is a diffeomorphism. To compute its derivative, let $u$ be a vector tangent to $F_0$ at $p$, $\gamma : I \to F_0$ a curve with $\dot{\gamma}(0) = u$. For simplicity of notation, identify $u$ with the vector in $M_p$ that is related to $u$ via the derivative of the inclusion $F_0 \subset M$. There is a unique basic vector field $X$ along $\gamma$ with $\pi_* X = x := \dot{c}(0)$. If $V : [0, a] \times I \to M$ denotes the variation by geodesics given by $V(t, s) = \exp(tX(s))$, then $h(\gamma(s)) = V(a, s)$ by definition. Thus,

$$h_* u = h_* \dot{\gamma}(0) = V_* D_2(a, 0) = Y(a),$$

where $Y$ is the variational vector field of $V$.

Now suppose $M$ above is a complete, noncompact manifold with curvature $K \geq 0$, $\pi : M \to S$ the Riemannian submersion onto the soul $S$ from Examples and Remarks 5.1(v). Let $p \in S$, $c : [0, a] \to S$ as above. Since $S$ is horizontal, it is an integral manifold of the horizontal distribution, and the $A$ tensor is zero along $S$. If $Z$ is any horizontal vector field along $c$, then

$$\langle Y', Z \rangle = \langle Y, Z \rangle' - \langle Y, Z' \rangle = -\langle Y, Z'^v \rangle = -\langle Y, A_{\dot{c}} Z \rangle = 0,$$

so that $Y'$ is vertical. On the other hand,

$$Y'(0) = \nabla_{D_1(0,0)} V_* D_2 = \nabla_{D_2(0,0)} V_* D_1 = \nabla_u X = -S_x u,$$

where $S$ is the second fundamental tensor field of $F_0$ at $p$ (here too, $S_x u$ is identified with a vector in $TM$). Since the fiber is totally geodesic at $p$, $Y'(0) = 0$. The same argument shows that $Y'(t) = 0$ for all $t$; i.e., $Y$ is parallel. Thus, $R(Y, \dot{c})\dot{c} = -Y'' = 0$.

Summarizing, we have that at any point $p$ in a soul, the curvature of the plane spanned by $x \in S_p$ and $u \perp S_p$ is zero. In particular, if $M$ has strictly positive curvature, then a soul must have trivial tangent space, and therefore consists of a single point. Since the normal bundle of a point $p$ is $M_p$, $M^n$ is then diffeomorphic to $\mathbb{R}^n$. The above argument can actually be refined to show that the conclusion is still valid if one only assumes that all sectional curvatures are positive at *some* point of $M$. This remarkable fact was conjectured by Cheeger and Gromoll [11], and proved more than twenty years later by Perelman [30].

Another famous question of Cheeger and Gromoll is whether *every* vector bundle over $S^n$ admits a metric of nonnegative curvature. At the time of writing, the answer to this question is not known, although all bundles over spheres of dimension $\leq 4$ have been shown to admit such metrics, as does the tangent bundle of the $n$-dimensional sphere, see Exercise 114. This is no longer true if one allows flat souls instead of positively curved ones: Among the plane bundles over the torus $S^1 \times S^1$, only the trivial one admits a metric of nonnegative curvature [29].

EXERCISE 126. Let $c : [0, b] \to M$ be a geodesic. $t_0 \in [0, b]$ is said to be a *conjugate point* of $c$ if there exists a nontrivial Jacobi field $Y$ along $c$ with $Y(0) = 0$, $Y(t_0) = 0$. Prove that $t_0$ is a conjugate point of $c$ iff $\exp_{c(0)}$ does not have maximal rank at $t_0 \dot{c}(0)$. Show, furthermore, that the nullity of $\exp_{*c(0)}$ equals the dimension of the vector space of Jacobi fields $Y$ that vanish at 0 and at $t_0$.

EXERCISE 127. (a) Use Example 6.1 to show that if $c$ is a normal (i.e., unit-speed) geodesic in $S^n$, then every Jacobi field $Y$ along $c$, orthogonal to $\dot{c}$, with $Y(0) = 0$ vanishes at $\pi$. Thus, $\exp_p$ has minimal rank 1 at any $v \in M_p$ with $|v| = \pi$ by Exercise 126.

(b) Let $M$ be a Riemannian manifold with sectional curvature $K \leq 0$, $c$ a geodesic in $M$. Prove that if $Y$ is Jacobi along $c$, then the function $|Y|^2$ is convex. Why does this imply that in a manifold of nonpositive curvature, geodesics have no conjugate points?

EXERCISE 128. Let $X$ be a Killing field on $M$ (cf. Exercise 111), $c$ a geodesic. Prove that $X \circ c$ is Jacobi along $c$. *Hint:* Recall that the flow $\Phi_t$ of $X$ consists of isometries. Consider the variation $(t, s) \mapsto \Phi_s(c(t))$ of $c$.

EXERCISE 129. Let $\pi : M \rightarrow B$ denote a Riemannian submersion with $M$ complete, $c : [0, a] \rightarrow B$ a geodesic, and $F_0$, $F_a$ the fibers of $\pi$ over the endpoints of $c$. Consider the diffeomorphism $h : F_0 \rightarrow F_a$ given by $h(p) = c_p(a)$, where $c_p$ is the horizontal lift of $c$ starting at $p$. Prove that for $u \in (F_0)_p \subset M_p$, $h_* u = Y(a)$, where $Y$ is the Jacobi field along $c_p$ with initial conditions

$$Y(0) = u, \qquad Y'(0) = -S_x u - A_x^* u.$$

Conclude that if the fibers of $\pi$ are totally geodesic, then they are all isometric to one another.

## 7. Length-Minimizing Properties of Geodesics

In this section, we will see that geodesics are the shortest curves between two points, provided the latter are "close enough." The curves we consider are assumed to be piecewise-smooth, but thanks to the following lemma, we only need to deal with differentiable ones:

LEMMA 7.1. *Any piecewise-smooth curve $c : [a, b] \rightarrow M$ admits a differentiable reparametrization $\tilde{c} : [a, b] \rightarrow M$.*

PROOF. By assumption, there exist numbers $a = t_0 < t_1 < \cdots < t_k = b$ such that each portion $c_i := c|_{[t_{i-1}, t_i]}$ is differentiable. Let $\phi_i : [a, b] \rightarrow \mathbb{R}$ be a smooth function such that $\phi_i(t) = 0$ if $t \leq t_{i-1}$, $\phi_i(t) = 1$ if $t \geq t_i$, and $\phi_i$ is strictly increasing on $[t_{i-1}, t_i]$, cf. Lemma 1.1 in Chapter 1. Then the function $\phi := t_0 + \sum_{i=1}^{k} (t_i - t_{i-1})\phi_i$ is differentiable, strictly increasing on $[a, b]$, and satisfies $\phi(t_i) = t_i$, $\phi^{(m)}(t_i) = 0$, for all $i = 0, \ldots, k$ and $m \in \mathbb{N}$. It follows that $\tilde{c} := c \circ \phi$ is a smooth reparametrization of $c$. $\square$

Consider a curve $\gamma : I \rightarrow M_p$ in the tangent space at $p \in M$, where $M_p$ is endowed with its canonical Riemannian structure. If $\gamma(t) \neq 0$ for all $t$, we may write $\dot{\gamma} = \dot{\gamma}_r + \dot{\gamma}_\theta$, with

$$\dot{\gamma}_r = \frac{1}{|\gamma|^2} \langle \dot{\gamma}, \mathcal{J}_\gamma \gamma \rangle \mathcal{J}_\gamma \gamma \quad \text{and} \quad \dot{\gamma}_\theta = \dot{\gamma} - \dot{\gamma}_r.$$

$\dot{\gamma}_r$ and $\dot{\gamma}_\theta$ are called the *radial* and *polar* components of $\dot{\gamma}$ respectively. Notice that they are mutually orthogonal. When $\gamma$ is a ray, i.e., $\gamma(t) = tv$ for some $v \in M_p$, then $\dot{\gamma} = \dot{\gamma}_r$, and $\gamma$ is length-minimizing. The following lemma says that this property is preserved under the exponential map:

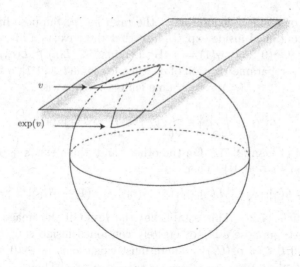

FIGURE 2

LEMMA 7.2. *Let $p \in M$, and consider a vector $v$ in the domain $\tilde{M}_p$ of $\exp_p$. Denote by $\gamma_v : [0,1] \to \tilde{M}_p$ the ray from 0 to $v$, $\gamma_v(t) = tv$. If $\gamma : [0,1] \to \tilde{M}_p$ is any piecewise-smooth curve with $\gamma(0) = \gamma_v(0) = 0$ and $\gamma(1) = \gamma_v(1) = v$, then*

$$L(\exp \circ \gamma) \geq L(\exp \circ \gamma_v).$$

*Inequality is strict if there is some $t_0 \in [0,1]$ for which the polar component of $\dot{\gamma}$ at $t_0$ does not vanish under the exponential map; i.e., if $\exp_{p*} \dot{\gamma}_\theta(t_0) \neq 0$.*

PROOF. We may assume, by Lemma 7.1, that $\gamma$ is differentiable, and that both $v$ and $\gamma(t)$ are nonzero for $t \in [0,1]$. By the Gauss Lemma,

$$(7.1) \qquad |\exp_{p*} \dot{\gamma}|^2 = |\exp_{p*} \dot{\gamma}_r|^2 + |\exp_{p*} \dot{\gamma}_\theta|^2 \geq |\exp_{p*} \dot{\gamma}_r|^2 = |\dot{\gamma}_r|^2.$$

Now, $|\gamma|' = |\dot{\gamma}_r|$, because

$$|\gamma|' = \langle \gamma, \gamma \rangle^{1/2'} = \frac{\langle \gamma, \gamma' \rangle}{|\gamma|} = \frac{\langle J_\gamma \gamma, \dot{\gamma} \rangle}{|\gamma|} = |\dot{\gamma}_r|.$$

Thus,

$$L(\exp_p \circ \gamma) = \int_0^1 |\exp_{p*} \dot{\gamma}| \geq \int_0^1 |\dot{\gamma}_r| = \int_0^1 |\gamma|' = |\gamma(1)| = |v| = L(\exp_p \circ \gamma_v).$$

The last assertion of the lemma is clear, since the inequality in (7.1) is strict if $\exp_{p*} \dot{\gamma}_\theta \neq 0$. $\qquad \square$

THEOREM 7.1. *Let $p \in M$, and choose $\epsilon > 0$ so that $\exp_p : U_\epsilon \to \exp_p(U_\epsilon)$ is a diffeomorphism, where $U_\epsilon = \{v \in M_p \mid |v| < \epsilon\}$ is the open ball of radius $\epsilon$ centered at $0 \in M_p$. For $v \in U_\epsilon$, denote by $c_v : [0,1] \to M$ the geodesic in direction $v$, $c_v(t) = \exp_p(tv)$. Then for any piecewise-smooth curve $c : [0,1] \to M$ with $c(0) = c_v(0) = p$ and $c(1) = c_v(1) = q$, the length of $c$ is at least as great as that of $c_v$, and is strictly greater unless $c$ equals $c_v$ up to reparametrization.*

PROOF. For $u \in U_\epsilon$, denote by $\gamma_u$ the ray $t \mapsto tu$. Suppose first that the image of $c$ is contained inside $\exp_p(U_\epsilon)$, so that there exists a lift $\gamma$ of $c$ in $U_\epsilon$; i.e., $c = \exp_p \circ \gamma$, $\gamma(0) = 0$, $\gamma(1) = v$. By Lemma 7.2, $L(c) \geq L(c_v)$. We claim that if $c$ is not a reparametrization of $c_v$, then for some $t \in [0,1]$, $\dot\gamma$ is not radial (and therefore $L(c) > L(c_v)$ by the same lemma): Otherwise,

$$(7.2) \qquad L(\gamma) = \int_0^1 |\dot\gamma| = \int_0^1 |\dot\gamma_r| = \int_0^1 |\gamma|' = |\gamma(1)|$$

as in the proof of Lemma 7.2. On the other hand, there exists $t_0 \in (0,1)$ such that $\gamma(t_0) \notin \{sv \mid s \in [0,1]\}$. Then

$$L(\gamma) = L(\gamma|_{[0,t_0]}) + L(\gamma|_{[t_0,1]}) \geq |\gamma(t_0)| + |\gamma(1) - \gamma(t_0)| > |\gamma(1)|,$$

which contradicts (7.2). This establishes the result if the image of $c$ lies in $\exp_p(U_\epsilon)$. Next, suppose $c$ is not entirely contained inside $\exp_p(U_\epsilon)$, and let $b = \sup\{t \mid c[0,t] \subset \exp_p(U_\epsilon)\}$. There must exist some $t_0 \in (0,b)$ such that $v_0 := (\exp_p|_{U_\epsilon})^{-1} c(t_0)$ has norm greater than that of $v$. Then

$$L(c) \geq L(c|_{[0,t_0]}) \geq |v_0| > |v| = L(c_v).$$

$\square$

Recall that a distance function on a set $M$ is a function $d : M \times M \to \mathbb{R}$ satisfying

  (1)  $d(p,q) = d(q,p)$,
  (2)  $d(p,q) \geq 0$, and $= 0$ iff $p = q$, and
  (3)  $d(p,q) \leq d(p,r) + d(r,q)$,    (the triangle inequality)

for all $p, q, r \in M$. When $d$ is a distance function on a topological space $M$, $(M,d)$ is called a *metric space* if the open balls $B_\epsilon(p) = \{q \in M \mid d(p,q) < \epsilon\}$ of radius $\epsilon > 0$ around $p \in M$ form a basis for the topology of $M$.

Suppose $M$ is a connected Riemannian manifold, and define a map $d : M \times M \to \mathbb{R}$ by

$$(7.3) \qquad d(p,q) = \inf\{L(c) \mid c : [0,1] \to M, c(0) = p, c(1) = q\},$$

where $c$ is assumed to be piecewise-smooth.

PROPOSITION 7.1. *Let $M$ be a connected Riemannian manifold. If $d$ is the function defined in (7.3), then $(M,d)$ is a metric space.*

PROOF. We first check that $d$ is a distance function on $M$. The only axiom that requires some work is: If $d(p,q) = 0$, then $p = q$. So consider the ball $U_\epsilon$ around $0$ in $M_p$, where $\epsilon > 0$ is small enough so that $\exp_p : U_\epsilon \to \exp_p(U_\epsilon)$ is a diffeomorphism. Then $q \in \exp_p(U_\epsilon)$, since any curve from $p$ which is not contained in $\exp_p(U_\epsilon)$ has length greater than or equal to $\epsilon$ by the proof of Theorem 7.1. By that same theorem,

$$(7.4) \qquad d(p,r) = |(\exp_p|_{U_\epsilon})^{-1} r|, \qquad r \in \exp_p(U_\epsilon).$$

Thus, $q = \exp_p 0 = p$.

We show next that the distance function is continuous on $M \times M$: If $p_n \to p$, then $d(p,p_n) \to 0$ by (7.4) and the continuity of $|(\exp_p|_{U_\epsilon})^{-1}|$. Now, if $(p_n, q_n) \to (p, q) \in M \times M$, then

$$d(p, q) - d(p, p_n) - d(q, q_n) \le d(p_n, q_n) \le d(p_n, p) + d(p, q) + d(q, q_n)$$

by the triangle inequality, and $d(p_n, q_n) \to d(p, q)$. Since $d$ is continuous, each $\epsilon$-ball $B_\epsilon(p)$ around $p \in M$ is open. It remains to show that any neighborhood of $p$ contains such a ball. This in turn will follow once we establish that

(7.5)                                    $$B_\epsilon(p) = \exp_p(U_\epsilon),$$

where, as usual, $\epsilon$ is chosen so that $\exp_p |_{U_\epsilon}$ is an imbedding. Now, $\exp_p(U_\epsilon) \subset B_\epsilon(p)$ by (7.4). Furthermore, $\exp_p(U_\epsilon)$ is open. In order to prove (7.5), it therefore remains to show that $\exp_p(U_\epsilon)$ is closed in $B_\epsilon(p)$, since the latter, being path-connected, is also connected. So let $q_n \in \exp_p(U_\epsilon)$, $q_n \to q \in B_\epsilon(p)$, $v_n := (\exp_p |_{U_\epsilon})^{-1}(q_n)$. $\{v_n\}$ is a bounded sequence in the vector space $M_p$, and contains a subsequence that converges to, say, $v$. Denoting the subsequence by $v_n$ again, we have $|v| = \lim |v_n| = \lim d(p, q_n) = d(p, q)$ by continuity of $d$, so that $|v| < \epsilon$. But $q = \exp_p v$ by continuity of the exponential map, and we conclude that $q \in \exp_p(U_\epsilon)$; i.e., $\exp_p(U_\epsilon)$ is closed in $B_\epsilon(p)$.  $\square$

DEFINITION 7.1. The *injectivity radius* $\mathrm{inj}_p$ at $p \in M$ is the supremum of the set of all $\epsilon > 0$ for which $U_\epsilon$ lies in the domain of $\exp_p$, and $\exp_p |_{U_\epsilon}$ is injective. The injectivity radius of a subset $A \subset M$ is

$$\mathrm{inj}_A = \inf\{\mathrm{inj}_p \mid p \in A\}.$$

LEMMA 7.3. *If $A \subset M$ is compact, then $\mathrm{inj}_A > 0$.*

PROOF. By Theorem 4.1 in Chapter 4, there exists an open neighborhood $U$ of the zero section $\{0_p \mid p \in A\}$ in $TM|_A$ on which $(\pi, \exp)$ is an imbedding. By compactness of $A$, there exist $p_1, \ldots, p_k \in A$ and $\epsilon_1, \ldots, \epsilon_k > 0$ such that the sets $B_{\epsilon_i/3}(p_i)$ cover $A$ and $B_{\epsilon_i}(p_i) \times B_{\epsilon_i}(p_i) \subset (\pi, \exp)(U)$. If $q$ is an arbitrary point of $A$, then $q \in B_{\epsilon_i/3}(p_i)$ for some $i$. Then $\exp_q$ is invertible on $U_{2\epsilon_i/3}$, so that $\mathrm{inj}_q \ge 2\epsilon_i/3$. Thus, $\mathrm{inj}_A \ge \epsilon$, where $\epsilon := \min_{1 \le i \le k}\{2\epsilon_i/3\} > 0$.  $\square$

A geodesic $c : [0, b] \to M$ is said to be *minimal* if $d(c(0), c(b)) = L(c)$. Recall that $U_\epsilon \subset M_p$ denotes the open ball of radius $\epsilon$ centered at $0_p \in M_p$.

LEMMA 7.4. *If $U_\epsilon$ is contained in the domain of $\exp_p$ for $\epsilon > 0$, then $p$ can be joined to any $q \in B_\epsilon(p)$ by a minimal geodesic. In particular, $\exp_p : U_\epsilon \to B_\epsilon(p)$ is surjective. Furthermore, each open ball $B_\delta(p)$ in $M$ has compact closure for $0 < \delta < \epsilon$.*

PROOF. For each $\delta \in (0, \epsilon)$, denote by $C_\delta$ the subset of $\overline{B_\delta(p)}$ consisting of all points $q$ that can be joined to $p$ by a minimal geodesic. The lemma will follow once we establish that

(1) $C_\delta$ is compact, and
(2) $C_\delta = \overline{B_\delta(p)}$.

For (1), we only need to show that $C_\delta$ is closed, since it is contained in the compact set $\exp_p(\overline{U_\delta})$. So let $q$ belong to the closure of $C_\delta$, and consider a sequence $\{q_n\}$ in $C_\delta$ converging to $q$. If $v_n$ denotes a vector in $\overline{U_\delta}$ with $|v_n| = d(p, q_n)$ and $\exp v_n = q_n$, $n \in \mathbb{N}$, then $\{v_n\}$ contains a subsequence that converges to some

$v$ in $\overline{U_\delta}$. By continuity of $\exp_p$, $\exp_p v = q$, and by continuity of $d$, $|v| = d(p, q)$. Thus, $q \in C_\delta$, and $C_\delta$ is closed.

For (2), consider the set $I = \{\delta \in (0, \epsilon) \mid C_\delta = \overline{B_\delta(p)}\}$. $I$ contains a neighborhood of 0 in $(0, \epsilon)$ by Theorem 7.1. We will show that $I$ is both open and closed in $(0, \epsilon)$, so that $I = (0, \epsilon)$, thereby establishing (2). To see that $I$ is closed, let $\delta_0 \in (0, \epsilon)$ be such that $\delta \in I$ for all $\delta < \delta_0$. Then $C_\delta = \overline{B_\delta(p)}$ for $\delta < \delta_0$, and thus, $B_{\delta_0}(p) \subset C_{\delta_0}$. Since $C_{\delta_0}$ is compact, $\overline{B_{\delta_0}(p)} \subset C_{\delta_0}$. The reverse inclusion is always true by definition, so that $I$ is closed. To see that $I$ is open, let $\delta_0 \in I$. By Lemma 7.3, there exists $\alpha > 0$ such that $\mathrm{inj}_{C_{\delta_0}} = \alpha$. Let $\alpha' = \min\{\alpha, \epsilon - \delta_0\} > 0$. We claim that $\delta_0 + \alpha' \in I$ (so that $I$ being an interval, is open). It actually suffices to show that $B_{\delta_0 + \alpha'}(p) \subset C_{\delta_0 + \alpha'}$, since the latter set is closed. Furthermore, we know that $\overline{B_{\delta_0}(p)} \subset C_{\delta_0 + \alpha'}$, so we only need to establish that

$$B_{\delta_0 + \alpha'}(p) \setminus \overline{B_{\delta_0}(p)} \subset C_{\delta_0 + \alpha'}.$$

Consider to this end a point $q$ belonging to the left side, and a sequence of curves $c_n : [a, b] \to M$ from $p$ to $q$ such that $L(c_n) < d(p, q) + 1/n$. By the intermediate value theorem, there exists $t_n \in (a, b)$ with $d(p, c_n(t_n)) = \delta_0$, and the sequence $r_n := c_n(t_n)$ subconverges to some $r$ at distance $\delta_0$ from $p$. Then

$$(7.6) \qquad\qquad d(p, q) = d(p, r) + d(r, q).$$

Indeed, $L(c_n) \geq d(p, r_n) + d(r_n, q)$, so that by construction of $c_n$,

$$\frac{1}{n} + d(p, q) \geq d(p, r_n) + d(r_n, q).$$

Furthermore, for each $n \in \mathbb{N}$, there exists $k > n$ such that $d(r_k, r) < 1/n$. Then

$$\begin{aligned}
\frac{3}{n} + d(p, q) &\geq \frac{2}{n} + d(p, r_k) + d(r_k, q) \\
&> d(p, r_k) + d(r_k, r) + d(r, r_k) + d(r_k, q) \\
&\geq d(p, r) + d(r, q).
\end{aligned}$$

Since this holds for arbitrary $n$, $d(p, q) \geq d(p, r) + d(r, q)$, which proves (7.6). Now, $r \in \overline{B_{\delta_0(p)}}$, so there exists a minimal geodesic $c_0$ from $p$ to $r$. There is also a minimal geodesic $c_1$ joining $r$ to $q$ because $d(r, q) < \alpha'$ by (7.6). The composition $c_0 * c_1$ is then a piecewise-smooth curve from $p$ to $q$ realizing the distance between them. By Theorem 7.1, $c_0 * c_1$ is locally (and hence globally), up to reparametrization, a normal minimal geodesic from $p$ to $q$. $\qquad\square$

We defined earlier a complete Riemannian manifold $M$ to be one with complete Levi-Civita connection, so that geodesics of $M$ are defined on all of $\mathbb{R}$. On the other hand, $(M, d)$ is a metric space, and such a space is said to be *complete* if every Cauchy sequence in $M$ converges (a sequence $\{p_n\}$ is said to be a Cauchy sequence if for any $\epsilon > 0$, there exists some natural number $N$ such that $d(p_n, p_m) < \epsilon$ whenever $n, m > N$). The next fundamental result implies that both concepts of completeness agree:

THEOREM 7.2 (Hopf-Rinow). *Let $M$ be a connected Riemannian manifold with distance function $d$. The following statements are equivalent:*

(1) *$M$ is complete as a metric space.*

(2) *For each $p \in M$, $\exp_p$ is defined on all of $M_p$.*

(3) *Any bounded set $B \subset M$ is relatively compact; i.e., $B$ has compact closure.*

*Furthermore, any one of the above conditions implies:*

(4) *Any two points in $M$ are joined by a minimal geodesic.*

PROOF. (3) $\Rightarrow$ (1) Since a Cauchy sequence is bounded, it lies inside a compact set, and therefore converges.

(1) $\Rightarrow$ (2) Let $v \in TM$, and consider the maximal integral curve $\gamma : I \to TM$ of the geodesic spray $S$ with $\gamma(0) = v$, so that $\exp(tv) = (\pi \circ \gamma)(t)$. We claim that $[0, \infty) \subset I$. Now $I$ is open, contains a neighborhood of 0, so it suffices to show that $I$ is closed relative to $[0, \infty)$; i.e., if $[0, s) \subset I$, then $[0, s] \subset I$. By Exercise 18 in Chapter 1, we only need to consider a sequence $\{t_n\}$ in $[0, s)$ that converges to $s$, and establish that $v_n := \gamma(t_n)$ subconverges in $TM$; but the sequence $p_n := \pi(v_n)$ is Cauchy, because

$$d(p_n, p_m) = d(\exp(t_n v), \exp(t_m v)) \leq |t_n - t_m| v.$$

Thus, $\{p_n\}$ converges to some $p \in M$. Since $|v_n| = |v|$, the sequence $\{v_n\}$ lies inside the total space of some sphere bundle $\pi : S_r(TM) \to M$ of radius $r = |v|$ over $M$. In fact, since $\pi(v_n) \to p$, $\{\pi(v_n)\}$ lies inside some compact neighborhood $C$ of $p$. Then $\{v_n\}$ is contained inside the compact set $\pi^{-1}(C)$, and must have a convergent subsequence.

(2) $\Rightarrow$ (3) This follows from Lemma 7.4, since any bounded set $B$ must be contained in a metric ball $B_r(p)$ for large enough $r$. If $B_r(p)$ has compact closure, so does $B$.

(2) $\Rightarrow$ (4) Immediate from Lemma 7.4.                      $\square$

EXERCISE 130. Let $M_1$ be a Riemannian submanifold of $M_2$, and denote by $d_i$ the distance function on $M_i$.

(a) Prove that $d_1(p, q) \geq d_2(p, q)$ for $p, q$ in $M_1$, and give an example where strict inequality occurs.

(b) Show that $M_1$ is totally geodesic in $M_2$ iff $d_1 = d_2$ locally.

EXERCISE 131. Determine $\text{inj}_M$ and $\text{inj}_p$ for $p \in M$, if
(a) $M = \mathbb{R}^n$,      (b) $M = \mathbb{R}^n \setminus \{0\}$,      (c) $M = S^n$.

EXERCISE 132. Let $M$ be complete.
(a) Show that a closed Riemannian submanifold of $M$ is complete.
(b) Given $p \in M$, $A \subset M$, define

$$d(p, A) := \inf\{d(p, q) \mid q \in A\}.$$

If $d(p, A) = 0$, does it necessarily follow that $p \in A$? Show that if $A$ is closed, then there exists a point $q \in A$ such that $d(p, q) = d(p, A)$.

EXERCISE 133. Let $\pi : M \to B$ be a Riemannian submersion.
(a) By Exercise 118, if $M$ is complete, then so is $B$. Is the converse always true?
(b) Given $B, C \subset M$, define

$$d(B, C) := \inf\{d(p, q) \mid p \in B, q \in C\}.$$

Prove that if $M$ is complete, then the fibers $\pi^{-1}(b_1)$, $\pi^{-1}(b_2)$ over any two points $b_1$, $b_2$ in $B$ are equidistant; i.e., for $p_i \in \pi^{-1}(b_i)$,

$$d(\pi^{-1}(b_1), \pi^{-1}(b_2)) = d(p_1, \pi^{-1}(b_2)) = d(p_2, \pi^{-1}(b_1)).$$

EXERCISE 134. Show that statement (2) in Theorem 7.2 is equivalent to:
(2') There exists some $p \in M$ for which $\exp_p$ is defined an all of $M_p$. *Hint:* Show that (2') implies completeness of $M$: Given a Cauchy sequence $\{q_n\}$ in $M$, let $v_n \in M_p$ so that $\exp_p v_n = q_n$, and $|v_n| = d(p, q_n)$. Observe that $\{v_n\}$ is bounded.

## 8. First and Second Variation of Arc-Length

Consider a normal (i.e., unit-speed) geodesic $c : [0, a] \to M$ in a Riemannian manifold $M$, and a variation $V : [0, a] \times J \to M$ of $c$, where $J$ is an interval containing 0. We assume that $c = V \circ \imath_0$, where $\imath_s : [0, a] \to [0, a] \times J$ maps $t$ to $(t, s)$ for $s \in J$. The *length function* $L : J \to \mathbb{R}$ of $V$ is defined by

$$L(s) = L(V \circ \imath_s) = \int_0^a |V_* D_1(t, s)| \, dt;$$

i.e., $L(s)$ is the length of the curve $t \mapsto V(t, s)$. Let $Y$ be the variational vector field along $c$ given by $Y(t) = V_* D_2(t, 0)$, $Y_\perp = Y - \langle Y, \dot{c} \rangle$ its component orthogonal to $c$. The derivative $\nabla_D Y$ will be denoted $Y'$. The following theorem will enable us to compare the length of $c$ with that of nearby curves:

THEOREM 8.1. *With notation as above,*
(1) $L'(0) = \langle Y, \dot{c} \rangle(t)|_0^a$, *and*
(2) $L''(0) = (\int_0^a |Y_\perp'|^2 - \langle R(Y_\perp, \dot{c})\dot{c}, Y_\perp \rangle) + \langle \nabla_{D_2} V_* D_2(t, 0), \dot{c}(t) \rangle|_0^a.$

PROOF. By continuity of $|V_* D_1|$ and compactness of $[0, a]$, there exists an interval $I \subset J$ around 0 such that $V_* D_1(t, s) \neq 0$ for all $(t, s) \in [0, a] \times I$. Then $L$ is differentiable on $I$, and for $s \in I$,

$$L'(s) = \int_0^a D_2 \langle V_* D_1, V_* D_1 \rangle^{1/2}(t, s) \, dt = \int_0^a \frac{\langle \nabla_{D_2} V_* D_1, V_* D_1 \rangle}{|V_* D_1|}(t, s) \, dt.$$

Since $\nabla_{D_2} V_* D_1 = \nabla_{D_1} V_* D_2$,

$$(8.1) \qquad L'(s) = \int_0^a \frac{\langle \nabla_{D_1} V_* D_2, V_* D_1 \rangle}{|V_* D_1|}(t, s) \, dt.$$

When $s = 0$, $|V_* D_1| = |\dot{c}| = 1$, and

$$\langle \nabla_{D_1} V_* D_2, V_* D_1 \rangle = \langle Y', \dot{c} \rangle = \langle Y, \dot{c} \rangle' - \langle Y, \nabla_D \dot{c} \rangle = \langle Y, \dot{c} \rangle',$$

which establishes (1).
Differentiating (8.1) yields

$$L''(s) = \int_0^a D_2 \frac{\langle \nabla_{D_1} V_* D_2, V_* D_1 \rangle}{|V_* D_1|}(t, s) \, dt,$$

and the integrand may be rewritten as

$$(|V_*D_1|D_2\langle\nabla_{D_1}V_*D_2, V_*D_1\rangle - \langle\nabla_{D_1}V_*D_2, V_*D_1\rangle D_2|V_*D_1|)/|V_*D_1|^2$$

$$= \frac{\langle\nabla_{D_2}\nabla_{D_1}V_*D_2, V_*D_1\rangle + \langle\nabla_{D_1}V_*D_2, \nabla_{D_2}V_*D_1\rangle}{|V_*D_1|}$$

$$- \frac{\langle\nabla_{D_1}V_*D_2, V_*D_1\rangle\langle\nabla_{D_2}V_*D_1, V_*D_1\rangle}{|V_*D_1|^3}$$

$$= \frac{\langle\nabla_{D_2}\nabla_{D_1}V_*D_2, V_*D_1\rangle + |\nabla_{D_1}V_*D_2|^2}{|V_*D_1|} - \frac{\langle\nabla_{D_1}V_*D_2, V_*D_1\rangle^2}{|V_*D_1|^3}.$$

When $s = 0$, $Y(t) = V_*D_2(t, 0)$. $Y'_\perp = Y' - \langle Y', \dot{c}\rangle\dot{c}$, and

(8.2)
$$|Y'_\perp|^2(t) = (|Y'|^2 - \langle Y', \dot{c}\rangle^2)(t) = |\nabla_{D_1}V_*D_2|^2(t, 0) - \langle\nabla_{D_1}V_*D_2, V_*D_1\rangle^2(t, 0).$$

Furthermore,

$$\langle R(Y_\perp, \dot{c})\dot{c}, Y_\perp\rangle(t) = \langle R(\dot{c}, Y)Y, \dot{c}\rangle(t)$$

$$= ((\langle\nabla_{D_1}\nabla_{D_2}V_*D_2, V_*D_1\rangle - \langle\nabla_{D_2}\nabla_{D_1}V_*D_2, V_*D_1\rangle))(t, 0)$$

$$= (D_1\langle\nabla_{D_2}V_*D_2, V_*D_1\rangle - \langle\nabla_{D_2}\nabla_{D_1}V_*D_2, V_*D_1\rangle)(t, 0),$$

so that

$$\langle\nabla_{D_2}\nabla_{D_1}V_*D_2, V_*D_1\rangle(t, 0) = D_1(t, 0)\langle\nabla_{D_2}V_*D_2, V_*D_1\rangle - \langle R(Y_\perp, \dot{c})\dot{c}, Y_\perp\rangle(t).$$

Together with (8.2), and the fact that $|V_*D_1|(t, 0) = 1$, this implies

$$L''(0) = \int_0^a (D_1(t, 0)\langle\nabla_{D_2}V_*D_2, V_*D_1\rangle) - \langle R(Y_\perp, \dot{c})\dot{c}, Y_\perp\rangle(t) + |Y'_\perp|^2(t)\, dt,$$

which establishes (2).                                                         □

The following is an immediate consequence:

COROLLARY 8.1. *Suppose* $V : [0, a] \times J \to M$ *is a variation of* $c$ *with fixed endpoints; i.e.,* $V(0, s) = c(0)$ *and* $V(a, s) = c(a)$ *for all* $s \in J$. *Then*

$$L'(0) = 0, \qquad L''(0) = \int_0^a |Y'_\perp|^2 - \langle R(Y_\perp, \dot{c})\dot{c}, Y_\perp\rangle.$$

EXAMPLES AND REMARKS 8.1. (i) Theorem 8.1 generalizes to a larger class of variations: A continuous map $V : [0, a] \times J \to M$ is said to be a *piecewise-smooth variation of* $c$ if there exists a partition $t_0 = 0 < t_1 < \cdots < t_k = a$ of $[0, a]$ such that each $V_i := V_{|[0,a]\times[t_{i-1},t_i]}$ is a variation of $c_{[t_{i-1},t_i]}$. Define a continuous vector field $V_*D_2$ along $V$ by $V_*D_2(t, s) = V_{i*}D_2(t, s)$ for $t \in [t_{i-1}, t_i]$. Then $t \mapsto Y(t) := V_*D_2(t, 0)$ is a piecewise-smooth vector field along $c$, and $Y'$ a (not necessarily continuous) vector field along $c$, if we define $Y'(t_i) = \nabla_{D_2(t_i,0)}V_{i-1*}D_2$. Let $Y_i := Y_{[t_{i-1},t_i]}$, $Y_{i\perp} := Y_i - \langle Y_i, \dot{c}\rangle\dot{c}$. By Theorem 8.1,

$$L'(0) = \sum_{i=1}^k \langle Y_i, \dot{c}\rangle(t)|_{t_{i-1}}^{t_i} = \langle Y, \dot{c}\rangle(t)|_0^a,$$

since $Y_i(t_i) = Y_{i+1}(t_i)$, and

$$L''(0) = \sum_{i=1}^{k} \left( \int_{t_{i-1}}^{t_i} |Y'_{i\perp}|^2 - \langle R(Y_{i\perp}, \dot c)\dot c, Y_{i\perp} \rangle \right) + \langle \nabla_{D_2} V_{i*} D_2(t,0), \dot c(t) \rangle|_{t_{i-1}}^{t_i}$$

$$= \left( \int_0^a |Y'_\perp|^2 - \langle R(Y_\perp, \dot c)\dot c, Y_\perp \rangle \right) + \langle \nabla_{D_2} V_{k*} D_2(a,0), \dot c(a) \rangle$$

$$- \langle \nabla_{D_2} V_{1*} D_2(0,0), \dot c(0) \rangle.$$

(ii) We saw in (i) that a piecewise-smooth variation $V$ of $c$ induces a (like-wise piecewise smooth) variational vector field $t \mapsto Y(t) := V_* D_2(t,0)$ along $c$. Conversely, any piecewise-differentiable vector field $Y$ along $c$ induces a variation $V$ of $c$ with $Y(t) = V_* D_2(t,0)$: By continuity of $Y$ and compactness of $[0,a]$, there exists an interval $J$ around 0 such that $sY(t)$ lies in the domain of exp for all $(t,s) \in [0,a] \times J$. Define $V : [0,a] \times J \to M$ by $V(t,s) = \exp(sY(t))$.

(iii) Let $V$ be a variation with fixed endpoints of a geodesic $c$ in a space of nonpositive curvature $K \leq 0$. By Corollary 8.1,

$$L'(0) = 0, \qquad L''(0) = \int_0^a \cdot |Y'_\perp|^2 - K_{\dot c, Y_\perp} |Y_\perp|^2 \geq 0,$$

so that all nearby variational curves (i.e., $V \circ \imath_s$ for sufficiently small $s$) are at least as long as $c$.

(iv) Recall from Exercise 126 that $t_0 \in [0,a]$ is said to be a conjugate point of a geodesic $c : [0,a] \to M$ if there exists a nontrivial Jacobi field $Y$ along $c$ that vanishes at 0 and $t_0$. Suppose $c : [0,a] \to M$ is a normal geodesic without conjugate points. We claim that for any piecewise-smooth variation $V : [0,a] \times J \to M$ of $c$ with fixed endpoints, all nearby curves are longer than $c$; more precisely, there is an interval $I \subset J$ around 0 such that $L(0) \leq L(s)$ for all $s \in I$, and $L(0) < L(s)$ if the curves $V_s := V \circ \imath_s$ and $V_0 = c$ are not reparametrizations of each other. To see this, let $p = c(0)$. By Exercise 126, $\exp_p$ has maximal rank on the compact set $C = \{ t\dot c(0) \mid 0 \leq t \leq a \}$ in $M_p$, so that $\exp_p$ is a local diffeomorphism from a neighborhood of $C$ in $M_p$ onto a neighborhood $U$ of the image of $c$ in $M$. Choose an interval $I \subset J$ around 0 so that $V([0,a] \times I)$ is contained in $U$. Then for each $s \in I$, there exists a curve $\gamma_s$ in $M_p$ from 0 to $a\dot c(0)$ such that $\exp \circ \gamma_s = V_s$. The claim now follows from Lemma 7.2.

When, in addition, $c$ is injective, then $\exp_p$ maps a neighborhood of $C$ diffeomorphically onto a neighborhood $U$ of the image of $c$, so that any curve in $U$ from $p$ to $c(a)$ which is not a reparametrization of $c$ must be longer than $c$. There may, however, be curves from $p$ to $c(a)$ that leave $U$ and are shorter than $c$; this occurs for example on a flat torus or on a cylinder.

Denote by $\Gamma_c$ the space of piecewise-smooth vector fields $Y$ along $c$ such that $\langle Y, \dot c \rangle = 0$.

DEFINITION 8.1. The *index form* of a normal geodesic $c : [0,a] \to M$ is the symmetric bilinear form $I : \Gamma_c \times \Gamma_c \to \mathbb{R}$ given by

$$I(X,Y) = \int_0^a \langle X', Y' \rangle - \langle R(X, \dot c)\dot c, Y \rangle.$$

Notice that if $V : [0,a] \to M$ is the canonical variation of $c$ induced by some $Y \in \Gamma_c$ from Examples and Remarks 8.1, then $L'(0) = 0$, and $L''(0) = I(Y,Y)$, since each curve $s \mapsto V(t,s) = \exp(sY(t))$ is a geodesic, so that $\nabla_{D_2} V_* D_2 = 0$.

Let us denote by $\Gamma_c^0$ the subspace consisting of those $Y$ in $\Gamma_c$ that vanish at $0$ and $a$. The index form is, in general, degenerate on both spaces. In fact, the next lemma characterizes the space of Jacobi fields as the degenerate subspace of the index form on $\Gamma_c^0$:

LEMMA 8.1. For $Y \in \Gamma_c^0$, $Y$ is Jacobi if and only if $I(Y,X) = 0$ for all $X \in \Gamma_c^0$.

PROOF. If $Y$ is Jacobi, then

$$I(Y,X) = \int_0^a \langle Y', X' \rangle - \langle R(Y,\dot{c})\dot{c}, X \rangle = \int_0^a \langle Y', X' \rangle + \langle Y'', X \rangle$$
$$= \int_0^a \langle Y', X \rangle' = \langle Y', X \rangle(t)|_0^a = 0.$$

Conversely, suppose $Y \in \Gamma_c^0$ satisfies $I(Y,X) = 0$ for all $X \in \Gamma_c^0$. Assuming first that $Y$ is differentiable, we have

$$0 = I(Y,X) = \int_0^a \langle Y', X' \rangle - \langle R(Y,\dot{c})\dot{c}, X \rangle$$
$$= \langle Y', X \rangle(t)|_0^a - \int_0^a \langle Y'' + R(Y,\dot{c})\dot{c}\, X \rangle$$
$$= -\int_0^a \langle Y'' + R(Y,\dot{c})\dot{c}, X \rangle.$$

Let $f$ be a smooth function on $[0,a]$, with $f$ positive on $(0,a)$, and $f(0) = f(a) = 0$. Then $X := f(Y'' + R(Y,\dot{c})\dot{c}) \in \Gamma_c^0$, and

$$0 = I(Y,X) = -\int_0^a f|Y'' + R(Y,\dot{c})\dot{c}|^2,$$

so that $Y'' + R(Y,\dot{c})\dot{c} = 0$, and $Y$ is a Jacobi field. If $Y$ is only assumed to be piecewise-smooth, then by a similar argument, there exists a partition $0 = t_0 < t_1 < \cdots < t_n = a$ such that each $Y_i := Y_{|[t_{i-1},t_i]}$ is Jacobi. Fix $j \in \{1,\ldots,n-1\}$, and consider a vector field $X_j \in \Gamma_c^0$ such that $X_j(t_j) = Y'_{j+1}(t_j) - Y'_j(t_j)$, and $X_j(t_i) = 0$ if $i \neq j$. Since each $Y_i$ is Jacobi,

$$0 = I(Y, X_j) = \sum_{i=0}^k \langle Y'_i, X_j \rangle(t)|_{t_{i-1}}^{t_i} = -\langle Y'_{j+1}, X_j \rangle(t_j) + \langle Y'_j, X_j \rangle(t_j)$$
$$= -\langle Y'_{j+1} - Y'_j, X_j \rangle(t_j) = -|Y'_{j+1} - Y'_j|^2(t_j).$$

Thus, $Y'_j(t_j) = Y'_{j+1}(t_j)$. Since Jacobi fields are determined by their value and that of their derivative at one point, $Y$ is smooth on $[0,a]$. $\square$

THEOREM 8.2. Let $c : [0,a] \to M$ be a normal geodesic. The index form of $c$ is positive definite on $\Gamma_c^0$ if and only if $c$ has no conjugate points.

PROOF. Suppose $c$ has no conjugate points, and let $Y \in \Gamma_c^0$. By Examples and Remarks 8.1(iv), the length function of the canonical variation induced by $Y$ has a minimum at $0$, so that $I(Y,Y) = L''(0) \geq 0$. Suppose $I(Y,Y) = 0$. For any $c \in \mathbb{R}$ and $Z \in \Gamma_c^0$, $0 \leq I(Y - cZ, Y - cZ) = -2cI(Y, Z) + c^2 I(Y, Z)$. This can only happen if $I(Y, Z) = 0$ for all $Z \in \Gamma_c^0$, and by Lemma 8.1, $Y$ is a Jacobi field that vanishes at 0 and $a$. Since $c$ has no conjugate points, $Y \equiv 0$.

For the converse, suppose $c$ has a conjugate point $t_0 \in (0, a]$. Let $Y$ be a nontrivial Jacobi field that vanishes at 0 and $t_0$, and define a piecewise-smooth vector field $X$ along $c$ by $X(t) = Y(t)$ if $t \leq t_0$, $X(t) = 0$ if $t \geq t_0$. Then $I(X, X) = 0$, and $I$ is not positive definite on $\Gamma_c^0$. $\square$

COROLLARY 8.2. *If $c : [0, a] \to M$ is a normal geodesic with a conjugate point $t_0 < a$, then $c$ is not minimal.*

PROOF. It suffices to construct a variation $V$ with fixed endpoints of $c$ such that $L(s) < L(0)$ for small $s$. By hypothesis, there exists a nontrivial Jacobi field $Y$ along $c$ with $Y(0) = 0$, $Y(t_0) = 0$. Observe that $Y \in \Gamma_c$, that is, $Y$ is orthogonal to $c$: By the remark following Definition 6.1, $\langle Y, \dot{c} \rangle'$ is constant. Thus, $\langle Y, \dot{c} \rangle$ is a linear function that vanishes at 0 and at $t_0$, and must be identically zero. Now, $Y'(t_0) \neq 0$, for otherwise $Y$ would be trivial; so consider the parallel vector field $E$ along $c$ with $E(t_0) = -Y'(t_0)$, and define $X \in \Gamma_c^0$ by $X = fE$, where $f$ is a function satisfying $f(0) = f(a) = 0$, $f(t_0) = 1$. Finally, for $\alpha > 0$, let $Y_\alpha \in \Gamma_c^0$ be given by

$$Y_\alpha(t) = \begin{cases} Y(t) + \alpha X(t), & \text{if } t \leq t_0, \\ \alpha X(t), & \text{if } t \geq t_0. \end{cases}$$

Then

$$I(Y_\alpha, Y_\alpha) = \int_0^{t_0} \langle Y' + \alpha X', Y' + \alpha X' \rangle - \langle R(Y + \alpha X, \dot{c})\dot{c}, Y + \alpha X \rangle$$
$$+ \int_{t_0}^a \langle \alpha X', \alpha X' \rangle - \langle R(\alpha X, \dot{c})\dot{c}, \alpha X \rangle$$
$$= \langle Y, Y' \rangle(t)|_0^{t_0} + 2\alpha \langle X, Y' \rangle(t)|_0^{t_0} + \alpha^2 I(X, X),$$

since $Y$ is Jacobi. Thus, $I(Y_\alpha, Y_\alpha) = \alpha^2 I(X, X) - 2\alpha \langle Y', Y' \rangle(t_0) < 0$ if $\alpha$ is small enough. The canonical variation of $c$ induced by $Y_\alpha$ has 0 as a strict maximum of its length function. $\square$

EXERCISE 135. Suppose $V : [0, a] \times J \to M$ is a variation with fixed endpoints of $c$. Show that if all curves $V_s = V \circ \imath_s$ are geodesics, then they have the same length, and $a$ is a conjugate point of each and everyone of them.

EXERCISE 136. Let $c : [0, a] \to M^n$ be a geodesic with no conjugate points.

(a) Show that for any $X \in \Gamma_c$, there exists a unique Jacobi field $Y$ with $Y(0) = X(0)$ and $Y(a) = X(a)$. *Hint:* Recall that the space of Jacobi fields along $c$ is $2n$-dimensional.

(b) Prove that for $X$ and $Y$ as in (a), $I(X, X) > I(Y, Y)$ if $X \neq Y$. *Hint:* $I(X - Y, X - Y) > 0$.

EXERCISE 137. Let $c : [0, a] \to S_\rho^n$ be a normal geodesic on the $n$-sphere of radius $\rho$, and consider Jacobi fields $Y_1$, $Y_2$ along $c$ with $Y_i(0) = 0$, $Y_i'(0) \perp \dot{c}(0)$.

(a) Show that $I(Y_1, Y_2) = (\rho/2) \sin(2a/\rho) \langle Y_1'(0), Y_2'(0) \rangle$.

(b) Conclude that if $a < \pi\rho/2$, then $I$ is positive definite on the subspace of all $X \in \Gamma_c$ with $X(0) = 0$.

EXERCISE 138. Show that the converse of Corollary 8.2 does not hold; i.e., give examples of geodesics that have no conjugate points, but are not minimal.


## 9. Curvature and Topology

As an application of the index form, we shall discuss two results illustrating how the knowledge of the sectional curvatures at every point can sometimes lead to an understanding of the large-scale structure of the space. The first one states that a manifold with curvature $K \leq 0$ has Euclidean space as its universal cover. The second is that a manifold with curvature bounded from below by a positive constant has compact universal cover. For further results in this direction, the reader is invited to consult a book that deals more exclusively with Riemannian geometry. Reference [31], for example, is a modern and thorough account of the subject.

By Example 6.1, a normal geodesic in a space of constant curvature $\kappa$ has $\pi/\sqrt{\kappa}$ as conjugate point if $\kappa > 0$, and no conjugate point if $\kappa \leq 0$. This seems to suggest that the more curved the space, the earlier conjugate points occur. This is indeed the case, and it can be phrased more precisely in terms of the index form: Let $M$, $\tilde{M}$ be Riemannian manifolds of the same dimension $n \geq 2$, $p \in M$, $\tilde{p} \in \tilde{M}$. Given unit vectors $u$, $\tilde{u}$ in $M_p$, $\tilde{M}_{\tilde{p}}$ respectively, choose a linear isometry $\imath : M_p \to \tilde{M}_{\tilde{p}}$ with $\imath u = \tilde{u}$, and denote by $c : [0, a] \to M$ (resp. $\tilde{c} : [0, a] \to \tilde{M}$) the geodesic given by $c(t) = \exp_p(tu)$ (resp. $\tilde{c}(t) = \exp_{\tilde{p}}(t\tilde{u})$).

Next, we construct an isomorphism $L : \Gamma_c \to \Gamma_{\tilde{c}}$ as follows: For $t \in [0, a]$, let $L_t : M_{c(t)} \to \tilde{M}_{\tilde{c}(t)}$ denote the isometry $\tilde{P}_{0,t} \circ \imath \circ P_{t,0}$, where $P_{t,0}$ is parallel translation along $c$ from $M_{c(t)}$ to $M_{c(0)}$, and $\tilde{P}_{0,t}$ parallel translation along $\tilde{c}$ from $\tilde{M}_{\tilde{c}(0)}$ to $\tilde{M}_{\tilde{c}(t)}$.

$$
\begin{array}{ccc}
M_{c(t)} & \xrightarrow{\ L_t\ } & \tilde{M}_{\tilde{c}(t)} \\
\scriptstyle P_{t,0} \downarrow & & \uparrow \scriptstyle \tilde{P}_{0,t} \\
M_p & \xrightarrow{\ \ \imath\ \ } & \tilde{M}_{\tilde{p}}
\end{array}
$$

For $X \in \Gamma_c$, define $LX \in \Gamma_{\tilde{c}}$ by $LX(t) = L_t X(t)$. To see that $L$ is an isomorphism, consider an orthonormal basis $u_1, \ldots, u_{n-1}$ of $u^\perp$, and denote by $E_i$ the parallel vector field along $c$ with $E_i(0) = u_i$, $1 \leq i \leq n - 1$. Similarly, let $\tilde{E}_i$ be the parallel field along $\tilde{c}$ with $\tilde{E}_i(0) = \imath u_i$. Any $X \in \Gamma_c$ can be written as $X = \sum_i f_i E_i$, where $f_i = \langle X, E_i \rangle$, and by definition of $L$,

(9.1) $$LX = \sum_i f_i \tilde{E}_i.$$

It follows that $L$ is an isomorphism.

LEMMA 9.1. *With notation as above, suppose that $K_{\dot{c}(t), v} \leq K_{\dot{\tilde{c}}(t), L_t v}$ for all $t \in [0, a]$, $v \in M_{c(t)}$. Then $I(X, X) \geq I(LX, LX)$ for any $X \in \Gamma_c$.*

PROOF. By (9.1), $\langle LX, LY \rangle = \langle X, Y \rangle$, and $(LX)' = L(X')$ for all $X$, $Y \in \Gamma_c$. If $X(t) \neq 0$, then

$$\langle R(X, \dot{c})\dot{c}, X \rangle(t) = K_{\dot{c}(t), X(t)} \langle X, X \rangle(t) \leq K_{\dot{\tilde{c}}(t), LX(t)} \langle LX, LX \rangle(t)$$
$$= \langle \tilde{R}(LX, \dot{\tilde{c}})\dot{\tilde{c}}, LX \rangle(t),$$

whereas both sides are zero if $X(t) = 0$. Thus,

$$I(X, X) = \int_0^a \langle X', X' \rangle - \langle R(X, \dot{c})\dot{c}, X \rangle \geq \int_0^a \langle (LX)', (LX)' \rangle - \langle \tilde{R}(LX, \dot{\tilde{c}})\dot{\tilde{c}}, LX \rangle$$
$$= I(LX, LX).$$

$\square$

THEOREM 9.1. *Let* $c : [0, a] \to M$ *be a normal geodesic in a Riemannian manifold with sectional curvature* $K$.

(1) *If* $K \leq 0$, *then* $c$ *has no conjugate points.*
(2) *If* $K \leq \kappa$, *where* $\kappa > 0$, *and* $L(c) < \pi/\sqrt{\kappa}$, *then* $c$ *has no conjugate points.*
(3) *If* $K \geq \kappa > 0$, *and* $L(c) \geq \pi/\sqrt{\kappa}$, *then* $c$ *has a conjugate point.*

PROOF. Statement (1) follows from (2) by taking $\kappa > 0$ arbitrarily small; see also Exercise 127. For (2), let $\tilde{M}$ denote the sphere of constant curvature $\kappa$ in Lemma 9.1. For $X \in \Gamma_c^0$, $LX \in \Gamma_{\tilde{c}}^0$, and $I(X, X) \geq I(LX, LX)$. Since $\tilde{c}$ has no conjugate points, the claim follows from Theorem 8.2. Statement (3) is argued as in (2), after interchanging the roles of $M$ and $\tilde{M}$. $\square$

THEOREM 9.2 (Hadamard, Cartan). *Let* $M^n$ *be a complete Riemannian manifold with sectional curvature* $K \leq 0$. *Then the universal cover of* $M$ *is diffeomorphic to* $\mathbb{R}^n$.

PROOF. The statement follows once we show that $\exp_p : M_p \to M$ is a covering map for any $p \in M$. Since $(M, g)$ has nonpositive curvature, its geodesics have no conjugate points, and $\exp_p$ has maximal rank everywhere. Let $\tilde{M}$ denote the Riemannian manifold $M_p$ together with the metric $\exp_p^* g$. Notice that for $v \in M_p$, the ray $t \mapsto tv$ through the origin is a geodesic, since $\exp_p$ is isometric. By Exercise 134, $\tilde{M}$ is complete. Given $q \in M$ and $\epsilon \in (0, \text{inj}_q)$, $\exp_p$ maps the open ball $B_\epsilon(u)$ of radius $\epsilon$ centered at any $u \in \exp_p^{-1}(q)$ onto $B_\epsilon(q)$, because $\exp_p$ is isometric and $\tilde{M}$ is complete. Furthermore it is one-to-one on this ball, since $\epsilon < \text{inj}_q$. Thus, $B_\epsilon(q)$ is evenly covered by $\exp_p$, and the latter is a covering map. $\square$

THEOREM 9.3 (Bonnet, Myers). *Let* $M^n$ *be a complete Riemannian manifold with sectional curvature* $K \geq \kappa > 0$. *Then* $d(p, q) \leq \pi/\sqrt{\kappa}$ *for all* $p, q \in M$. *In particular,* $M$ *is compact and has finite fundamental group.*

PROOF. By Theorem 9.1, any geodesic in $M$ of length greater that $\pi/\sqrt{\kappa}$ has a conjugate point $t_0 \in (0, \pi/\sqrt{\kappa})$, and cannot be minimal by Corollary 8.2. On the other hand, any two points of $M$ are joined by a minimal geodesic because $M$ is complete. This show that the diameter $\text{diam}(M) = \sup\{d(p, q)$ |

$p, q \in M\}$ of $M$ cannot exceed $\pi/\sqrt{\kappa}$. By the Hopf-Rinow theorem, $M$ is compact.

If $\rho : \tilde{M} \to M$ denotes the universal covering map, endow $\tilde{M}$ with the unique differentiable structure for which $\rho$ becomes smooth. The above argument may now be applied to the complete Riemannian manifold $(\tilde{M}, \rho^*g)$, where $g$ is the metric of $M$. Thus, $\tilde{M}$ is compact, and the fiber $\pi_1(M)$ of $\rho$ is finite. $\qquad\square$

EXERCISE 139. Generalize Myer's theorem to Riemannian manifolds whose Ricci curvature satisfies $\text{Ric}(x) \geq (n-1)\kappa > 0$ for all $|x| = 1$, as follows: Let $c : [0, \pi/\sqrt{\kappa}] \to M$ be a normal geodesic, and consider parallel orthonormal vector fields $E_1, \ldots, E_{n-1}$ along $c$, with $\langle E_i, \dot{c} \rangle = 0$. Define $X_i \in \Gamma_c^0$ by $X_i(t) = \sin(\sqrt{\kappa}t)E_i(t)$. Prove that $\sum_i I(X_i, X_i) = 0$, and conclude that $c$ has a conjugate point.

EXERCISE 140. Show by means of an example that the conclusion of Myers' theorem no longer holds if one only requires that $K > 0$.

EXERCISE 141. A normal geodesic $c : \mathbb{R} \to M$ is said to be a *line* if $d(c(t), c(t')) = |t - t'|$ for all $t, t' \in \mathbb{R}$. Prove that in a complete, simply connected manifold of nonpositive curvature, every normal geodesic is a line.

EXERCISE 142. A normal geodesic $c : [0, \infty) \to M$ is said to be a *ray* if $d(c(0), c(t)) = t$ for all $t \geq 0$. Show that if $M$ is complete and noncompact, then for any $p \in M$, there exists a ray $c$ with $c(0) = p$. Is the statement necessarily true if one replaces "ray" by "line"?

## 10. Actions of Compact Lie Groups

In this section, we will prove (a somewhat stronger version of) Theorem 14.2 in Chapter 1. Let $G$ be a compact Lie group acting on $M$ via $\mu : G \times M \to M$. When there is no risk of confusion, we write $g(p)$ instead of $\mu(g, p)$. Recall that the *orbit* $G(p)$ of $p \in M$ is the set $\{g(p) \mid g \in G\}$, and the *isotropy group* $G_p$ of $p$ is the subgroup consisting of all $g \in G$ such that $g(p) = p$. Notice that an isotropy group is necessarily closed in $G$.

LEMMA 10.1. *Let $p \in M$, and set $H := G_p$. Then the map $f : G/H \to M$ given by $f(gH) = g(p)$ is an imbedding onto $G(p)$.*

PROOF. The map $f$ is clearly well-defined and bijective onto the orbit of $p$. Since $G/H$ is compact and $M$ is Hausdorff, $f$ is a topological imbedding. It remains to show that $f$ has maximal rank everywhere. By equivariance of $f$, it suffices to do so at $eH$; equivalently, we claim that if $\pi : G \to G/H$ is projection, then $x \in G_e$ belongs to $H_e$ whenever $x \in \ker(f \circ \pi)_{*e}$. To see this, consider the vector field $X \in \mathfrak{g}$ with $X(e) = x$, and the curve $t \mapsto c(t) := (f \circ \pi)(\exp tx)$. Equivariance of $f$ implies that $f \circ \pi = \mu_g \circ f \circ \pi \circ L_{g^{-1}}$ for any $g \in G$. Thus,

$$\dot{c}(t) = f_* \circ \pi_*(X(\exp tx)) = \mu_{\exp tx*} \circ f_* \circ \pi_* \circ L_{\exp -tx*}X(\exp tx)$$

$$= \mu_{\exp tx*} \circ f_* \circ \pi_* x = 0.$$

$\exp tx$ therefore belongs to $H$ for all $t$, and $x \in H_e$. $\qquad\square$

LEMMA 10.2. *There exists a Riemannian metric on M for which the action of G is by isometries.*

PROOF. Given a Riemannian metric $\langle,\rangle$ on $M$, define a new metric $\widetilde{\langle,\rangle}$ by $\widetilde{\langle x,y\rangle} := \int_G f$, where $f(g) = \langle g_*x, g_*y\rangle$. For $h \in G$, $\widetilde{\langle h_*x, h_*y\rangle} = \int_G \tilde{f}$, with $\tilde{f}(g) = \langle g_*h_*x, g_*h_*y\rangle = f(gh)$. Thus, $\widetilde{\langle h_*x, h_*y\rangle} = \int_G f \circ R_h = \int_G f = \widetilde{\langle x,y\rangle}$ by Proposition 14.1 in Chapter 1. $\qquad\square$

In view of Lemma 10.2, we will from now on assume that $G$ acts by isometries. The following version of the tubular neighborhood theorem will also be needed:

PROPOSITION 10.1. *Let $N$ be a compact submanifold of a Riemannian manifold $M$ with normal bundle $\nu$ in $M$. There exists $\epsilon > 0$ such that exp :* $E(\nu^\epsilon) \to B_\epsilon(N)$ *is a diffeomorphism of the total space $E(\nu^\epsilon) = \{u \in E(\nu) \mid |u| < \epsilon\}$ of the disk bundle $\nu^\epsilon$ onto the $\epsilon$-neighborhood $B_\epsilon(N)$ of $N$ in $M$.*

PROOF. Since exp has maximal rank on the zero section $s(N)$ of $\nu$ and is injective on $s(N)$, there exists, by Lemma 1.1 in Chapter 3, a neighborhood $U$ of $s(N)$ in $\nu$ such that $\exp : U \to M$ is a diffeomorphism onto its image. By compactness of $N$, $U$ contains (the total space of) some $\epsilon$-disk bundle $\nu^\epsilon$, and $\exp(\nu^\epsilon) \subset B_\epsilon(N)$. It remains to show that $\exp(\nu^\epsilon)$ contains $B_\epsilon(N)$. Consider a point $q \in B_\epsilon(N)$. Choosing a smaller $\epsilon > 0$ if necessary, we may assume that $B_\epsilon(N)$ has injectivity radius larger than $\epsilon$. By Exercise 132, there is a point $p \in N$ such that $d(q, N) = d(q, p)$, and thus a minimal normal geodesic $c : [0, a] \to M$ from $q$ to $p$. It must be shown that $\dot{c}(a) \perp N_p$; if not, choose a curve $\gamma : (-\delta, \delta) \to N$ with $\gamma(0) = p$ and $\langle \dot{\gamma}(0), \dot{c}(a)\rangle < 0$. For small enough $\delta$, the image of $\gamma$ lies inside the $\epsilon$-neighborhood of $q$, so that $\gamma$ may be lifted via $\exp_q$ to a curve $\tilde{\gamma}$ in $M_q$.

Consider the variation $V : [0, a] \times (-\delta, \delta) \to M$ of $c$ by geodesics, $V(t, s) = \exp_q(t\tilde{\gamma}(s))$, and denote by $L(s)$ the length of the geodesic $t \mapsto V(t, s)$. By assumption, $L : (-\delta, \delta) \to \mathbb{R}^+$ has a minimum at 0. Denoting by $Y$ the variational vector field of $V$, we have by Theorem 8.1

$$L'(0) = \langle Y, \dot{c}\rangle(t)|_0^a = \langle \dot{\gamma}(0), \dot{c}(a)\rangle < 0,$$

contradicting the fact that $L$ has a minimum at 0. $\qquad\square$

The above proposition can be used to explicitly describe tubular neighborhoods of orbits:

PROPOSITION 10.2. *Given $p \in M$, denote by $H$ the isotropy group $G_p$ at $p$, by $\nu_p$ the normal bundle in $M$ of the orbit $G(p) = G/H$ of $p$, and by $\nu_p^\epsilon$ the corresponding disk bundle of radius $\epsilon$. There exists $\epsilon > 0$ such that the tubular neighborhood $B_\epsilon(G(p))$ of radius $\epsilon$ about the orbit is equivariantly diffeomorphic to $G \times_H U$, where $U$ is the fiber of $\nu_p^\epsilon$ over $p$, and $H$ acts on $U$ by $h(u) = h_*u$. (The diffeomorphism is equivariant with respect to the action of $G$ on $G \times_H U$ given by $g[a, u] = [ga, u]$.)*

PROOF. Choose $\epsilon > 0$ so that Proposition 10.1 holds for $N = G(p)$, and define $F : G \times_H U \to B_\epsilon(G(p))$ by $F[g, u] = g \circ \exp u$. $G$ is assumed to act by isometries, so that $h \circ \exp = \exp \circ h_*$ for $h \in H$, and $F$ is well-defined. It is clearly equivariant. Furthermore, $F$ is the composition

$$F : G \times_H U \xrightarrow{\phi} E(\nu_p^\epsilon) \xrightarrow{\exp} B_\epsilon(G(p)),$$

where $\phi[g, u] = g_* u$. The statement then follows from Proposition 10.1, once we establish that $\phi$ is a diffeomorphism. We claim more, namely that $\phi$ is a bundle map covering $f : G/H \to G(p)$, with $f$ as in Lemma 10.1. It is clear that $\phi$ covers $f$, and therefore induces a map $G \times_H U \to E(f^* \nu_p^\epsilon)$. The latter is given by $[g, u] \mapsto (gH, g_* u)$, which is smooth, covers the identity, and has smooth inverse $(gH, u) \mapsto [g, g_*^{-1} u]$; it is therefore an equivalence, so that $\phi$ is a bundle map. □

Recall from Section 14 in Chapter 1 that two orbits are said to have the *same type* if there exists an equivariant bijection between them. When all orbits have the same type, $G$ is said to act by *principal orbits*. Theorem 14.2 (2) in Chapter 1 states that in this case, the orbit space $M/G$ is a differentiable manifold, and the projection $\pi : M \to M/G$ is a submersion. We are now in a position to deduce a stronger result, namely that $M \to M/G$ is a fiber bundle:

THEOREM 10.1. *Let $G$ be a compact Lie group acting on $M$, $H$ a closed subgroup of $G$. If all orbits have type $G/H$, then the projection $\pi : M \to M/G$ is a fiber bundle with fiber $G/H$ and group $N(H)/H$, where $N(H)$ denotes the normalizer $\{g \in G \mid gH = Hg\}$ of $H$.*

PROOF. The previous proposition implies that any point $p$ of $M$ has a $G$-invariant neighborhood equivariantly diffeomorphic to $G \times_H U$, where $U$ is a metric ball around the origin in the subspace of $M_p$ that is orthogonal to the orbit through $p$. Consider a point $u$ in $U$. By hypothesis on the orbit type, there exists an equivariant diffeomorphism $f : G/G_{[e,u]} \to G/H$. Let $aH := f(G_{[e,u]})$. Since $f$ is equivariant, $f(gG_{[e,u]}) = gaH$ for any $g \in G$. If we now choose $g$ to lie in $G_{[e,u]}$, then $gaH$ must equal $aH$; i.e., $a^{-1}ga \in H$ for all $g \in G_{[e,u]}$, and $G_{[e,u]}$ is conjugate to some subgroup of $H$. Arguing in a similar fashion with $f^{-1}$, we conclude that $G_{[e,u]}$ is conjugate to $H$. But if $g[e, u] = [e, u]$, then $[g, u] = [e, u]$, so that $g \in H$, and $G_{[e,u]} = H$. The latter implies that $[e, u] = [g, u] = [e, gu]$; thus, $H$ acts trivially on $U$, and $G \times_H U$ is equivariantly diffeomorphic to $(G/H) \times U$. Composing this with the projection onto the second factor yields a homeomorphism of a neighborhood of $G(p)$ in the orbit space (endowed with the quotient topology) with $U$, which is an open set in a vector space. Thus, $M/G$ is a topological manifold. Consider two such homeomorphisms originating from the construction in the previous proposition by taking $p_1$, $p_2$ in $M$ and the fibers $U_i$ of $\nu_{p_i}^{\epsilon_i}$ over $p_i$. The transition function is $\exp_{\nu_{p_2}}^{-1} \circ g \circ \exp_{\nu_{p_1}}$ for some $g \in G$, which is differentiable. We therefore obtain a differentiable structure on $M/G$ for which the projection is a submersion.

Let $V_i = \pi(\exp(U_i)$. The construction above induces equivariant bundle charts $(\pi, \phi_i) : \pi^{-1}(V_i) \to V_i \times G/H$ whose transition function at any point is a $G$-equivariant diffeomorphism of $G/H$. It remains to show that the group $\text{Diff}_G(G/H)$ of such diffeomorphisms is isomorphic to $N(H)/H$. To see this, observe that if $f \in \text{Diff}_G(G/H)$, then $f(gH) = gaH$ for some $a \in G$ with

$a^{-1}Ha \subset H$ by the argument at the beginning of the proof. We claim that $a^{-1}Ha = H$; i.e., $a \in N(H)$: Indeed, if $A = \{a^n \mid n = 0, 1, 2, \dots\}$, then by Lemma 10.3 below, the closure $\bar{A}$ of $A$ contains $a^{-1}$. Now, the map $F : G \times G \to G$ sending $(b, c)$ to $b^{-1}cb$ is continuous, and by hypothesis, $F(A \times H) \subset H$. Since $H$ is closed, $F(\bar{A} \times H) \subset H$. Thus, $aHa^{-1} \subset H$, so that $H \subset a^{-1}Ha$ as claimed. Summarizing, any $f \in \text{Diff}_G(G/H)$ has the form $f(gH) = ga^{-1}H = (gH)a$ for some $a$ in the normalizer of $H$. It now easily follows that $\text{Diff}_G(G/H)$ is isomorphic to $N(H)/H$ acting by right multiplication on $G/H$.     $\square$

LEMMA 10.3. *For any $a \in G$, the closure of the set $A = \{a^n \mid n = 0, 1, 2, \dots\}$ is a subgroup of $G$.*

PROOF. Notice first of all that the closure of a subgroup is again a subgroup by continuity of $(a, b) \mapsto ab^{-1}$. It suffices therefore to show that $a^{-1} \in \bar{A}$, or equivalently, that any neighborhood of $a^{-1}$ intersects $A$. Consider the subgroup $\langle a \rangle$ generated by $a$. If $e$ is an isolated point of $\overline{\langle a \rangle}$, then $\overline{\langle a \rangle}$ is discrete, and being compact, must be finite, so that $a^n = e$ for some $n \in \mathbb{N}$. If $n = 1$, then $a^{-1} = e \in A$, and otherwise, $a^{-1} = a^{n-1} \in A$. So assume that $e$ is not isolated. If $U$ is a neighborhood of $e$, then so is $V = U \cap U^{-1}$, where $U^{-1} := \{g^{-1} \mid g \in U\}$. It must therefore contain $a^n$ for some positive $n$, so that $a^{n-1} \in L_{a^{-1}}(V) \cap A$. In other words, if $U$ is any neighborhood of $e$, then $L_{a^{-1}}(U)$ intersects $A$. But then any neighborhood $W$ of $a^{-1}$ intersects $A$, because $L_a(W)$ is a neighborhood of $e$, so that $W \cap A = L_{a^{-1}}(L_a(W)) \cap A \neq \emptyset$.     $\square$

COROLLARY 10.1. *If $G$ is a compact Lie group acting freely on $M$, then $\pi : M \to M/G$ is a principal $G$-bundle.*

The corollary yields an alternative immediate proof of the fact that the Hopf fibrations, as well as the projections $G \to G/H$ for $G$ compact and $H$ closed, are principal bundles. Neither the theorem nor its corollary remain true when $G$ is no longer compact: Consider for example the $\mathbb{R}$-action on the torus $S^1 \times S^1$ given by $t(z_1, z_2) = (e^{it}z_1, e^{i\alpha t}z_2)$ with $\alpha$ irrational. There do exist criteria guaranteeing that certain maps $M \to M/G$ are fibrations for noncompact $G$. One such (see [14]) is the condition that $G$ act freely and *properly*; i.e., any two points that do not lie in the same orbit can be separated by open sets $U$, $V$, with the property that $g(U) \cap V = \emptyset$ for all $g \in G$; equivalently, the orbit space $M/G$ is Hausdorff in the quotient topology.

# CHAPTER 6

# Characteristic Classes

Let $\xi = \pi : E \to M$ denote a rank $n$ bundle over $M$ with connection $\nabla$ and curvature $R$. The Bianchi identity $d^\nabla R = 0$ from Exercise 94 implies that certain polynomial functions in $R$ are closed differential forms on $M$, and thus represent cohomology classes in $H^*(M, \mathbb{R})$. These classes are called *characteristic classes* of $\xi$, and turn out to be independent of the choice of connection.

Since the algebraic machinery needed to establish this is fairly involved, we illustrate the process by discussing a simple example: Recall that the trace function $\operatorname{tr} : \mathfrak{gl}(n) \to \mathbb{R}$ is invariant under the adjoint action of $GL(n)$: Given $A \in \mathfrak{gl}(n)$, $B \in GL(n)$, $\operatorname{tr}(\operatorname{Ad}_B A) = \operatorname{tr}(BAB^{-1}) = \operatorname{tr} A$ (throughout the chapter, we identify $\mathfrak{gl}(n)$ with the space of $n \times n$ matrices). This elementary fact implies that the trace operator induces a parallel section $\operatorname{Tr}$ of the bundle $\operatorname{End}(\xi)^*$, if the latter is given the connection induced by $\nabla$: Given $p \in M$, the fiber of $\operatorname{End}(\xi)$ over $p$ is $\mathfrak{gl}(E_p)$. Now choose a basis of $E_p$, that is, an isomorphism $b : \mathbb{R}^n \to E_p$, and define, for $L \in \mathfrak{gl}(E_p)$,

$$\operatorname{Tr}(p)(L) := \operatorname{tr}(b^{-1} \circ L \circ b).$$

There is no ambiguity here, for if $\tilde{b}$ is a different basis of $E_p$, then $M := \tilde{b}^{-1} \circ b \in GL(n)$, and $\tilde{b}^{-1} \circ L \circ \tilde{b} = M(b^{-1} \circ L \circ b)M^{-1}$ has trace equal to that of $b^{-1} \circ L \circ b$. Thus, $\operatorname{Tr}$ is a section of $\operatorname{End}(\xi)^*$. To see that it is parallel, consider a curve $c$ in $M$ and a basis $\beta$ of parallel sections of $\xi$ along $c$. If $X$ is a parallel section of $\operatorname{End}(\xi)$ along $c$, then $t \mapsto \beta(t)^{-1} \circ X(t) \circ \beta(t)$ is a constant curve in $\mathfrak{gl}(n)$, and $\operatorname{Tr} \circ c(X)$ is a constant function. Thus, $\operatorname{Tr}$ is parallel along any curve, as claimed.

In general, given a vector bundle $\eta$ over $M$ with connection $\nabla$, a section $L$ of $\eta^*$ assigns to each $\eta$-valued $r$-form $\omega \in A_r(M, \eta)$ on $M$ an ordinary differential form $L\omega \in A_r(M)$ on $M$ of the same degree. For vector fields $X_i$ on $M$, we have

$$d(L\omega)(X_0, \ldots, X_r) = \sum_i (-1)^i X_i((L\omega)(X_0, \ldots, \widehat{X}_i, \ldots, X_r))$$
$$+ \sum_{i<j} (-1)^{i+j} (L\omega)([X_i, X_j], X_0, \ldots, \widehat{X}_i, \ldots, \widehat{X}_j, \ldots, X_r).$$

On the other hand,

$$X_i((L\omega)(X_0, \ldots, \widehat{X}_i, \ldots, X_r)) = (\nabla_{X_i} L)(\omega(X_0, \ldots, \widehat{X}_i, \ldots, X_r))$$
$$+ L\nabla_{X_i}(\omega(X_0, \ldots, \widehat{X}_i, \ldots, X_r)),$$

so that

$$d(L\omega)(X_0,\ldots,X_r) = \sum_i (-1)^i (\nabla_{X_i} L)(\omega(X_0,\ldots,\widehat{X}_i,\ldots,X_r))$$
$$+ L d^\nabla \omega(X_0,\ldots,X_r).$$

In particular, if $L$ is a parallel section of $\eta^*$, then

(0.1) $$d \circ L = L \circ d^\nabla.$$

Returning to our original vector bundle $\xi$, the curvature tensor $R$ of the connection is an $\mathrm{End}(\xi)$-valued 2-form on $M$. Then $\mathrm{Tr} \circ R$ is an ordinary 2-form on $M$, and by Equation (0.1) together with the Bianchi identity,

$$d \circ \mathrm{Tr} \circ R = \mathrm{Tr} \circ d^\nabla \circ R = 0.$$

In other words, $\mathrm{Tr} \circ R$ represents an element $w \in H^2(M)$.

PROPOSITION 0.3. *The element $w \in H^2(M)$ represented by $\mathrm{Tr} \circ R$ is independent of the choice of connection.*

PROOF. Consider connections $\mathcal{H}_i$ on $\xi$ with curvature $R_i$, $i = 1, 2$. Let $I = [0,1]$, and denote by $p : M \times I \to M$ and $t : M \times I \to I$ the respective projections. The bundle $p^*\xi$ then admits connections $p^*\mathcal{H}_0$ and $p^*\mathcal{H}_1$, with corresponding covariant derivatives $\nabla_0$ and $\nabla_1$. One easily checks that

$$\nabla := (1 - t)\nabla_0 + t\nabla_1$$

is a covariant derivative on $p^*\xi$, cf. also Exercise 90 in Chapter 4. Furthermore, if $\mathcal{H}$ and $R$ denote the corresponding connection and curvature, then $\imath_0^* \mathcal{H} = \mathcal{H}_0$, $\imath_1^* \mathcal{H} = \mathcal{H}_1$, and similar equations hold for the curvature tensors; here, as usual, $\imath_s : M \to M \times I$ maps $p$ to $(p, s)$. Now, by (0.1), the 2-form $\mathrm{Tr} \circ R$ on $M$ is closed. The Poincaré Lemma then implies that the form

$$\mathrm{Tr} \circ R_1 - \mathrm{Tr} \circ R_0 = (\imath_1^* - \imath_0^*)\, \mathrm{Tr} \circ R = d(I \circ \mathrm{Tr} \circ R)$$

is exact. $\square$

## 1. The Weil Homomorphism

In order to generalize the example discussed in the previous section, we need some algebraic preliminaries.

DEFINITION 1.1. A function $f : \mathbb{R}^n \to \mathbb{R}$ is said to be *symmetric* if for any permutation $\sigma$ of $\{1,\ldots,n\}$, $f(\lambda_{\sigma(1)},\ldots,\lambda_{\sigma(n)}) = f(\lambda_1,\ldots,\lambda_n)$ for all $\lambda_i \in \mathbb{R}$. The *elementary symmetric functions* $s_1,\ldots,s_n : \mathbb{R}^n \to \mathbb{R}$ are defined by

$$s_k(\lambda_1,\ldots,\lambda_n) = \sum_{i_1 < \cdots < i_k} \lambda_{i_1} \lambda_{i_2} \cdots \lambda_{i_k}, \qquad 1 \le k \le n.$$

For example, $s_1(\lambda_1,\ldots,\lambda_n) = \sum_i \lambda_i$, and $s_n(\lambda_1,\ldots,\lambda_n) = \prod_i \lambda_i$. A straightforward induction argument shows that

$$(x - \lambda_1) \cdots (x - \lambda_n) = x^n - s_1(\lambda_1,\ldots,\lambda_n)x^{n-1} + \cdots + (-1)^n s_n(\lambda_1,\ldots,\lambda_n).$$

The polynomials $s_i$ may be extended to $\mathbb{C}^n$. Notice that they are algebraically independent over the reals; i.e., if $p : \mathbb{R}^n \to \mathbb{R}$ is a real polynomial such that

$$p(s_1(\lambda_1,\ldots,\lambda_n),\ldots,s_n(\lambda_1,\ldots,\lambda_n)) = 0$$

for all $\lambda_i$ with $s_j(\lambda_1, \ldots, \lambda_n) \in \mathbb{R}$, then $p \equiv 0$. To see this, let $a_1, \ldots, a_n \in \mathbb{R}$, and denote by $\lambda_1, \ldots, \lambda_n \in \mathbb{C}$ the roots of the equation

$$(1.1) \qquad x^n - a_1 x^{n-1} + \cdots + (-1)^n a_n = 0.$$

Then $a_i = s_i(\lambda_1, \ldots, \lambda_n)$, and by assumption, $p(a_1, \ldots, a_n) = 0$.

DEFINITION 1.2. A *polynomial of degree $k$* on a vector space $V$ is a function $p : V \to \mathbb{R}$ such that if $\omega^1, \ldots, \omega^n$ is a basis of $V^*$, then there exist $a_{i_1 \cdots i_k} \in \mathbb{R}$ with $p(v) = \sum a_{i_1 \cdots i_k} \omega^{i_1}(v) \cdots \omega^{i_k}(v)$ for all $v \in V$.

The coefficients of $p$ may, and will, be assumed to be symmetric in the indices. The space of these polynomials will be denoted $P_k(V)$, and $P(V) := \oplus_{k=0}^{\infty} P_k(V)$ is then an algebra with the usual product of functions. For example, $s_k \in P_k(\mathbb{R}^n)$. In fact, any symmetric polynomial $f$ on $\mathbb{R}^n$ is a function of $s_1, \ldots, s_n$: Given $a_1, \ldots, a_n \in \mathbb{R}$, let $\lambda_1, \ldots, \lambda_n$ denote the corresponding roots of (1.1), and define $F : \mathbb{R}^n \to \mathbb{R}$ by $F(a_1, \ldots, a_n) = f(\lambda_1, \ldots, \lambda_n)$. Then

$$f(x_1, \ldots, x_n) = F(s_1(x_1, \ldots, x_n), \ldots, s_n(x_1, \ldots, x_n)).$$

It can be shown that $F$ may be chosen to be a polynomial.

When $V$ is the Lie algebra $\mathfrak{g}$ of a group $G$, a polynomial $p$ in $P(\mathfrak{g})$ is said to be *invariant* if $p(\mathrm{Ad}_g v) = p(v)$ for all $v \in V$, $g \in G$. The collection $P_G$ of invariant polynomials on $\mathfrak{g}$ is a subalgebra of $P(\mathfrak{g})$.

EXAMPLE 1.1. For $A \in \mathfrak{gl}(n)$, define $f_i(A) = s_i(\lambda_1, \ldots, \lambda_n)$, where $\lambda_1, \ldots, \lambda_n$ are the eigenvalues of $A$; equivalently, $f_i$ is determined by the equation

$$\det(xI - A) = x^n - f_1(A)x^{n-1} + \cdots + (-1)^n f_n(A).$$

Then $f_i$ is an invariant polynomial of degree $i$ on $\mathfrak{gl}(n)$.

Instead of working with $P_k(\mathfrak{g})$, it is often more convenient to deal with the space $S_k(\mathfrak{g})$ of symmetric tensors of type $(0, k)$ on $\mathfrak{g}$: The *polarization* of a polynomial $p = \sum a_{i_1 \cdots i_k} \omega^{i_1} \cdots \omega^{i_k} \in P_k(\mathfrak{g})$ is $\mathrm{pol}(p) := \sum a_{i_1 \cdots i_k} \omega^{i_1} \otimes \cdots \otimes \omega^{i_k} \in S_k(\mathfrak{g})$. $\mathrm{pol} : P_k(\mathfrak{g}) \to S_k(\mathfrak{g})$ has inverse $\mathrm{pol}^{-1}(T)(v) = T(v, \ldots, v)$ for $T \in S_k(\mathfrak{g})$, $v \in \mathfrak{g}$. If we define multiplication in $S(\mathfrak{g}) := \oplus_{k=0}^{\infty} S_k(\mathfrak{g})$ by

$$(ST)(v_1, \ldots, v_{k+l}) = \frac{1}{(k+l)!} \sum_{\sigma \in P_{k+l}} S(v_{\sigma(1)}, \ldots v_{\sigma(k)}) \cdot T(v_{\sigma(k+1)}, \ldots, v_{\sigma(k+l)})$$

for $S \in S_k(\mathfrak{g})$, $T \in S_l(\mathfrak{g})$, then the natural extension of pol to $P(\mathfrak{g})$ is an algebra isomorphism $\mathrm{pol} : P(\mathfrak{g}) \to S(\mathfrak{g})$.

$T \in S_k(\mathfrak{g})$ is said to be *invariant* if $T(\mathrm{Ad}_g v_1, \ldots, \mathrm{Ad}_g v_k) = T(v_1, \ldots, v_k)$ for $g \in G$, $v_i \in \mathfrak{g}$. The subalgebra $S_G$ of invariant symmetric tensors is the isomorphic image of $P_G$ via pol.

We are now ready to carry these concepts over to bundles: Recall that if $\xi = \pi : E \to M$ is a vector bundle over $M$, then $\mathrm{End}(\xi) = \mathrm{Hom}(\xi, \xi)$ is the bundle over $M$ with fiber $\mathfrak{gl}(E_p)$ over $p$. Let us denote by $\mathrm{End}_k(\xi)$ the $k$-fold tensor product $\mathrm{End}(\xi) \otimes \cdots \otimes \mathrm{End}(\xi)$, and set $\otimes \mathrm{End}(\xi) := \oplus_{k \geq 0} \mathrm{End}_k(\xi)$. Since the fiber of this bundle is an algebra, we may define the product $\alpha \otimes \beta \in A_{k+l}(M, \otimes \mathrm{End}(\xi))$ of $\otimes \mathrm{End}(\xi)$-valued forms $\alpha \in A_k(M, \otimes \mathrm{End}(\xi))$ and

$\beta \in A_l(M, \otimes \mathrm{End}(\xi))$ by

$$(\alpha \otimes \beta)(x_1, \ldots, x_{k+l}) = \frac{1}{k!l!} \sum_{\sigma \in P_{k+l}} \alpha(x_{\sigma(1)}, \ldots, x_{\sigma(k)}) \otimes \beta(x_{\sigma(k+1)}, \ldots, x_{\sigma(k+l)}).$$

By Examples and Remarks 2.1(v) in Chapter 4, a connection on $\xi$ induces one on $\otimes \mathrm{End}(\xi)$. Since multiplication in the algebra bundle is parallel, an argument similar to the one for the trivial line bundle $\epsilon^1$ over $M$ shows that the exterior covariant derivative operator satisfies

(1.2)                  $d^\nabla(\alpha \otimes \beta) = d^\nabla \alpha \otimes \beta + (-1)^k \alpha \otimes d^\nabla \beta$

for $\alpha$, $\beta$ as above.

PROPOSITION 1.1. *Let $T$ denote a symmetric $(0, k)$ tensor on $\mathfrak{gl}(n)$, and $\xi$ a rank $n$ bundle over $M$ with total space $E$ and connection $\nabla$. If $T$ is invariant, then it induces a parallel section $\overline{T}$ of $\mathrm{End}_k(\xi)^*$.*

PROOF. Given $p \in M$, choose an isomorphism $b : \mathbb{R}^n \to E_p$, and define

$$\overline{T}(p)(L_1 \otimes \cdots \otimes L_k) = T(b^{-1} \circ L_1 \circ b, \ldots, b^{-1} \circ L_k \circ b), \qquad L_i \in \mathfrak{gl}(E_p).$$

The argument used in the last section to show that Tr is a well-defined parallel section of $\mathrm{End}(\xi)^*$ applies equally well to $\overline{T}$.                    □

If $R$ is the curvature tensor of the connection on $\xi$ and $T \in S_k(\mathfrak{gl}(n))$ is invariant, then the $k$-fold product $R^k = R \otimes \cdots \otimes R \in A_{2k}(M, \mathrm{End}_k(\xi))$, and $\overline{T}(R^k)$ is an ordinary $2k$-form on $M$.

THEOREM 1.1 (Weil). *Let $\xi$ denote a rank $n$ bundle over $M$, and $R$ the curvature tensor of some connection on $\xi$. If $T \in S_k(\mathfrak{gl}(n))$ is invariant, then*

(1) *the $2k$-form $\overline{T}(R^k)$ is closed; and*
(2) *if $w(T)$ denotes the element of $H^{2k}(M)$ determined by $\overline{T}(R^k)$, then $w(T)$ is independent of the choice of connection, and $w : S_{GL(n)} \to H^*(M) = \oplus_{i \geq 0} H^i(M)$ is an algebra homomorphism.*

PROOF. By (0.1) and the Bianchi identity,

$$d\overline{T}(R^k) = \overline{T}d^\nabla(R^k) = T(d^\nabla R \otimes R \otimes \cdots \otimes R) + \cdots + T(R \otimes \cdots \otimes R \otimes d^\nabla R) = 0,$$

which establishes (1). For (2), replace Tr by $\overline{T}$ in the proof of Proposition 0.3. The fact that $w$ is a homomorphism is straightforward to prove.                    □

The map $w : S_{GL(n)} \to H^*(M)$ is called the *Weil homomorphism*. It is natural with respect to pull-backs:

PROPOSITION 1.2. *Let $\xi$ denote a vector bundle over $M$, $f : N \to M$. If $w$, $\tilde{w}$ denote the Weil homomorphisms associated to $\xi$, $f^*\xi$ respectively, then $\tilde{w} = f^* \circ w$.*

PROOF. Let $R$ denote the curvature tensor of some connection on $\xi$. By Cartan's structure equation, the induced connection on $f^*\xi$ is $f^*R$, so that for $T \in S_{GL(n)}$ of type $(0, k)$,

$$\tilde{w}(T) = [\overline{T}(f^*R)^k] = [f^*\overline{T}(R^k)] = f^* w(T),$$

where $[\alpha]$ denotes the cohomology class containing $\alpha$.                    □

EXERCISE 143. Recall that for an $n \times n$ matrix $A = (a_{ij})$, the determinant of $A$ equals $\sum_{\sigma \in P_n} (\mathrm{sgn}\,\sigma) a_{1\sigma(1)} \cdots a_{n\sigma(n)}$. Use this fact to show that if $f_k$ is the invariant polynomial of degree $k$ on $\mathfrak{gl}(n)$ from Example 1.1, then

$$f_k(A) = \sum_{1 \le h_1 < \cdots < h_k \le n} \det(a_{h_i h_j}).$$

EXERCISE 144. A linear transformation $L : V \to V$ on an $n$-dimensional vector space $V$ induces for each $k = 1, \ldots, n$ a linear map $L_k^* : A_k(V) \to A_k(V)$ given by

$$(L_k^* \omega)(v_1, \ldots, v_k) = \omega(Lv_1, \ldots, Lv_k), \qquad \omega \in A_k(V), \quad v_i \in V.$$

Since $f_k$ is an invariant polynomial on $\mathfrak{gl}(n)$, we may define $f_k(L) := f_k([L])$, where $[L]$ denotes the matrix of $L$ in some fixed basis of $V$.

(a) Use Exercise 143 to show that $f_k(L) = \mathrm{tr}\, L_k^*$. In particular, $L_n^* : A_n(V) \to A_n(V)$ is multiplication by $\det L$.

(b) Show that for $A, B \in \mathfrak{gl}(n)$, $f_k(AB) = f_k(BA)$, and use this to give another proof of the invariance of $f_k$.

EXERCISE 145. Let $R$ denote the curvature tensor of some connection on $E \to M$.

(a) Given $p \in M$, $x_i \in M_p$, express $R^2(x_1, \ldots, x_4) \in \mathrm{End}_2(E_p)$ explicitly in terms of $R(x_i, x_j) \in \mathrm{End}(E_p)$.

(b) Suppose $E$ has rank 4. If $g_2 = \mathrm{pol}\, f_2$, find an expression for $\bar{g}_2(R^2)$ in terms of an orthonormal basis of $E_p$.

## 2. Pontrjagin Classes

If $w : S_{GL(n)} \to H^*(M)$ is the Weil homomorphism associated to a vector bundle $\xi$ over $M$, the element $w(f) \in H^*(M)$, for $f \in S_{GL(n)}$, is called a *characteristic class* of $\xi$. By Proposition 1.2, equivalent bundles have the same characteristic classes. A natural question to ask is whether generators can be found for these classes. The problem is simplified by the fact that any rank $n$ bundle allows a reduction of its structure group to $O(n)$, and thus admits a Riemannian connection. The corresponding curvature tensor $R$ then belongs to $A_2(M, \mathfrak{o}(\xi))$, where $\mathfrak{o}(\xi) = \{L \in \mathrm{End}(\xi) \mid L + L^t = 0\}$. Since a characteristic class is independent of the choice of connection, we may restrict the Weil homomorphism to the group $O(n)$. In other words, we wish to find generators of $S_{O(n)}$, or equivalently, of the algebra $P_{O(n)}$ of invariant polynomials on $\mathfrak{o}(n)$.

Now, the polynomials $f_1, \ldots, f_n$ defined by

$$\det(xI - A) = x^n - f_1(A)x^{n-1} + \cdots + (-1)^n f_n(A), \qquad A \in \mathfrak{gl}(n),$$

are invariant polynomials on $\mathfrak{gl}(n)$. They are therefore also invariant (under $\mathrm{Ad}_{O(n)}$) as polynomials on $\mathfrak{o}(n)$. Furthermore, if $A \in \mathfrak{o}(n)$, then $\det(xI - A) = \det(xI - A)^t = \det(xI + A)$, so that $f_i(A) = f_i(-A)$. By Example 1.1, $f_i(A)$ must be zero for odd $i$.

THEOREM 2.1. *Let $n = 2k$ or $2k + 1$. The algebra $P_{O(n)}$ of invariant polynomials on $\mathfrak{o}(n)$ is generated by $f_2, f_4, \ldots, f_{2k}$.*

PROOF. Given $\lambda_1, \ldots, \lambda_k \in \mathbb{R}$, set

$$(\lambda_1 \ldots \lambda_k) = \begin{pmatrix} 0 & \lambda_1 & \cdots & 0 & 0 \\ -\lambda_1 & 0 & \cdots & 0 & 0 \\ \vdots & \vdots & \ddots & \vdots & \vdots \\ 0 & 0 & \cdots & 0 & \lambda_k \\ 0 & 0 & \cdots & -\lambda_k & 0 \end{pmatrix}$$

if $n = 2k$, and

$$(\lambda_1 \ldots \lambda_k) = \begin{pmatrix} 0 & \lambda_1 & \cdots & 0 & 0 & 0 \\ -\lambda_1 & 0 & \cdots & 0 & 0 & 0 \\ \vdots & \vdots & \ddots & \vdots & \vdots & \vdots \\ 0 & 0 & \cdots & 0 & \lambda_k & 0 \\ 0 & 0 & \cdots & -\lambda_k & 0 & 0 \\ 0 & 0 & \cdots & 0 & 0 & 0 \end{pmatrix}$$

if $n = 2k + 1$. By elementary linear algebra, for any $M \in \mathfrak{o}(n)$, there exists $A \in O(n)$ such that $AMA^{-1} = (\lambda_1 \ldots \lambda_k)$ for some $\lambda_i \in \mathbb{R}$. Consider $f \in P_{O(n)}$. Since both $f$ and $f_i$ are invariant, it suffices to establish that there exists a polynomial $p$ such that $f(\lambda_1 \ldots \lambda_k) = p(f_2(\lambda_1 \ldots \lambda_k), \ldots, f_{2k}(\lambda_1 \ldots \lambda_k))$ for all $\lambda_i \in \mathbb{R}$. Notice first of all that

$$(2.1) \qquad\qquad f_{2i}(\lambda_1 \ldots \lambda_k) = s_i(\lambda_1^2, \ldots, \lambda_k^2).$$

In fact, since the left side is independent of whether $n$ is even or odd, we may assume that $n = 2k$. The characteristic polynomial of $(\lambda_1 \ldots \lambda_k)$ is

$$\prod_{j=1}^{k}(x - i\lambda_j)(x + i\lambda_j) = \sum_{j=0}^{2k}(-1)^j f_j(\lambda_1 \ldots \lambda_k)x^{2k-j},$$

where $f_0 :\equiv 1$. It can also be written as

$$\prod_{j=1}^{k}(x^2 + \lambda_j^2) = \sum_{j=0}^{k}(-1)^j s_j(-\lambda_1^2, \ldots, -\lambda_k^2)(x^2)^{k-j} = \sum_{j=0}^{k} s_j(\lambda_1^2, \ldots, \lambda_k^2)x^{2k-2j},$$

with $s_0 :\equiv 1$. The claim follows by comparing coefficients.

In view of (2.1), it suffices to show that $f(\lambda_1 \ldots \lambda_k)$ may be expressed as a polynomial in $s_1(\lambda_1^2, \ldots, \lambda_k^2), \ldots, s_k(\lambda_1^2, \ldots, \lambda_k^2)$; i.e., that $f(\lambda_1 \ldots \lambda_k) = q(\lambda_1^2, \ldots, \lambda_k^2)$ for some symmetric polynomial $q$. To see this, let $p$ denote the polynomial given by $p(\lambda_1, \ldots, \lambda_k) = f(\lambda_1 \ldots \lambda_k)$. If

$$A = \begin{pmatrix} 0 & 1 & \\ 1 & 0 & \\ & & I_{n-2} \end{pmatrix} \in O(n),$$

then $A(\lambda_1 \ldots \lambda_k)A^{-1} = (-\lambda_1\lambda_2 \ldots \lambda_k)$. Thus, $p(\lambda_1, \ldots, \lambda_k) = p(-\lambda_1, \ldots, \lambda_k)$, and $p$ contains only even powers of $\lambda_1$. An obvious modification of $A$ shows that this is true for any $\lambda_i$, so that $p(\lambda_1, \ldots, \lambda_k) = q(\lambda_1^2, \ldots, \lambda_k^2)$ for some polynomial

$q$. It remains to show that $p$, and hence $q$, is symmetric. But if

$$B = \begin{pmatrix} 0 & I_2 & \\ I_2 & 0 & \\ & & I_{n-4} \end{pmatrix} \in SO(n),$$

then $B(\lambda_1 \ldots \lambda_k)B^{-1} = (\lambda_2 \lambda_1 \ldots \lambda_k)$. Similarly, any pair $(\lambda_i, \lambda_j)$ can be transposed by an appropriate $B \in SO(n)$. This concludes the argument.     □

Let $g_{2i}$ denote the polarization of $f_{2i}$. By Theorem 1.1, if $R$ is the curvature tensor of a Riemannian connection on a rank $n$ bundle $\xi$ over $M$, then $\bar{g}_{2i}(R^{2i})$ is a closed $4i$-form on $M$, and its cohomology class is independent of the choice of connection.

DEFINITION 2.1. The $i$-th *Pontrjagin class* of a rank $n$ bundle $\xi$ over $M$ is the element $p_i(\xi) \in H^{4i}(M)$ represented by the form

$$p_i = \frac{1}{(2\pi)^{2i}} \bar{g}_{2i}(R^{2i}), \qquad i = 1, \ldots, [n/2].$$

An explicit formula for $p_i$ can be given using Exercise 143: For example,

$$f_2(A) = \sum_{1 \le i < j \le n} \det \begin{pmatrix} a_{ii} & a_{ij} \\ a_{ji} & a_{jj} \end{pmatrix}.$$

Given $p \in M$ and an orthonormal basis $u_1, \ldots, u_n$ of $E_p$, define 2-forms $R^{ij}$ on $M_p$ by $R^{ij}(x, y) = \langle R(x, y)u_j, u_i \rangle$, $x, y \in M_p$. Since $R^{ii} = 0$ and $R^{ij} = -R^{ji}$,

$$p_1(p)(x_1, \ldots, x_4) = \frac{1}{(2\pi)^2} \sum_{i<j} \frac{1}{2!2!} \sum_{\sigma \in P_4} (\operatorname{sgn} \sigma) R^{ij}(x_{\sigma(1)}, x_{\sigma(2)}) \cdot R^{ij}(x_{\sigma(3)}, x_{\sigma(4)});$$

i.e., $p_1 = \frac{1}{(2\pi)^2} \sum_{i<j} R^{ij} \wedge R^{ij}$. More generally,

$$p_k = \frac{1}{(2\pi)^{2k}} \sum_{\substack{1 \le i_1 < \cdots < i_{2k} \le n \\ \sigma \in P\{i_1, \ldots, i_{2k}\}}} (\operatorname{sgn} \sigma) R^{i_1 \sigma(i_1)} \wedge \cdots \wedge R^{i_{2k} \sigma(i_{2k})}.$$

EXAMPLE 2.1 (Pontrjagin Classes of $S^n$). The Pontrjagin classes of a manifold are defined to be those of its tangent bundle. Consider the canonical connection on $\tau S^n$. By Examples and Remarks 3.1 in Chapter 4, its curvature tensor is given by

$$R(x, y)u = \langle y, u \rangle x - \langle x, u \rangle y.$$

Thus, if $\omega^1, \ldots, \omega^n$ is a (local) orthonormal basis of 1-forms on $S^n$, then $R^{ij} = \omega^i \wedge \omega^j$. The summands in $p_k$ are of the form $\omega^{i_1} \wedge \omega^{\sigma(i_1)} \wedge \cdots \wedge \omega^{i_{2k}} \wedge \omega^{\sigma(i_{2k})}$, where $\sigma$ is a permutation of $i_1, \ldots, i_k$; i.e., $p_k \equiv 0$.

A similar argument shows that any space of constant curvature has trivial Pontrjagin classes.

EXERCISE 146. Suppose $\xi$ is a rank $n = 2k$ bundle over $M$ that admits a nowhere-zero cross-section. Show that $p_k(\xi) = 0$. *Hint*: Choose a connection for which the cross-section is parallel.

EXERCISE 147. Introduce a Euclidean metric on the bundle $\mathfrak{o}(\xi)$ by setting

$$\langle A, B \rangle = \frac{1}{2}\operatorname{tr}(A^t B), \qquad A, B \in \mathfrak{o}(E_p), \quad p \in M.$$

(More precisely, $\langle A, B \rangle = (1/2)\operatorname{tr}(b^{-1} \circ A^t B \circ b)$ for some isomorphism $b : \mathbb{R}^n \to E_p$.) Given a Riemannian connection on $\xi$ with curvature tensor $R$, consider the 4-form $\alpha$ on $M$ given by

$$\alpha(x_1, \ldots, x_4) = \frac{1}{(2\pi)^2}\frac{1}{(2!)^2}\sum_{\sigma \in P_4}(\operatorname{sgn}\sigma)\langle R(x_{\sigma(1)}, x_{\sigma(2)}), R(x_{\sigma(3)}, x_{\sigma(4)})\rangle.$$

Show that $\alpha$ represents $p_1(\xi)$.

EXERCISE 148. Use Exercise 147 to determine the Pontrjagin class of the rank 4 bundle $S^7 \times_{S^3} \mathbb{R}^4 \to S^4$ associated to the Hopf fibration $S^7 \to S^4$.

## 3. The Euler Class

In this section, we investigate characteristic classes of oriented bundles of rank $n$; i.e., bundles the structure group of which reduces to $SO(n)$. Since any polynomial on $\mathfrak{o}(n)$ which is invariant under the adjoint action of $O(n)$ is also invariant under that of $SO(n)$, the algebra $P_{SO(n)}$ contains $P_{O(n)}$. When $n$ is odd, we will see that both algebras coincide. When $n$ is even, however, a new polynomial, the Pfaffian, arises, yielding an additional class called the Euler class.

THEOREM 3.1. If $n = 2k + 1$, then the algebra $P_{SO(n)}$ of invariant polynomials on $\mathfrak{o}(n)$ is generated by $f_2, f_4, \ldots, f_{2k}$.

PROOF. The argument in the proof of Theorem 2.1 goes through as before, with one modification: The matrix

$$A = \begin{pmatrix} 0 & 1 & \\ 1 & 0 & \\ & & I_{n-2} \end{pmatrix}$$

used in that proof does not lie in $SO(n)$. However, since $n$ is odd, the last diagonal entry in the matrix $(\lambda_1 \ldots \lambda_k)$ is 0. Instead of conjugating by $A$, conjugate by

$$\begin{pmatrix} 0 & 1 & & \\ 1 & 0 & & \\ & & I_{n-1} & \\ & & & -1 \end{pmatrix} \in SO(n).$$

$\square$

DEFINITION 3.1. For $n = 2k$, the *Pfaffian* $\operatorname{Pf}(A)$ of $A = (a_{ij}) \in \mathfrak{gl}(n)$ is

$$\operatorname{Pf}(A) = \frac{1}{2^k k!}\sum_{\sigma \in P_{2k}}(\operatorname{sgn}\sigma)a_{\sigma(1)\sigma(2)}\cdots a_{\sigma(2k-1)\sigma(2k)}.$$

Given $1 \le i_1, j_1, \ldots, i_k, j_k \le 2k$, define $\epsilon^{i_1 j_1 \ldots i_k j_k}$ to be 0 if the indices are not all distinct, and equal to the sign of the permutation $(i_1 j_1 \ldots i_k j_k)$ otherwise.

LEMMA 3.1. *For* $A \in \mathfrak{o}(2k)$,

$$\mathrm{Pf}(A) = \sum_{\{(i_1,j_1),\dots,(i_k,j_k)\} \in P} \epsilon^{i_1 j_1 \dots i_k j_k} a_{i_1 j_1} \cdots a_{i_k j_k},$$

*where* $P$ *denotes the collection of all sets* $\{(i_1, j_1), \dots, (i_k, j_k)\}$ *of* $k$ *pairs of integers* $(i_l, j_l)$ *with* $1 \le i_l < j_l \le 2k$.

For example, if $A \in \mathfrak{o}(4)$, then $\mathrm{Pf}(A) = a_{12}a_{34} - a_{13}a_{24} + a_{14}a_{23}$.

PROOF. Notice that the expression $(\mathrm{sgn}\,\sigma)a_{\sigma(1)\sigma(2)} \cdots a_{\sigma(2k-1)\sigma(2k)}$ remains unchanged when two pairs $(\sigma(2l-1), \sigma(2l))$ and $(\sigma(2m-1), \sigma(2m))$ are interchanged. It therefore remains unchanged under permutations of pairs. This means that

$$\mathrm{Pf}(A) = \frac{1}{2^k} \sum_{\{(i_1,j_1),\dots,(i_k,j_k)\} \in \tilde{P}} \epsilon^{i_1 j_1 \dots i_k j_k} a_{i_1 j_1} \cdots a_{i_k j_k},$$

where $\tilde{P}$ denotes the collection of all sets $\{(i_1, j_1), \dots, (i_k, j_k)\}$ of $k$ pairs of integers between 1 and $n$. Since $A$ is skew-adjoint, each summand in the above formula is unchanged when we interchange $i_l$ and $j_l$. The statement now follows. $\square$

Our next aim is to show that the Pfaffian is invariant under the adjoint action of $SO(n)$. Recall that for $B \in \mathfrak{gl}(n)$, $\det B = \sum_{\sigma \in P_n} (\mathrm{sgn}\,\sigma) b_{1\sigma(1)} \cdots b_{n\sigma(n)}$. Notice that for $\tau \in P_n$, $b_{\tau(1)\sigma(1)} = b_{k\,\sigma\circ\tau^{-1}(k)}$, where $k = \tau(1)$. Thus, given a permutation $\tau = (i_1, \dots, i_n)$,

$$\sum_{\sigma \in P_n} (\mathrm{sgn}\,\sigma) b_{i_1\sigma(1)} \cdots b_{i_n\sigma(n)} = \sum_{\sigma \in P_n} (\mathrm{sgn}\,\sigma) b_{1\,\sigma\circ\tau^{-1}(1)} \cdots b_{n\,\sigma\circ\tau^{-1}(n)}$$

$$= \sum_{\sigma\circ\tau^{-1} \in P_n} \epsilon^{i_1 \dots i_n} (\mathrm{sgn}\,\sigma \circ \tau^{-1}) b_{1\,\sigma\circ\tau^{-1}(1)}$$

$$\cdots b_{n\,\sigma\circ\tau^{-1}(n)}$$

$$= \epsilon^{i_1 \dots i_n} \det B.$$

PROPOSITION 3.1. *For* $n = 2k$ *and* $A, B \in \mathfrak{gl}(n)$, $\mathrm{Pf}(B^t AB) = (\det B)\,\mathrm{Pf}(A)$. *In particular, if* $B \in SO(n)$, *then* $\mathrm{Pf}(B^{-1}AB) = \mathrm{Pf}(B^t AB) = \mathrm{Pf}(A)$; *i.e., the Pfaffian is invariant under the adjoint action of* $SO(n)$.

Proof.

$$2^k k! \operatorname{Pf}(B^t A B) = \sum_{\sigma \in P_n} (\operatorname{sgn} \sigma)) \left( \sum_{i_1, i_2 = 1}^{n} b_{i_1 \sigma(1)} a_{i_1 i_2} b_{i_2 \sigma(2)} \right)$$

$$\cdots \left( \sum_{i_{n-1}, i_n = 1}^{n} b_{i_{n-1} \sigma(n-1)} a_{i_{n-1} i_n} b_{i_n \sigma(n)} \right)$$

$$= \sum_{i_1, \ldots, i_n = 1}^{n} \left( \sum_{\sigma \in P_n} (\operatorname{sgn} \sigma) b_{i_1 \sigma(1)} \cdots b_{i_n \sigma(n)} \right) a_{i_1 i_2} \cdots a_{i_{n-1} i_n}$$

$$= \sum_{i_1, \ldots, i_n = 1}^{n} \epsilon^{i_1 \ldots i_n} (\det B)) a_{i_1 i_2} \cdots a_{i_{n-1} i_n}$$

$$= 2^k k! (\det B) \operatorname{Pf}(A).$$

$\square$

COROLLARY 3.1. *For $A \in \mathfrak{o}(2k)$, $\det A = \operatorname{Pf}^2(A)$.*

PROOF. Choose $B \in O(n)$ such that $BAB^{-1} = (\lambda_1 \ldots \lambda_k)$. It follows from Lemma 3.1 that $\det(\lambda_1 \ldots \lambda_k) = \lambda_1 \cdots \lambda_k$. By Proposition 3.1,

$$\operatorname{Pf}(A) = (\det B) \lambda_1 \cdots \lambda_k = \pm \lambda_1 \cdots \lambda_k.$$

Thus, $\operatorname{Pf}^2(A) = \lambda_1^2 \cdots \lambda_k^2 = \det A$. $\square$

THEOREM 3.2. *When $n = 2k$, the algebra $P_{SO(n)}$ of invariant polynomials on $\mathfrak{o}(n)$ is generated by $f_2, f_4, \ldots, f_{n-2}$ and $\operatorname{Pf}$.*

PROOF. The modification of the matrix

$$A = \begin{pmatrix} 0 & 1 & \\ 1 & 0 & \\ & & I_{n-2} \end{pmatrix}$$

used in the proof of Theorem 3.1 no longer works in this case because $n$ is even. However, the matrix

$$\tilde{A} = \begin{pmatrix} 0 & 1 & & & \\ 1 & 0 & & & \\ & & 0 & 1 & \\ & & 1 & 0 & \\ & & & & I_{n-4} \end{pmatrix}$$

lies in $SO(n)$, and $\tilde{A}(\lambda_1 \ldots \lambda_k) \tilde{A}^{-1} = (-\lambda_1 - \lambda_2 \ldots \lambda_k)$. Similarly, conjugation by an appropriate element of $SO(n)$ changes the sign of any two elements of $(\lambda_1 \ldots \lambda_k)$. Thus, each monomial in the polynomial $p$ from the proof of Theorem 2.1 contains either even powers of $\lambda_i$ for all $i$, or odd powers of $\lambda_i$ for all $i$; write $p = p_0 + p_1$, where $p_0$ is the sum of monomials of the former type, and $p_1$ of the latter. Since $p$ is symmetric, so are $p_0$ and $p_1$, and

$$p_0(\lambda_1, \ldots, \lambda_k) = \bar{p}(s_1(\lambda_1^2, \ldots, \lambda_k^2), \ldots, s_k(\lambda_1^2, \ldots, \lambda_k^2))$$

for some $\bar{p}$. On the other hand, $p_1(\lambda_1, \ldots, \lambda_k) = \lambda_1 \cdots \lambda_k q(\lambda_1^2, \ldots, \lambda_k^2)$ for some symmetric $q$, so that

$$f(M) = \bar{p}(f_2(M), \ldots, f_{2k}(M)) + \mathrm{Pf}(M)\tilde{p}(f_2(M), \ldots, f_{2k}(M))$$

for some $\tilde{p}$. Moreover, $f_{2k} = f_n$ may be dispensed with, since it equals $\mathrm{Pf}^2$. $\square$

DEFINITION 3.2. Let $\xi$ be an oriented vector bundle of rank $n = 2k$. If pf denotes the polarization of Pf, and $\bar{\mathrm{pf}}$ the parallel section of $\mathrm{End}_k(\xi)^*$ induced by pf, then the *Euler class* of $\xi$ is the element $e(\xi)$ represented by

$$e = \frac{1}{(2\pi)^k}\bar{\mathrm{pf}}(R^k).$$

When $n$ is odd, $e(\xi)$ is defined to be 0. The form $e$ representing the Euler class will be called the *Euler form* of the bundle.

By Corollary 3.1,

(3.1) $$p_k(\xi) = e(\xi) \cup e(\xi),$$

where $\cup$ denotes the product in $H^*(M)$.

We now derive a local expression for $e$ which is convenient for computations. Let $U \subset M$ be an open set such that $\xi_{|U}$ is trivial, and consider a positively oriented orthonormal basis of sections $U_1, \ldots, U_{2k}$ of $\xi_{|U}$. Define 2-forms $R^{ij}$ on $U$ by $R^{ij}(X,Y) = \langle R(X,Y)U_j, U_i \rangle$ as before. Then the Euler form is the $2k$-form on $M$ with restriction to $U$ given by

$$e(X_1, \ldots, X_{2k}) = \frac{1}{(2\pi)^k 2^k k!} \sum_{i_1, \ldots, i_{2k}} \epsilon^{i_1 \ldots i_{2k}} \frac{1}{2^k} \sum_{\sigma \in P_{2k}} (\mathrm{sgn}\,\sigma) R^{i_1 i_2}(X_{\sigma(1)}, X_{\sigma(2)})$$
$$\cdots R^{i_{2k-1} i_{2k}}(X_{\sigma(2k-1)}, X_{\sigma(2k)}).$$

Thus,

$$e = \frac{1}{(2\pi)^k 2^k k!} \sum_{i_1, \ldots, i_{2k}} \epsilon^{i_1 \ldots i_{2k}} R^{i_1 i_2} \wedge \cdots \wedge R^{i_{2k-1} i_{2k}},$$

and by Lemma 3.1,

(3.2) $$e = \frac{1}{(2\pi)^k} \sum_{\{(i_1,j_1);\ldots,(i_k,j_k)\} \in P} \epsilon^{i_1 j_1 \ldots i_k j_k} R^{i_1 j_1} \wedge \cdots \wedge R^{i_k j_k},$$

where $P$ is as in the lemma.

The next proposition is an immediate consequence of (3.2), and its proof is left as an exercise.

PROPOSITION 3.2. *If the bundle $\xi$ admits a nowhere-zero section, then its Euler class vanishes.*

EXAMPLES AND REMARKS 3.1. (i) When $n = 2$, the Euler form is given by

$$e(p)(x,y) = \frac{1}{2\pi}\langle R(x,y)v, u \rangle, \qquad p \in M, \quad x, y \in M_p,$$

where $u, v$ is a positively oriented basis of $E_p$. In particular, if $\xi$ is the tangent bundle $\tau M$ of $M$, then $e = (1/2\pi)K\omega$, with $K$ and $\omega$ denoting the sectional curvature and the volume form respectively of the Riemannian manifold $M$.

(ii) When $n = 4$, the Euler form is given by

$$\frac{1}{(2\pi)^2}(R^{12} \wedge R^{34} - R^{13} \wedge R^{24} + R^{41} \wedge R^{23})$$

according to (3.2). It can alternatively be described as follows: View $R$ as an element of $A_2(M, \Lambda_2(\xi))$, so that for $x, y, z, w \in M_p$, $R(x,y) \wedge_\xi R(z,w) \in \Lambda_4 \xi$, with $\wedge_\xi$ denoting the wedge product in $\Lambda \xi$. Since $\xi$ is oriented, there exists a unique section $\omega_\xi$ of $\Lambda_4 \xi$ representing the orientation, with $\langle \omega_\xi, \omega_\xi \rangle = 1$; in fact, $\omega_\xi(p) = e_1 \wedge e_2 \wedge e_3 \wedge e_4$ for any positively oriented orthonormal basis $e_1, \ldots, e_4$ of $E_p$. Notice that

$$\langle R(x,y) \wedge_\xi R(z,w), \omega_\xi(p) \rangle = \sum_{\substack{i<j \\ k<l}} \langle R^{ij}(x,y) R^{kl}(z,w) e_i \wedge e_j \wedge e_k \wedge e_l, \omega_\xi(p) \rangle$$

$$= 2 \sum_{\{(i,j),(k,l)\} \in P} \epsilon^{ijkl} R^{ij}(x,y) R^{kl}(z,w),$$

with $P$ as in Lemma 3.1. Thus, if we define $R \wedge_\xi R \in A_4(M, \Lambda_4 \xi)$ by

$$R \wedge_\xi R(X_1, \ldots, X_4) = \frac{1}{2!2!} \sum_{\sigma \in P_4} (\text{sgn}\,\sigma) R(X_{\sigma(1)}, X_{\sigma(2)}) \wedge_\xi R(X_{\sigma(3)}, X_{\sigma(4)}),$$

then the Euler form is given by $(1/8\pi^2)\langle R \wedge_\xi R, \omega_\xi \rangle$.

In general, when $V$ is an oriented $n$-dimensional inner product space with volume form $\omega$, the *Hodge star operator* is the endormorphism $\star : \Lambda_k(V) \rightarrow \Lambda_{n-k}(V)$ defined by

$$\alpha \wedge \star\beta = \langle \alpha, \beta \rangle \omega, \qquad \alpha, \beta \in \Lambda_k(V).$$

It is easily seen that $\star$ is an isomorphism satisfying

$$\star \circ \star = (-1)^{k(n-k)} 1_{\Lambda_k(V)}, \qquad \star\omega = 1.$$

The Hodge operator extends naturally to oriented Euclidean bundles, and the resulting operator $\star_\xi$ is a parallel section of $\text{Hom}(\Lambda_k \xi, \Lambda_{n-k} \xi)$, since $\langle,\rangle$ and $\omega$ are both parallel. The Euler form of a rank 4 oriented bundle is then given by

$$(3.3) \qquad\qquad e = \frac{1}{8\pi^2} \star_\xi (R \wedge_\xi R).$$

More generally, when the rank of $\xi$ is $2k$, then

$$(3.4) \qquad\qquad e = \frac{1}{k!(2\pi)^k} \star_\xi R^k,$$

where $R^k$ denotes the $k$-fold wedge product in $\Lambda \xi$ of $R$ with itself.

(iii) Let $-\xi$ denote $\xi$ with the opposite orientation. It follows from (3.4) that $e(-\xi) = -e(\xi)$.

EXERCISE 149. Let $V$ be a $2k$-dimensional oriented inner product space, with $k$ even, so that $\star : \Lambda_k(V) \rightarrow \Lambda_k(V)$ equals its own inverse.

(a) Show that $\star$ is a self-adjoint operator, and that $\Lambda_k(V)$ splits into an orthogonal direct sum $\Lambda_k^+ \oplus \Lambda_k^-$ of the $\pm 1$-eigenspaces of $\star$. $\alpha \in \Lambda_k(V)$ is said to be *self-dual* (resp. *anti-self-dual*) if $\star\alpha = \alpha$ (resp. $\star\alpha = -\alpha$).

(b) If $e_1, \ldots, e_4$ is a positively oriented orthonormal basis of $V^4$, write down explicit orthonormal bases of $\Lambda_2^\pm(V)$.

EXERCISE 150. Use (3.2) to prove that $e(\xi) = 0$ whenever $\xi$ admits a nowhere-zero section. *Hint*: Choose a connection as in Exercise 146.

EXERCISE 151. (a) Show that for the canonical connection on $\tau S^{2k}$, each summand in (3.2) is just the volume form of $S^{2k}$.

(b) Prove that the set $P$ in that equation has $(2k-1)(2k-3)\cdots 3 = (2k)!/(2^k k!)$ elements.

(c) Use the fact that the volume of the $2k$-sphere equals $(\pi^k 2^{2k+1} k!)/(2k)!$ to prove that the Euler form $e$ in (3.2) of $\tau S^{2k}$ satisfies $\int_{S^{2k}} e = 2$. It turns out that the Euler (and Pontrjagin) classes are always integral cohomology classes, so that $\int_M e \in \mathbb{Z}$.

EXERCISE 152. Use the fact that $R : \Lambda_2 \tau S^{2k} \to \Lambda_2 \tau S^{2k}$ is the identity ($R$ being the curvature tensor of the canonical connection) to redo Exercise 151 using (3.4) instead.

## 4. The Whitney Sum Formula for Pontrjagin and Euler Classes

It is worth emphasizing one important property of the characteristic classes studied so far, which follows immediately from Proposition 1.2:

THEOREM 4.1. *Let $\xi$ denote a vector bundle over $M$, $p_k(\xi)$ its $k$-th Pontrjagin class, and $e(\xi)$ its Euler class if the bundle is oriented. Given $f : N \to M$,*

$$p_k(f^*\xi) = f^* p_k(\xi), \qquad e(f^*\xi) = f^* e(\xi).$$

Our next goal is to understand the relation between the classes of two bundles $\xi_i$ over $M$ and those of their Whitney sum $\xi_1 \oplus \xi_2$. We begin with the Euler class; extend the Pfaffian to all skew-adjoint matrices, by setting $\mathrm{Pf}(A) = 0$ is $A$ is odd-dimensional. This way, the relation $\det A = \mathrm{Pf}^2(A)$ for $A \in \mathfrak{o}(n)$ holds regardless of whether $n$ is odd or even, cf. Corollary 3.1. For $A \in \mathfrak{o}(n)$, $B \in \mathfrak{o}(m)$, set

$$A \circledast B = \begin{pmatrix} A & 0 \\ 0 & B \end{pmatrix} \in \mathfrak{o}(m+n).$$

LEMMA 4.1. $\mathrm{Pf}(A \circledast B) = \mathrm{Pf}(A)\,\mathrm{Pf}(B)$.

PROOF. The statement is clear if $n$ or $m$ is odd, since

$$\mathrm{Pf}^2(A \circledast B) = \det(A \circledast B) = \det A \cdot \det B = \mathrm{Pf}^2(A) \cdot \mathrm{Pf}^2(B),$$

and both sides vanish. If $n = 2k$ and $m = 2l$ are both even, then there exist orthogonal matrices $M_1$, $M_2$ such that $M_1^{-1} A M_1 = (\lambda_1 \ldots \lambda_k)$ and $M_2^{-1} B M_2 = (\mu_1 \ldots \mu_l)$. Thus,

$$\begin{aligned}
\mathrm{Pf}(A \circledast B) &= \mathrm{Pf}(M_1 \circledast M_2 \cdot (\lambda_1 \ldots \lambda_k \mu_1 \ldots \mu_l) \cdot M_1^{-1} \circledast M_2^{-1}) \\
&= (\det M_1)(\det M_2)\lambda_1 \cdots \lambda_k \cdot \mu_1 \cdots \mu_l \\
&= (\det M_1)\lambda_1 \cdots \lambda_k (\det M_2)\mu_1 \cdots \mu_l \\
&= \mathrm{Pf}(A)\,\mathrm{Pf}(B).
\end{aligned}$$

$\square$

LEMMA 4.2.

$$f_{2i}(A \circledast B) = \sum_{j=1}^{i} f_{2j}(A) f_{2i-2j}(B).$$

PROOF.

$$\det(xI_{n+m} - A \circledast B) = \det((xI_n - A) \circledast (xI_m - B))$$
$$= \det(xI_n - A)\det(xI_m - B),$$

so that

$$\sum_{i=0}^{k+l} x^{2(k+l-i)} f_{2i}(A \circledast B) = \sum_{j=1}^{k} x^{2(k-j)} f_{2j}(A) \cdot \sum_{r=1}^{l} x^{2(l-r)} f_{2r}(B).$$

The statement then follows by comparing coefficients.  $\square$

Given Riemannian connections on vector bundles $\xi_i$ of rank $n_i$ over $M$, $i = 1, 2$, the induced connection on $\xi_1 \oplus \xi_2$ is the pullback via the diagonal imbedding $\Delta : M \to M \times M$ of the product connection on $\xi_1 \times \xi_2$, cf. Examples and Remarks 2.1 in Chapter 4. Let $p \in M$, and consider an orthonormal basis $b : \mathbb{R}^{n_1+n_2} \to E(\xi_1 \oplus \xi_2)_p$ such that the restrictions $b_1 := b_{|\mathbb{R}^{n_1} \times 0}$ and $b_2 := b_{|0 \times \mathbb{R}^{n_2}}$ form orthonormal bases of $E(\xi_1)_p$ and $E(\xi_2)_p$ respectively. If $B : \mathfrak{o}(E(\xi_1 \oplus \xi_2)_p) \to \mathfrak{o}(n_1 + n_2)$ and $B_i : \mathfrak{o}(E(\xi_i)_p) \to \mathfrak{o}(n_i)$ denote the isomorphisms from Proposition 1.1 induced by $b$ and $b_i$, then the curvature tensors $R$, $R_i$ of $\xi_1 \oplus \xi_2$, $\xi_i$ satisfy

$$BR(p) = B_1 R_1(p) \circledast B_2 R_2(p).$$

THEOREM 4.2. *If $\xi_1$, $\xi_2$ are oriented vector bundles over $M$, then*

$$e(\xi_1 \oplus \xi_2) = e(\xi_1) \cup e(\xi_2).$$

PROOF. Denote by $n_i$ the rank of $\xi_i$. The inclusion $SO(n_1) \circledast SO(n_2) \subset SO(n_1+n_2)$ induces an orientation of $\xi_1 \oplus \xi_2$. Consider Riemannian connections on $\xi_i$, and the induced connection on $\xi_1 \oplus \xi_2$. With notation as above, $BR(p) = BR_1(p) \circledast BR_2(p)$ for all $p \in M$. By Lemma 4.1, if one of the bundles is odd-dimensional, then the Whitney sum has vanishing Euler class, and the statement is true. So assume $n_i = 2k_i$. By the same lemma, together with the fact that the product of polynomials corresponds to the wedge product of forms,

$$\overline{\mathrm{pf}}(R^{k_1+k_2}) = \overline{\mathrm{pf}}(R_1^{k_1}) \wedge \overline{\mathrm{pf}}(R_2^{k_2}),$$

which establishes the claim.  $\square$

In order to state the Whitney sum formula for Pontrjagin classes, denote by $p_0(\xi)$ the class in $H^0(M)$ containing the constant function 1 on $M$.

DEFINITION 4.1. The *total Pontrjagin class* of a rank $n$ bundle $\xi$ is

$$p(\xi) = p_0(\xi) + p_1(\xi) + \cdots + p_{[\frac{n}{2}]}(\xi) \in H^0(M) \oplus H^4(M) \oplus \cdots \subset H^*(M).$$

THEOREM 4.3. *If $\xi_1$, $\xi_2$ are vector bundles over $M$, then $p(\xi_1 \oplus \xi_2) = p(\xi_1) \cup p(\xi_2)$.*

PROOF. The statement means that $p_k(\xi_1 \oplus \xi_2) = \sum_{i=0}^{k} p_i(\xi_1) \cup p_{k-i}(\xi_2)$. This follows from Lemma 4.2 by an argument similar to the one used for the Euler class of a Whitney sum. $\qquad\square$

EXERCISE 153. Use the results from this section to reprove that an oriented bundle which admits a nowhere-zero section has vanishing Euler class.

EXERCISE 154. A bundle $\xi^n$ is said to be *stably trivial* if there exists a trivial bundle $\epsilon^k$ such that $\xi^n \oplus \epsilon^k = \epsilon^{n+k}$. For example, the tangent bundle of the sphere is stably trivial. Show that a stably trivial bundle has vanishing total Pontrjagin class.

## 5. Some Examples

In this section, we look at characteristic classes of vector bundles over low-dimensional spheres. It turns out that these classes determine the bundles up to "finite ambiguity." Since $H^k(S^n) = 0$ except when $k = 0$ or $n$, any characteristic class lives in $H^n(S^n)$; in fact, they can only exist when $n$ is even, so if $n \leq 4$, we are left with bundles over $S^2$ and $S^4$. This leaves out only one bundle, for any bundle over $S^3$ is trivial, and there is exactly one nontrivial bundle over $S^1$.

By Theorem 15.2 in Chapter 1, the map

$$H^n(S^n) \longrightarrow \mathbb{R},$$

$$[\omega] \longmapsto \int_{S^n} \omega$$

is an isomorphism, so that the Euler class and appropriate Pontrjagin class may be identified with numbers. As noted earlier, these numbers are actually integers, and are called the *Euler and Pontrjagin numbers* of the bundle.

Recall that equivalence classes of rank $k$ vector bundles over $S^n$ are in bijective correspondence with $\pi_{n-1}(SO(k))$.

LEMMA 5.1. *Let $\alpha$ denote the Euler or Pontrjagin form corresponding to rank $k$ bundles over $S^n$. Then the map*

$$\mathrm{Vect}_k(S^n) \cong \pi_{n-1}(SO(k)) \longrightarrow \mathbb{Z},$$

$$\xi \longmapsto \int_{S^n} \alpha(\xi)$$

*is a homomorphism.*

PROOF. Let $\tilde{G}_{k,l}$ be a classifying space for rank $k$ bundles over $S^n$. If $f : S^n \to \tilde{G}_{k,l}$ is a classifying map for $\xi$, then $\xi$ is equivalent to $f^*\tilde{\gamma}_{k,l}$, and by Theorem 4.1,

$$\int_{S^n} \alpha(\xi) = \int_{S^n} \alpha(f^*\tilde{\gamma}_{k,l}) = \int_{S^n} f^*\alpha(\tilde{\gamma}_{k,l}) = (\deg f) \int_{\tilde{G}_{k,l}} \alpha(\tilde{\gamma}_{k,l}).$$

But $\deg : \pi_n(\tilde{G}_{k,l}) \cong \pi_{n-1}(SO(k)) \to \mathbb{Z}$ is a homomorphism (cf. Examples and Remarks 2.1 in Chapter 3), and the statement follows. $\qquad\square$

**5.1. Bundles over $S^2$.** Rank $k$ bundles over $S^2$ are classified by $\pi_1(SO(k))$. When $k = 2$, there exists, for each $n \in \mathbb{Z}$, precisely one bundle $\xi_n$ with $C(\xi_n) = n$, according to the discussion in Section 5 of Chapter 3.

Letting $R$ denote the curvature tensor of some Riemannian connection on $\xi_n$, the 2-form $e_n$ on $S^2$ given by

$$e_n(p)(x,y) = \frac{1}{2\pi}\langle R(x,y)v, u\rangle$$

(for $x$, $y \in S_p^2$ and a positively oriented orthonormal basis $u, v$ of $E(\xi_n)_p$) represents the Euler class of $\xi_n$ by Examples and Remarks 3.1(i). In the case of the tangent bundle $\xi_2$ of the 2-sphere together with the canonical connection,

$$e_2(p)(x,y) = \frac{1}{2\pi}\langle R(x,y)y, x\rangle = \frac{1}{2\pi}$$

if $x, y$ is a positively oriented orthonormal basis of $S_p^2$. Thus, $e_2$ equals $(1/2\pi)$ times the volume form $\omega$ of $S^2$, and the Euler number of $\xi_2$ is $\frac{1}{2\pi}\int_{S^2}\omega = 2$. By Lemma 5.1, the Euler number of $\xi_n$ is $n$, and thus determines the bundle.

When $k > 2$, there is exactly one nontrivial rank $k$ bundle over $S^2$; it cannot be distinguished from the trivial one by any characteristic class.

**5.2. Bundles over $S^4$.** Rank $k$ bundles over $S^4$ are classified by $\pi_3(SO(k))$. When $k < 3$, any such bundle is trivial. For $k = 3$, there is one and only one rank 3 bundle $\xi_n^3$ over $S^4$ with $C(\xi_n^3) = n$ for each $n \in \mathbb{Z}$. We will shortly see that the rank 4 bundle $\epsilon^1 \oplus \xi_n^3$ has first Pontrjagin number $-4n$. Assuming this for the moment, we have

$$p(\epsilon^1 \oplus \xi_n^3) = p(\epsilon^1) \cup p(\xi_n^3) = p(\xi_n^3),$$

so that $\xi_3^n$ is determined by its first Pontrjagin number $-4n$.

Rank 4 bundles over $S^4$ are classified by $\pi_3(SO(4)) \cong \pi_3(S^3) \oplus \pi_3(SO(3)) \cong \mathbb{Z} \oplus \mathbb{Z}$. Denote by $\xi_{m,n}^4$ the bundle corresponding to $(m[1_{S^n}], n[\rho]) \in \pi_3(S^3) \oplus \pi_3(SO(3))$, where $\rho : S^3 \to SO(3)$ is the covering homomorphism from Chapter 3. $\xi_{m,0}^4$ has structure group reducible to $S^3$, and $\xi_{0,n}^4$ to $SO(3)$. In fact, $\xi_{0,n}^4 \cong \epsilon^1 \oplus \xi_n^3$.

Insofar as the Pontrjagin class is concerned, we shall work in a slightly more general setting: Let $M$ denote a 4-dimensional compact, oriented Riemannian manifold with volume form $\omega$, and Hodge operator $\star : \Lambda_2^* M \to \Lambda_2^* M$, cf. Section 3. For $\alpha \in \Lambda_2^* M$, the identity $\alpha = \frac{1}{2}(\alpha + \star\alpha) + \frac{1}{2}(\alpha - \star\alpha)$ decomposes $\Lambda_2^* M$ into a direct sum $\Lambda^+ \oplus \Lambda^-$ of the $+1$ and $-1$ eigenspaces of $\star$. This decomposition is orthogonal, since $\star$ is norm-preserving:

$$\langle \star\alpha, \star\alpha\rangle\omega = \star\alpha \wedge \alpha = \alpha \wedge \star\alpha = \langle\alpha,\alpha\rangle\omega.$$

Furthermore, $\omega$ is parallel, so that $\star$ is a parallel section of the bundle $\operatorname{End}(\Lambda_2^* M)$. There is a corresponding decomposition of the space $A_2(M) = A_2^+(M) \oplus A_2^-(M)$ of 2-forms on $M$. $\alpha \in A_2(M)$ is said be *self-dual* if $\star\alpha = \alpha$, *anti-self-dual* if $\star\alpha = -\alpha$.

For a vector bundle $\xi$ over $M$, we have, as above, a splitting $A_2(M, \operatorname{End}\xi) = A_2^+(M, \operatorname{End}\xi) \oplus A_2^-(M, \operatorname{End}\xi)$. The curvature tensor $R$ of a connection on $\xi$ decomposes as $R = R^+ + R^-$, and we say $R$ is *self-dual* if $R = R^+$, *anti-self-dual* if $R = R^-$.

When the bundle $\xi$ is Euclidean, the inner product on $\Lambda_k(M_p^*)$ extends to $\mathrm{Hom}(\Lambda_k(M_p), E(\xi)_p) = \Lambda_k(M_p^*) \otimes E(\xi)_p$ by defining

$$\langle \alpha, \beta \rangle = \sum_{i_1 < \cdots < i_k} \langle \alpha(x_{i_1}, \ldots, x_{i_k}), \beta(x_{i_1}, \ldots, x_{i_k}) \rangle,$$

where $x_i$ is an orthonormal basis of $M_p$. (When $\xi$ is the trivial line bundle over $M$ with the standard inner product on the $\mathbb{R}$-factor, this inner product coincides with the one on $\Lambda_k(M_p^*)$. In general, it induces a pointwise inner product on $A_k(M, \xi)$, which, when integrated over $M$, yields one on all of $A_k(M, \xi)$.)

In order to apply this to $R \in A_2(M, \mathfrak{o}(\xi))$, we introduce an inner product on $\mathfrak{o}(\xi)$ by defining

$$(5.1) \qquad\qquad \langle A, B \rangle = \frac{1}{2} \mathrm{tr}(A^t \cdot B).$$

This inner product is in fact the one for which the equivalence

$$L : \Lambda_2(\xi) \longrightarrow \mathfrak{o}(\xi),$$
$$u \wedge v \longmapsto (w \mapsto \langle v, w \rangle u - \langle u, w \rangle v).$$

becomes a linear isometry, if $\Lambda_2(\xi)$ is endowed with the Euclidean metric induced by the one on $\xi$: It is straightforward to check that if $u_i$ is an orthonormal basis of $E_p$, then $\{u_i \wedge u_j \mid i < j\}$ is one for $\Lambda_2(E_p)$.

PROPOSITION 5.1. *Let $\xi$ be a Euclidean bundle over $M^4$ with curvature tensor $R$. The first Pontrjagin form of $\xi$ is given by*

$$p_1 = \frac{1}{(2\pi)^2}(|R^+|^2 - |R^-|^2)\omega.$$

PROOF. Let $R^{ij}$ denote as before the local 2-form on $M$ given by $R^{ij}(p)(x, y) = \langle R(x, y)U_j(p), U_i(p) \rangle$, where $\{U_i\}$ is a local orthonormal basis of sections of the bundle. Given an orthonormal basis $x_i$ of $M_p$, we have

$$|R|^2(p) = \sum_{k<l} \frac{1}{2} \mathrm{tr}\, R^t(x_k, x_l) \cdot R(x_k, x_l) = \frac{1}{2} \sum_{i<j, k<l} R^{ij2}(x_k, x_l) = \sum_{i<j} |R^{ij}|^2(p).$$

In particular, $|R^\pm|^2 = \sum_{i<j} |R^{ij\pm}|^2$. Now, $R^{ij+} \wedge R^{ij+} = R^{ij+} \wedge \star R^{ij+} = |R^{ij+}|^2 \omega$, whereas $R^{ij+} \wedge R^{ij-} = -R^{ij+} \wedge \star R^{ij-} = -\langle R^{ij+}, R^{ij-} \rangle \omega = 0$. Similarly, $R^{ij-} \wedge R^{ij-} = -R^{ij-} \wedge \star R^{ij-} = -|R^{ij-}|^2 \omega$. Thus,

$$p_1 = \frac{1}{(2\pi)^2} \sum_{i<j} R^{ij} \wedge R^{ij} = \frac{1}{(2\pi)^2} \sum_{i<j} (|R^{ij+}|^2 - |R^{ij-}|^2)\omega$$

$$= \frac{1}{(2\pi)^2}(|R^+|^2 - |R^-|^2)\omega.$$

$\square$

Next, we derive an analogous formula for the Euler form of $\xi$, where $\xi$ is now assumed to be oriented, of rank 4. The Hodge operator $\star_\xi : \Lambda_2(\xi) \to \Lambda_2(\xi)$ is a parallel section of $\mathrm{End}(\Lambda_2(\xi))$, and induces an orthogonal, parallel splitting $\Lambda_2(\xi) = \Lambda_2^+(\xi) \oplus \Lambda_2^-(\xi)$ of $\Lambda_2$ into a direct sum of the $\pm 1$-eigenspaces of $\star_\xi$. Given $p \in M$, $x, y \in M_p$, $R(x, y) \in \Lambda_2(E_p)$, and we write $R = R_+ + R_-$ for the corresponding decomposition of the curvature tensor. By (3.3) and arguing as above, the Euler form $e$ of $\xi$ is given by

$$e = \frac{1}{8\pi^2}\langle R_+ \wedge_\xi R_+, \omega_\xi\rangle + \langle R_- \wedge_\xi R_-, \omega_\xi\rangle.$$

Now,

$$R_\pm(x,y) \wedge_\xi R_\pm(z,w) = \pm\langle R_\pm(x,y), R_\pm(z,w)\rangle\omega_\xi,$$

so that

$$\begin{aligned}
e(x_1,\ldots,x_4) &= \frac{1}{(8\pi^2)}\frac{1}{4}\sum_{\sigma\in P_4}(\operatorname{sgn}\sigma)\frac{1}{2}(\operatorname{tr} R_+^t(x_{\sigma(1)},x_{\sigma(2)})R_+(x_{\sigma(3)},x_{\sigma(4)}) \\
&\quad - \operatorname{tr} R_-^t(x_{\sigma(1)},x_{\sigma(2)})R_-^t(x_{\sigma(3)},x_{\sigma(4)})) \\
&= \frac{1}{8\pi^2}\frac{1}{4}\sum_{\sigma\in P_4}(\operatorname{sgn}\sigma)\sum_{i<j}(R_+^{ij}(x_{\sigma(1)},x_{\sigma(2)})R_+^{ij}(x_{\sigma(3)},x_{\sigma(4)}) \\
&\quad - R_-^{ij}(x_{\sigma(1)},x_{\sigma(2)})R_-^{ij}(x_{\sigma(3)},x_{\sigma(4)})) \\
&= \frac{1}{8\pi^2}\sum_{i<j}(R_+^{ij}\wedge R_+^{ij} - R_-^{ij}\wedge R_-^{ij})(x_1,\ldots,x_4).
\end{aligned}$$

This may be rewritten as

$$\begin{aligned}
e &= \frac{1}{8\pi^2}\sum_{i<j}R_+^{ij+}\wedge R_+^{ij+} + R_+^{ij-}\wedge R_+^{ij-} - R_-^{ij+}\wedge R_-^{ij+} - R_-^{ij-}\wedge R_-^{ij-} \\
&= \frac{1}{8\pi^2}\sum_{i<j}(|R_+^{ij+}|^2 - |R_+^{ij-}|^2 - |R_-^{ij+}|^2 + |R_-^{ij-}|^2)\omega.
\end{aligned}$$

Summarizing, we have proved:

PROPOSITION 5.2. *Let $\xi$ be an oriented rank 4 Euclidean bundle over $M^4$. The Euler form $e$ of $\xi$ is given by*

$$e = \frac{1}{8\pi^2}(|R_+^+|^2 - |R_+^-|^2 - |R_-^+|^2 + |R_-^-|^2)\omega.$$

It turns out that the rank 4 bundle $\xi$ is determined by the two rank 3 bundles $\Lambda_2^\pm(\xi)$: Let $\phi : S^3 \times S^3 \to SO(4)$ denote the covering homomorphism given by $\phi(q_1,q_2)u = q_1 u q_2^{-1}$, $q_i \in S^3$, $u \in \mathbb{H} = \mathbb{R}^4$. Denote by $S_+^3$ (resp. $S_-^3$) the subgroup $\phi(S^3 \times 1)$ (resp. $\phi(1 \times S^3)$) of $SO(4)$. Since these are normal subgroups, Exercise 155 below implies that the bundles

$$P_\mp := P \times_{SO(4)}(SO(4)/S_\pm^3) = P/S_\pm^3 \longrightarrow M$$

associated to the orthonormal frame bundle $SO(\xi) = P \to M$ of $\xi$ are in fact principal bundles over $M$ with group $SO_\mp(3) = SO(4)/S_\pm^3$ isomorphic to $SO(3)$.

LEMMA 5.2. *$P_\pm \to M$ is the principal $SO(3)$-bundle of $\Lambda_2^\pm(\xi)$.*

PROOF. An orthonormal basis $e_1,\ldots,e_4$ of $\mathbb{R}^4$ induces an orthonormal basis $\frac{1}{\sqrt{2}}(e_1 \wedge e_2 + e_3 \wedge e_4)$, $\frac{1}{\sqrt{2}}(e_1 \wedge e_3 + e_4 \wedge e_2)$, $\frac{1}{\sqrt{2}}(e_1 \wedge e_4 + e_2 \wedge e_3)$ of $\Lambda_2^+(\mathbb{R}^4)$ and a corresponding one for $\Lambda_2^-(\mathbb{R}^4)$ (obtained by changing the sign of the second term in each basis element of $\Lambda^+$). Any $g \in SO(4)$ extends to a linear isometry (also denoted by) $g : \Lambda_2(\mathbb{R}^4) \to \Lambda^2(\mathbb{R}^4)$, by setting $g(u \wedge v) := gu \wedge gv$

and extending linearly. This action leaves $\Lambda_2^\pm$ invariant since $(ge_1, \ldots, ge_4)$ is positively oriented whenever $(e_1, \ldots, e_4)$ is. Given $q \in S^3$,

$$(1 \wedge i + j \wedge k)q = q \wedge iq + jq \wedge kq = 1 \wedge i + j \wedge k,$$

and the same is true for the other basis elements of $\Lambda_2^+$. Thus, $S_-^3$ is the kernel of the representation $SO(4) \to SO(\Lambda_2^+ \mathbb{R}^4)$, and similarly, $S_+^3$ is the kernel of $SO(4) \to SO(\Lambda_2^- \mathbb{R}^4)$; i.e., the special orthogonal group $SO(\Lambda_2^\pm \mathbb{R}^4)$ is $SO_\pm(3)$. Since the map $P_\pm \to SO(\Lambda_2^\pm \xi)$ which takes $bS_\mp^3$ (where $b : \mathbb{R}^4 \to E_p$ is a linear isometry) to the orthonormal basis $\frac{1}{\sqrt{2}}b(e_1 \wedge e_2 \pm e_3 \wedge e_4), \frac{1}{\sqrt{2}}b(e_1 \wedge e_3 \pm e_4 \wedge e_2), \frac{1}{\sqrt{2}}b(e_1 \wedge e_4 \pm e_2 \wedge e_3)$ of $\Lambda_2^\pm(E_p)$ is $SO_\pm(3)$-equivariant, it is an equivalence by Theorem 3.1 in Chapter 2. $\qquad\square$

Our next objective is to relate the first Pontrjagin numbers $p_\pm$ of $\Lambda_2^\pm(\xi)$ to the Euler and first Pontrjagin numbers of $\xi$. Recall that if $\nabla$ is a covariant derivative operator on $\xi$, then the one induced on $\Lambda_2(\xi)$ is given by

$$\tilde{\nabla}_x(U_1 \wedge U_2) = (\nabla_x U_1) \wedge U_2(p) + U_1(p) \wedge (\nabla_x U_2), \quad p \in M, \quad x \in M_p, \quad U_i \in \Gamma\xi.$$

This implies that the corresponding curvature tensor $\tilde{R}$ of $\Lambda_2(\xi)$ is related to the one on $\xi$ by

$$\tilde{R}(x, y)(u_1 \wedge u_2) = (R(x, y)u_1) \wedge u_2 + u_1 \wedge (R(x, y)u_2).$$

The equivalence $L : \Lambda_2(\xi) \to \mathfrak{o}(\xi)$ induces a Lie algebra structure on each fiber of $\Lambda_2(\xi)$, see also Exercise 156 below; the corresponding Lie bracket is the one used in the following:

**PROPOSITION 5.3.** *The subbundles $\Lambda_2^\pm(\xi)$ are parallel under the induced connection; i.e., $\tilde{R} = \tilde{R}_+ + \tilde{R}_-$, with $\tilde{R}_\pm \in A_2(M, \operatorname{End} \Lambda_2^\pm(\xi))$. Given $x, y \in M_p$, and $\alpha, \beta \in \Lambda_2^\pm(E_p)$,*

$$\langle \tilde{R}(x, y)\alpha, \beta \rangle = \langle \tilde{R}_\pm(x, y)\alpha, \beta \rangle = \langle R_\pm(x, y), [\alpha, \beta] \rangle,$$

*where $R(x, y) = R_+(x, y) + R_-(x, y) \in \Lambda_2^+(E_p) \oplus \Lambda_2^-(E_p)$.*

**PROOF.** By Exercise 157, $\tilde{R}(x, y)\alpha = [R(x, y), \alpha]$. Since $\Lambda_2(E_p)$ is a direct sum of the ideals $\Lambda_2^\pm(E_p)$, the first statement is clear. The inner product on $\mathfrak{o}(E_p) = \Lambda_2(E_p)$ is Ad-invariant, so that ad is skew-adjoint. Thus,

$$\langle \tilde{R}_\pm(x, y)\alpha, \beta \rangle = \langle [R_\pm(x, y), \alpha], \beta \rangle = -\langle \operatorname{ad}_\alpha R_\pm(x, y), \beta \rangle = \langle R_\pm(x, y), \operatorname{ad}_\alpha \beta \rangle$$
$$= \langle R_\pm(x, y), [\alpha, \beta] \rangle.$$

$\qquad\square$

**PROPOSITION 5.4.** *Let $p_\pm$, $p_1$ denote the first Pontrjagin forms of $\Lambda_2^\pm(\xi)$, $\xi$, and $e$ the Euler form of $\xi$. Then*

(1) $p_+ = \frac{2}{(2\pi)^2}(|R_+^+|^2 - |R_+^-|^2)\omega, \quad p_- = \frac{2}{(2\pi)^2}(|R_-^+|^2 - |R_-^-|^2)\omega$; *and*

(2) $2p_1 = p_+ + p_-, \quad 4e = p_+ - p_-.$

**PROOF.** Consider a positively oriented orthonormal basis $u_1, \ldots, u_4$ of $E_q$, and denote by $I^\pm$, $J^\pm$, $K^\pm$ the induced orthonormal bases of $\Lambda_2^\pm(E_q)$; i.e.,

$I^{\pm} = (1/\sqrt{2})(u_1 \wedge u_2 \pm u_3 \wedge u_4)$, etc. The first Pontrjagin form of $\Lambda_2^+(\xi)$ is given at $q$ by

$$\frac{1}{(2\pi)^2}(\tilde{R}_+^{IJ} \wedge \tilde{R}_+^{IJ} + \tilde{R}_+^{JK} \wedge \tilde{R}_+^{JK} + \tilde{R}_+^{IK} \wedge \tilde{R}_+^{IK}),$$

where we have omitted the superscripts in $I$, $J$, $K$ for simplicity of notation. By Proposition 5.3,

$$\tilde{R}_+^{IJ}(x,y) = \langle R_+(x,y), [J,I]\rangle = -\sqrt{2}\langle R_+(x,y), K\rangle,$$

and similarly, $\tilde{R}_+^{JK}(x,y) = -\sqrt{2}\langle R_+(x,y), I\rangle$, $\tilde{R}_+^{IK}(x,y) = \sqrt{2}\langle R_+(x,y), J\rangle$. Thus,

$$p_+ = \frac{1}{(2\pi^2)}\frac{1}{4}\sum_{\sigma \in P_4}(\operatorname{sgn}\sigma)\sum_{\alpha \in \{I,J,K\}} 2\langle R_+(x_{\sigma(1)}, x_{\sigma(2)}), \alpha\rangle\langle R_+(x_{\sigma(3)}, x_{\sigma(4)}), \alpha\rangle$$

$$= \frac{1}{(2\pi^2)}\frac{1}{4}\sum_{\sigma \in P_4}(\operatorname{sgn}\sigma)2\langle R_+(x_{\sigma(1)}, x_{\sigma(2)}), R_+(x_{\sigma(3)}, x_{\sigma(4)})\rangle.$$

Referring to Exercise 147, we see that $p_+$ equals two times the first Pontrjagin form of $\xi$ with $R$ replaced by $R_+$. Proposition 5.1 then implies part (1). Comparing the expressions in (1) with Propositions 5.1 and 5.2 yields (2). $\quad\square$

Thus, for example, if the structure group of the bundle reduces to $SO(3)$, then $e = 0$, and $p_+ = p_-$. In order to see what happens when the group reduces to $S_\pm^3$, we use the following:

LEMMA 5.3. *A principal $G$-bundle $P \to M$ admits a reduction to a subgroup $H$ of $G$ iff the associated bundle $P \times_G (G/H) \to M$ with fiber $G/H$ admits a cross-section.*

PROOF. Suppose $\pi_Q : Q \to M$ is an $H$-reduction of $\pi_P : P \to M$. Then there exists a fiber-preserving diffeomorphism $F : Q \times_H (G/H) \to P \times_G (G/H)$ between the associated bundles with fiber $G/H$. Define $s : M \to P \times_G (G/H)$ by $s(m) := F[q, H]$, where $q$ is any point in $\pi_Q^{-1}(m)$. $s$ is a well-defined section, since $[qh, H] = [q, H]$ for $h \in H$.

Conversely, let $s : M = P/G \to P/H = P \times_G (G/H)$ be a section; i.e., for $m \in M$, $s(m)$ equals the $H$-orbit of some $p \in \pi_P^{-1}(m)$. Define $Q := \cup_{m \in M} s(m) \subset P$. $H$ acts on $Q$ by restriction, and $\pi_Q : Q \to M$ is a principal $H$-bundle equivalent to $s^*(P \to P/H)$. It is also clearly a subbundle of $\pi_P$. $\quad\square$

Notice that Lemma 5.3 generalizes Theorem 4.2 in Chapter 2: When $H = \{e\}$, the statement says that a principal bundle is trivial if and only if it admits a cross section.

COROLLARY 5.1. *An oriented rank 4 bundle $\xi$ over $S^4$ admits a reduction to $S_\pm^3$ iff $\Lambda_2^\pm(\xi)$ is trivial; i.e., iff $p_1 = \pm 2e$.*

PROOF. The first assertion is an immediate consequence of the lemma, since in our case, $H$ is normal in $G$, so that $P \times_G (G/H) \to M$ is principal by Exercise 155 below, and thus admits a section iff it is trivial. The second assertion follows from Proposition 5.4(2), together with the fact (which will be proved shortly) that a rank 3 bundle over $S^4$ is determined by its Pontrjagin class, so that $\Lambda_2^\pm(\xi)$ is trivial iff $p_\pm = 0$. $\quad\square$

The Hopf bundle, for example, is a principal $S^3$-bundle. As such, it is the reduction of a principal $SO(4)$-bundle to a subgroup isomorphic to $S^3$. In order to determine which subgroup, we use the following:

LEMMA 5.4. *If $Q \to M$ is a principal $H$-bundle, where $H$ is a subgroup of $G$, then $Q \times_H G \to M$ is a principal $G$-bundle which reduces to the original $H$-bundle $Q \to M$.*

PROOF. There is a well-defined action by right multiplication of $G$ on $Q \times_H G$, $[q, g]a := [q, ga]$ for $a \in G$, and the quotient is $M$. In order to exhibit a principal bundle atlas, consider a principal bundle chart $\phi : \pi_Q^{-1}(U) \to H$ of $\pi_Q : Q \to M$. By the proof of Theorem 2.1 in Chapter 2, the induced chart $\bar\phi : \pi^{-1}(U) \to G$ on the associated bundle $\pi : Q \times_H G \to M$ is given by $\bar\phi : \pi^{-1}(U) \to G$, where $\bar\phi[q, g] = \phi(q)g$. But then for $a \in G$,

$$\bar\phi([q, g]a) = \bar\phi[q, ga] = \phi(q)ga = (\bar\phi[q, g])a,$$

so that $\bar\phi$ is a principal bundle chart. Clearly, $Q = Q \times_H H \to M$ is a reduction of $Q \times_H G \to M$. $\quad\square$

Consider the subgroup $S_-^3$ of $SO(4)$. It acts on $\mathbb{H}$ *from the left* via $\mu : S_-^3 \times \mathbb{H} \to \mathbb{H}$, where $\mu(q, u) = uq^{-1}$ for $q \in S_-^3$, $u \in \mathbb{H}$. Define a right action $\bar\mu$ of $S_-^3$ on $\mathbb{H}$ by $\bar\mu(u, q) = \mu(q^{-1}, u) = uq$. This action extends to $\mathbb{H} \times \mathbb{H}$, and its restriction to $S^7$ is the Hopf fibration. By the above lemma, the bundle $S^7 \times_{S^3} SO(4) \to S^4$, with $S_-^3$ acting on $S^7$ via $\bar\mu$, is a principal $SO(4)$-bundle which reduces to the Hopf fibration with group $S_-^3$. Corollary 5.1 then implies that the associated rank 4 bundle $\xi_{-1,0}$ has first Pontrjagin form $p_1 = -2e$.

On the other hand, the tangent bundle $\xi_{2,-1}$ of $S^4$ has Euler number 2 (see Exercise 151), and by Lemma 5.1, $e(\xi_{2,-1}) = e(\xi_{2,0}) + e(\xi_{0,-1}) = 2$. But $\xi_{0,-1}$ admits a nowhere-zero section, so that its Euler number vanishes, and $e(\xi_{2,0}) = 2$, or more generally, $e(\xi_{k,0}) = k$. For $k = -1$, this implies that the Hopf bundle has Pontrjagin number $p_1(\xi_{-1,0}) = -2e(\xi_{-1,0}) = 2$. More generally, $p_1(\xi_{k,0}) = -2k$. Finally, since $p_1(\xi_{2,-1}) = 0$, $p_1(\xi_{2,0}) = p_1(\xi_{0,1})$, and $p_1(\xi_{0,k}) = p_1(\xi_{2k,0}) = -4k$. Summarizing, we have:

THEOREM 5.1. *Oriented rank 4 bundles over $S^4$ are determined by their Pontrjagin and Euler numbers. Specifically, $p_1(\xi_{m,n}) = -2m - 4n$, $e(\xi_{m,n}) = m$.*

Bundles over $S^n$ with rank larger than $n$ are in general not classified by their characteristic numbers. This can clearly be seen in the cases $n = 2$ and $n = 4$ that we discussed: According to Proposition 5.1 in Chapter 3, such a bundle is equivalent to a Whitney sum of a rank $n$ bundle with a trivial bundle, so that by Theorem 4.2, it must have zero Euler class.

As a final application, consider a vector bundle $\xi$ over a compact manifold $M$, with structure group a compact subgroup $G$ of $GL(k)$. If $\mathcal{C}$ denotes the affine space of connections $\nabla$ on $\xi$ with holonomy group $G$, the *Yang-Mills functional* on $\mathcal{C}$ is defined by

$$\mathcal{YM}(\nabla) = \frac{1}{2} \int_M |R|^2,$$

where $R$ is the curvature tensor of $\nabla$. A critical point of this functional is called a *Yang-Mills connection*. Such a connection is said to be *stable* if it is a local minimum of the functional.

Simons [7] showed that there are no stable Yang-Mills connections on bundles over $S^n$ if $n > 4$. Bourguignon and Lawson studied the four-dimensional case, which turns out to be quite different: Let $\xi$ be an oriented rank 4 bundle over $S^4$ with Pontrjagin number $p$ and Euler number $e$. By Propositions 5.1 and 5.2, the Yang-Mills functional for $SO(4)$-connections satisfies

$$\mathcal{YM}(\nabla) \geq 2\pi^2|p|, 4\pi^2|e|.$$

For example, when $\xi$ is the tangent bundle of the 4-sphere, the curvature $R$ of the canonical connection $\nabla$ is the identity on $\Lambda_2$, so that $R^+_- = R^-_+ = 0$. Thus, $\mathcal{YM}(\nabla) = 4\pi^2 e$, and since the tangent bundle has vanishing Pontrjagin class, the Levi-Civita connection is an absolute minimum of the Yang-Mills functional. Similarly, for a bundle with structure group $S^3_-$,

$$p = \frac{1}{4\pi^2} \int_{S^4} |R^+_+|^2 - |R^-_+|^2.$$

It is know that such a bundle admits connections the curvature tensor of which is self-dual or anti-self-dual depending on whether $p$ is positive or negative. Any such connection is therefore stable. The reader is referred to [6] for further details.

EXERCISE 155. Consider a principal $G$-bundle $P \to M$. Show that if $H$ is a normal subgroup of $G$, then the associated bundle with fiber $G/H$ is principal. (Identify $P/H$ with $P \times_G (G/H)$ via $pH \mapsto [p, H]$. The action of $G/H$ on $P/H$ is then given by $(pH)(aH) := paH$, for $p \in P$, $a \in G$).

EXERCISE 156. Let $\phi : S^3 \times S^3 \to SO(4)$ denote the covering homomorphism, $\phi(q_1, q_2)u = q_1 u q_2^{-1}$, $q_i \in S^3$, $u \in \mathbb{H} = \mathbb{R}^4$. Define $\phi_\pm : S^3 \to SO(4)$ by $\phi_\pm = \phi \circ \imath_\pm$, where $\imath_\pm : S^3 \to S^3 \times S^3$ are the inclusion homomorphisms $\imath_+(q) = (q, 1)$, $\imath_-(q) = (1, q)$.

(a) Prove that the Lie algebra $\mathfrak{o}(4)$ is isomorphic to $\phi_{+*}\mathfrak{o}(3) \times \phi_{-*}\mathfrak{o}(3)$ (recall that the Lie algebra of $S^3$ is isomorphic to $\mathfrak{o}(3)$).

(b) Let $L^{-1} : \mathfrak{o}(4) \to \Lambda_2(\mathbb{R}^4) = \Lambda_2^+ \oplus \Lambda_2^-$ denote the usual isometry. Show that $L^{-1} \circ \phi_{\pm*}$ maps the Lie algebra $\mathfrak{o}(3)$ isomorphically onto $\Lambda_2^\pm$.

EXERCISE 157. Given $A \in \mathfrak{o}(n)$, define $\tilde{A} : \Lambda_2(\mathbb{R}^n) \to \Lambda_2(\mathbb{R}^n)$ by

$$\tilde{A}(v \wedge w) = (Av) \wedge w + v \wedge (Aw)$$

on decomposable elements, and extending linearly.

(a) Prove that $\tilde{A}(v \wedge w) = [L^{-1}A, v \wedge w]$, where $L : \Lambda_2(\mathbb{R}^n) \to \mathfrak{o}(n)$ is the canonical isomorphism.

(b) Let $R$ be the curvature tensor of some connection on the bundle $\xi$ over $M$, and $\tilde{R}$ the induced curvature tensor of $\Lambda_2\xi$. Show that for $\alpha \in \Lambda_2(E(\xi)_p)$,

$$\tilde{R}(x, y)\alpha = [R(x, y), \alpha], \qquad x, y \in M_p,$$

after identifying $R(x, y)$ with an element of $\Lambda_2(E(\xi)_p)$ via $L^{-1}$.

## 6. The Unit Sphere Bundle and the Euler Class

Consider an oriented rank $n = 2k$ Euclidean bundle $\xi = \pi : E \to M$ and its unit sphere bundle $\xi^1 = \pi_{|E^1} : E^1 \to M$, where $E^1 = \{u \in E \mid |u| = 1\}$. Our goal in this section is to show that the pullback of the Euler form of $\xi$ to $E^1$ is exact, a fact that will be needed in the proof of the generalized Gauss-Bonnet theorem in the next section.

Recall that for $u \in E$, $\mathcal{J}_u$ denotes the canonical isomorphism of the fiber $E_{\pi(u)}$ of $\xi$ through $u$ with its tangent space at $u$. For convenience of notation, the latter will be identified with a subspace of $T_u E$ via the derivative of the inclusion $E_{\pi(u)} \hookrightarrow E$, so that $\mathcal{J}_u : E_{\pi(u)} \to (\mathcal{V}E)_u \subset T_u E$.

LEMMA 6.1. *There is a canonical isomorphism $\mathcal{J} : \Gamma_\pi \xi \to \Gamma \mathcal{V}\xi$ of the space $\Gamma_\pi \xi$ of sections of $\xi$ along $\pi$ with the space $\Gamma \mathcal{V}\xi$ of sections of the vertical bundle $\mathcal{V}\xi$ over $E$. A Riemannian connection $\tilde{\nabla}$ on $\xi$ induces a Riemannian connection $\nabla$ on $\mathcal{V}\xi$ given by*

$$\nabla_x \mathcal{J}U = \mathcal{J}_u \tilde{\nabla}_x U, \qquad U \in \Gamma_\pi \xi, \quad x \in T_u E, \quad u \in E.$$

*($\tilde{\nabla}$ in the above identity denotes the covariant derivative along $\pi : E \to M$.)*

PROOF. The equivalence

$$\pi^* \xi \longrightarrow \mathcal{V}\xi,$$
$$(u, v) \longmapsto \mathcal{J}_u v,$$

induces an isomorphism between $\Gamma \pi^* \xi$ and $\Gamma \mathcal{V}\xi$. On the other hand, the map $\Gamma \pi^* \xi \to \Gamma_\pi \xi$ that takes $U$ to $\pi_2 U$ is an isomorphism with inverse $V \mapsto (1_E, V)$, where $(1_E, V)(u) = (u, V(u))$. Combining the two, we obtain an isomorphism $\mathcal{J} : \Gamma_\pi \xi \to \Gamma \mathcal{V}\xi$ given by $(\mathcal{J}U)(v) = \mathcal{J}_v U(v)$. This establishes the first part of the lemma.

A Riemannian connection $\tilde{\nabla}$ on $\xi$ induces a connection $\bar{\nabla}$ on $\pi^* \xi$, where

$$\bar{\nabla}_x (1_E, U) = (u, \tilde{\nabla}_x U), \qquad U \in \Gamma_\pi \xi, \quad x \in T_u E, \quad u \in E.$$

The above equivalence $\pi^* \xi \cong \mathcal{V}\xi$ then yields a connection $\nabla$ on $\mathcal{V}\xi$, and $\nabla_x \mathcal{J}U = \mathcal{J}_u \tilde{\nabla}_x U$, as claimed; $\nabla$ is Riemannian because $\tilde{\nabla}$ is, and because $\mathcal{J}_u$ is isometric. $\qquad \square$

Denoting by $\tilde{R}$, $R$ the corresponding curvature tensors, the structure equation (Lemma 3.1 in Chapter 4) implies

(6.1)
$$R(x, y) \mathcal{J}_u v = \mathcal{J}_u \tilde{R}(\pi_* x, \pi_* y) v, \qquad x, y \in T_u E, \quad u, v \in E, \quad \pi(u) = \pi(v).$$

(Equivalently, in the bundle $\pi^* \xi$, $\bar{R}(x, y)(u, v) = (u, \tilde{R}(\pi_* x, \pi_* y) v)$.)

There is a canonical section of $\xi$ along $\pi$, namely the identity $1_E$. Under the isomorphism of Lemma 6.1, it corresponds to the position vector field $P$ on the manifold $TE$; i.e., $P$ is the section of $\mathcal{V}\xi$ defined by $P(u) = \mathcal{J}_u u$ for $u \in E$. Notice that

(6.2)
$$d^\nabla P(x) = \nabla_x P = x^v, \qquad x \in TE.$$

To see this, observe that if $\kappa$ denotes the connection map of $\nabla$ and $x \in T_u E$, then
$$\nabla_x P = \nabla_x \mathcal{J} 1_E = \mathcal{J}_u \tilde{\nabla}_x 1_E = \mathcal{J}_u \kappa 1_{E*} x = \mathcal{J}_u \kappa x = x^v.$$
By Theorem 3.1 in Chapter 4, $d^{\nabla^2} P(x,y) = R(x,y)P$, which together with (6.1) implies

(6.3)     $d^{\nabla^2} P(x,y) = \mathcal{J}_u \tilde{R}(\pi_* x, \pi_* y)u, \qquad x,y \in T_u E, \quad u \in E.$

The following observations will be used throughout the section:

REMARK 6.1. (i) The wedge product from (3.4) extends to all of $A(M, \Lambda\xi)$: For $\alpha \in A_p(M, \Lambda_q \xi)$, $\beta \in A_r(M, \Lambda_s \xi)$, define $\alpha \wedge_\xi \beta \in A_{p+r}(M, \Lambda_{q+s}\xi)$ by

$$(\alpha \wedge_\xi \beta)(X_1, \ldots, X_{p+r}) = \frac{1}{p!r!} \sum_{\sigma \in P_{p+r}} (\operatorname{sgn} \sigma) \alpha(X_{\sigma(1)}, \ldots, X_{\sigma(p)})$$
$$\wedge \beta(X_{\sigma(p+1)}, \ldots, X_{\sigma(p+r)}).$$

(The wedge product in the right side is of course the one in $\Lambda\xi$). Then $\alpha \wedge_\xi \beta = (-1)^{pr+qs} \beta \wedge_\xi \alpha$, and $d^\nabla(\alpha \wedge_\xi \beta) = (d^\nabla \alpha) \wedge_\xi \beta + (-1)^p \alpha \wedge_\xi d^\nabla \beta$. Notice that $\tilde{R} \in A_2(M, \Lambda_2 \xi)$ commutes with any other $\Lambda\xi$-valued form.

(ii) Since $\star_\xi$ is parallel, (0.1) implies
$$d(\star_\xi \alpha) = \star_\xi d^\nabla \alpha, \qquad \alpha \in A(M, \Lambda_n \xi).$$

(iii) If $u_i$ is an orthonormal basis of $E_{\pi(u)}$, then $\mathcal{J}_u u_i$ is an orthonormal basis of $(\mathcal{V}E)_u$. Thus, by (6.1), $\star_\mathcal{V} R^k = \pi^* \star_\xi \tilde{R}^k$.

(iv) Let $U_1, \ldots, U_n$ denote a local orthonormal basis of sections of $\xi$. If $\alpha$, $\beta$ are sections of $\Lambda_p \xi$, $\Lambda_{n-p}\xi$, respectively, then locally,

$$\alpha = \sum_{i_1 < \cdots < i_p} \langle \alpha, U_{i_1} \wedge \cdots \wedge U_{i_p} \rangle U_{i_1} \wedge \cdots \wedge U_{i_p}.$$

A similar expression holds for $\beta$, so that

$$\langle \alpha \wedge \beta, U_1 \wedge \cdots \wedge U_n \rangle = \sum_{\substack{i_1 < \cdots < i_p \\ j_1 < \cdots < j_{n-p}}} \epsilon^{i_1 \ldots i_p j_1 \ldots j_{n-p}} \langle \alpha, U_{i_1} \wedge \cdots \wedge U_{i_p} \rangle$$
$$\langle \beta, U_{j_1} \wedge \cdots \wedge U_{j_{n-p}} \rangle$$
$$= \frac{1}{p!} \frac{1}{(n-p)!} \sum_{\sigma \in P_n} (\operatorname{sgn} \sigma) \langle \alpha, U_{\sigma(1)} \wedge \cdots \wedge U_{\sigma(p)} \rangle$$
$$\langle \beta, U_{\sigma(p+1)} \wedge \cdots \wedge U_{\sigma(n)} \rangle.$$

This identity also holds when $\alpha$, $\beta$ are $\Lambda\xi$-valued forms on M as in (i).

From now on we will work in $E^1$, so let $\mathcal{V}\xi$ denote the restriction $\imath^* \mathcal{V}\xi$ of the vertical bundle to $E^1$, where $\imath : E^1 \hookrightarrow E$ is inclusion. Similarly, $P$ will denote the restriction $P \circ \imath$ of the position vector field, $R$ the pullback $\imath^* R \in A_2(E^1, \Lambda_2 \mathcal{V}\xi)$ of $R$, and $\pi : E^1 \to M$ the projection. For $i = 1, \ldots, k$, define $\omega_i \in A_{n-1}(E^1, \Lambda_n \mathcal{V}\xi)$ by

$$\omega_i = P \wedge_\mathcal{V} (d^\nabla P)^{2i-1} \wedge_\mathcal{V} R^{k-i},$$

with the wedge product as defined in Remark 6.1(i).

LEMMA 6.2.
$$d^\nabla \omega_i = (d^\nabla P)^{2i} \wedge_\nu R^{k-i} - \frac{2i-1}{k-i+1}(d^\nabla P)^{2i-2} \wedge_\nu R^{k-i+1}.$$

PROOF. By the Bianchi identity,
$$d^\nabla \omega_i = (d^\nabla P)^{2i} \wedge_\nu R^{k-i} + (2i-1)P \wedge_\nu d^{\nabla 2}P \wedge_\nu (d^\nabla P)^{2i-2} \wedge R^{k-i}$$
$$= (d^\nabla P)^{2i} \wedge_\nu R^{k-i} + (2i-1)(d^\nabla P)^{2i-2} \wedge_\nu P \wedge_\nu d^{\nabla 2}P \wedge_\nu R^{k-i}.$$

In order to evaluate the second summand, consider a positively oriented orthonormal basis $U_j$ of local sections of $\mathcal{V}\xi$ with $U_{2i-1} = P$. Then

$$\langle (d^\nabla P)^{2i-2} \wedge_\nu P \wedge_\nu d^{\nabla 2}P \wedge_\nu R^{k-i}, U_1 \wedge \cdots \wedge U_n \rangle$$

$$= \frac{1}{2^{k-i}} \sum_{\sigma \in P_n} (\operatorname{sgn}\sigma)\langle d^\nabla P, U_{\sigma(1)}\rangle \wedge \cdots \wedge \langle d^\nabla P, U_{\sigma(2i-2)}\rangle \wedge \langle P, U_{\sigma(2i-1)}\rangle$$

$$\wedge \langle d^{\nabla 2}P, U_{\sigma(2i)}\rangle \wedge \langle R, U_{\sigma(2i+1)} \wedge U_{\sigma(2i+2)}\rangle \wedge \cdots$$

$$\wedge \langle R, U_{\sigma(n-1)} \wedge U_{\sigma(n)}\rangle$$

$$= -\frac{1}{2^{k-i}} \sum_{\{\sigma | \sigma(2i-1)=2i-1\}} (\operatorname{sgn}\sigma)\langle d^\nabla P, U_{\sigma(1)}\rangle \wedge \cdots \wedge \langle d^\nabla P, U_{\sigma(2i-2)}\rangle$$

$$\wedge \langle R, U_{\sigma(2i-1)} \wedge U_{\sigma(2i)}\rangle \wedge \cdots \wedge \langle R, U_{\sigma(n-1)} \wedge U_{\sigma(n)}\rangle$$

by (6.3) and (6.1). Fix any $\sigma \in P_n$ with $\sigma(2i-1) \neq 2i-1$, so that $P = U_{\sigma(l)}$ for some $l \neq 2i-1$. If $l < 2i-1$, the corresponding expression in the last equality vanishes, because $\langle d^\nabla P, U_{\sigma(l)}\rangle = \langle d^\nabla P, P\rangle$, and $\langle d^\nabla P(x), P\rangle = \langle \nabla_x P, P\rangle = \frac{1}{2}x\langle P, P\rangle = 0$ on $E^1$ where $P$ has constant norm 1. If $l > 2i-1$, then the corresponding expression is the same as in the case $\sigma(2i-1) = 2i-1$: In fact, $(\operatorname{sgn}\sigma)\langle R, U_{\sigma(2i-1)} \wedge U_{\sigma(2i)}\rangle \wedge \cdots \wedge \langle R, U_{\sigma(n-1)} \wedge U_{\sigma(n)}\rangle$ remains unchanged when switching pairs $(\sigma(2p-1), \sigma(2p))$ and $(\sigma(2q-1), \sigma(2q))$. Similarly, this expression undergoes a sign change twice when interchanging $\sigma(2p-1)$ and $\sigma(2p)$ (once in $\langle R, U_{\sigma(2p-1)} \wedge U_{\sigma(2p)}\rangle$ and again in $(\operatorname{sgn}\sigma)$). Thus,

$$\langle (d^\nabla P)^{2i-2} \wedge_\nu P \wedge_\nu d^{\nabla 2}P \wedge_\nu R^{k-i}, U_1 \wedge \cdots \wedge U_n \rangle$$

$$= -\frac{1}{2^{k-i}2(k-i+1)} \sum_{\sigma \in P_n} (\operatorname{sgn}\sigma)\langle d^\nabla P, U_{\sigma(1)}\rangle \wedge \cdots \wedge \langle d^\nabla P, U_{\sigma(2i-2)}\rangle$$

$$\wedge \langle R, U_{\sigma(2i-1)} \wedge U_{\sigma(2i)}\rangle \wedge \cdots \wedge \langle R, U_{\sigma(n-1)} \wedge U_{\sigma(n)}\rangle$$

$$= -\frac{2^{k-i+1}}{2^{k-i}2(k-i+1)} \langle (d^\nabla P)^{2i-2} \wedge_\nu R^{k-i+1}, U_1 \wedge \cdots \wedge U_n \rangle,$$

so that

$$(d^\nabla P)^{2i-2} \wedge_\nu P \wedge_\nu d^{\nabla 2}P \wedge_\nu R^{k-i} = -\frac{1}{k-i+1}(d^\nabla P)^{2i-2} \wedge_\nu R^{k-i+1}.$$

Substituting into the original expression for $d^\nabla \omega_i$ then yields the result. $\qquad\square$

THEOREM 6.1. *Consider an oriented Euclidean bundle* $\xi = E \to M$ *of rank* $n = 2k$, *with Riemannian connection and corresponding Euler form* $e$. *If* $\xi^1 = \pi : E^1 \to M$ *denotes the unit sphere bundle of* $\xi$, *then the pullback* $\pi^*e \in A_n(E^1)$ *of the Euler form is exact. Specifically, there exists* $\Omega \in A_{n-1}(E^1)$ *such that*

(1) $\pi^* e = d\Omega$; and

(2) for $p \in M$, $\int_{(E^1)_p} \jmath^*\Omega = -1$, where $\jmath : (E^1)_p \hookrightarrow E^1$ denotes inclusion.

PROOF. Set $\Omega_i := \star_\mathcal{V}\omega_i \in A_{n-1}(E^1)$. We seek constants $a_i \in \mathbb{R}$ so that $\Omega = \sum a_i \Omega_i$ satisfies (1) and (2). We begin with (2): Given $u \in (E^1)_p$, $p \in M$, consider a local positively oriented orthonormal basis $U_i$ of sections of $\mathcal{V}\xi$ in a neighborhood of $u$ with $U_1 = P$. Since $P$ is orthogonal to the unit sphere $S^{n-1} = (E^1)_p$, the volume form of $(E^1)_p$ is given locally by $(U_2 \wedge \cdots \wedge U_n)^\flat \circ \jmath$. Now, $\jmath^*\Omega$ is vertical, whereas $R$ is horizontal by (6.1), so that

$$\jmath^*\Omega = a_k \jmath^*\Omega_k = a_k \langle \jmath^*\omega_k, U_1 \wedge \cdots \wedge U_n \rangle$$

$$= a_k \sum_{\sigma \in P_n} (\text{sgn}\,\sigma)\jmath^*(\langle P, U_{\sigma(1)} \rangle \wedge \langle d^\nabla P, U_{\sigma(2)} \rangle \wedge \cdots \wedge \langle d^\nabla P, U_{\sigma(n)} \rangle)$$

$$= a_k \sum_{\{\sigma | \sigma(1)=1\}} (\text{sgn}\,\sigma)\jmath^*(\langle d^\nabla P, U_{\sigma(2)} \rangle \wedge \cdots \wedge \langle d^\nabla P, U_{\sigma(n)} \rangle)$$

$$= a_k \sum_{\tau \in P_{n-1}} (\text{sgn}\,\tau)\jmath^*(\langle d^\nabla P, U_{1+\tau(1)} \rangle \wedge \cdots \wedge \langle d^\nabla P, U_{1+\tau(n-1)} \rangle)$$

$$= a_k(n-1)! \jmath^*(\langle d^\nabla P, U_2 \rangle \wedge \cdots \wedge \langle d^\nabla P, U_n \rangle).$$

Now, for $x_i$ in the tangent space at $u$ of the fiber $(E^1)_p$ over $p$, $\jmath_* x_i$ is vertical, and $\jmath^* d^\nabla P(x_i) = \jmath_* x_i$ by (6.2). Thus,

$$\jmath^*\Omega(x_1, \ldots, x_{n-1}) = a_k(n-1)! \sum_{\sigma \in P_{n-1}} (\text{sgn}\,\sigma)\langle \jmath_* x_{\sigma(1)}, U_2(u) \rangle \cdots$$

$$\langle \jmath_* x_{\sigma(n-1)}, U_n(u) \rangle$$

$$= a_k(n-1)! \det\langle \jmath_* x_i, U_j(u) \rangle$$

$$= a_k(n-1)! \langle \jmath_* x_1 \wedge \cdots \wedge \jmath_* x_{n-1}, U_2(u) \wedge \cdots \wedge U_n(u) \rangle$$

$$= a_k(n-1)! \,\omega(x_1, \ldots, x_{n-1}),$$

where $\omega$ denotes the volume form of $(E^1)_p$. Since the latter is isometric to $S^{n-1}$,

$$\int_{(E^1)_p} \jmath^*\Omega = a_k(n-1)! \,\text{vol}(S^{n-1}) = a_k(n-1)! \frac{2\pi^k}{(k-1)!} = -1,$$

if we set $a_k = -(k-1)!/2\pi^k(2k-1)!$.

Next, we determine $a_i$ for $i < k$ so that $\Omega$ satisfies (1):

$$d^\nabla \left( \sum a_i \omega_i \right) = \sum a_i \left( (d^\nabla P)^{2i} \wedge_\mathcal{V} R^{k-i} - \frac{2i-1}{k-i+1}(d^\nabla P)^{2i-2} \wedge_\mathcal{V} R^{k-i+1} \right)$$

$$= -\frac{a_1}{k} R^k + a_k(d^\nabla P)^{2k} + \sum_{i=1}^{k-1} \left( a_i - \frac{2i+1}{k-i} a_{i+1} \right) (d^\nabla P)^{2i}$$

$$\wedge_\mathcal{V} R^{k-i}.$$

With $a_k$ as above, define $a_i = (2i+1/k-i)a_{i+1}$ inductively; then

$$a_i = -\frac{(i-1)!}{2^{k-i+1}\pi^k(k-i)!(2i-1)!},$$

and

$$d^\nabla \left( \sum a_i \omega_i \right) = \frac{1}{(2\pi)^k} R^k + a_k (d^\nabla P)^{2k},$$

so that

$$d\Omega = d \left( \star_V \sum a_i \omega_i \right) = \star_V d^\nabla \left( \sum a_i \omega_i \right) = \frac{1}{(2\pi)^k} \star_V R^k + a_k \star_V (d^\nabla P)^{2k}.$$

Finally, $\star_V (d^\nabla P)^{2k} = 0$ by (6.2), and $\star_V R^k = \pi^* \star_\xi \tilde{R}^k$. Thus, $d\Omega = \pi^* e$, as claimed. $\qquad\qquad\square$

EXERCISE 158. Use Theorem 6.1 to show once again that the Euler class of a vector bundle vanishes if the bundle admits a nowhere-zero section.

## 7. The Generalized Gauss-Bonnet Theorem

Let $M$ be a compact, oriented, $n$-dimensional manifold. The *Euler characteristic* of $M$ is defined to be $\chi(M) = \sum_{k=0}^{n} (-1)^k \dim H^k(M)$. It turns out that this number can be computed by looking at the behavior of any vector field $X$ on $M$ with finitely many zeros. We shall only explain the procedure and concepts involved. For a proof, the reader is referred to [**25**]. Let $p$ be a zero of $X$, choose $\epsilon \in (0, \mathrm{inj}_p)$ (for some Riemannian metric on $M$) so that the ball $B_\epsilon(p)$ of radius $\epsilon$ centered at $p$ contains no other zeros of $X$, and denote by $S^{n-1}$ the unit sphere centered at 0 in $M_p$. Consider the maps $\imath_t : S^{n-1} \to S^{n-1} \times [0, \epsilon]$, $t \in [0, \epsilon]$, $\imath_t(v) = (v, t)$, and $H : S^{n-1} \times [0, \epsilon] \to M$, $H(v, t) = \exp_p(tv)$. The *index* $\mathrm{ind}_p X$ of $X$ at $p$ is the degree of the map $f_\epsilon := \Phi_\epsilon \circ (X/|X|) \circ H \circ \imath_\epsilon : S^{n-1} \to S^{n-1}$. Here, $\Phi_\epsilon : T^1 M_{\partial B_\epsilon(p)} \to S^{n-1}$ maps $u \in T^1 M \cap M_{\exp \epsilon v}$ to the parallel translate of $u$ along the geodesic $t \mapsto \exp_p((\epsilon - t)v)$, $0 \le t \le \epsilon$.

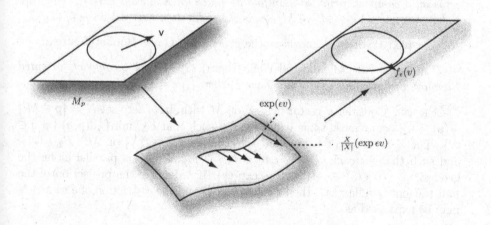

FIGURE 1

Thus, to obtain the value of $f_\epsilon$ at a point $v \in S^{n-1}$, one goes out at distance $\epsilon$ along the geodesic in direction $v$, evaluates the normalized vector field $X/|X|$ at that point, and parallel translates it back to $p$ along the same geodesic. The index of $X$ at $p$ is well-defined, for if $\delta \in (0, \mathrm{inj}_p)$, then $f_\delta$ and $f_\epsilon$ are homotopic via $(v, t) \mapsto f_{t\delta + (1-t)\epsilon}(v)$.

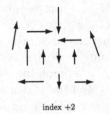

FIGURE 2

The proof of the following theorem can be found, for example, in [25]:

THEOREM 7.1 (Poincaré-Hopf). *Let $X$ be a vector field with finitely many zeros on a compact, oriented manifold $M$ (such an $X$ always exists). Then the Euler characteristic $\chi(M)$ of $M$ equals the index sum over all zeros of $X$.*

The next theorem is known as the generalized Gauss-Bonnet theorem:

THEOREM 7.2 (Allendoerfel-Weil, Chern). *If $M^{2k}$ is a compact, oriented Riemannian manifold with Euler form $e$, then $\int_M e = \chi(M)$.*

PROOF. Consider a vector field $X$ on $M$ with finite zero set $N = \{p \in M \mid X(p) = 0\}$, and choose some $0 < \epsilon < \mathrm{inj}_M$ such that $\epsilon < \min\{\frac{1}{2}d(p, q) \mid p, q \in N\}$. Let $Z$ be a vector field on $M \setminus N$ that equals $X/|X|$ on $M \setminus B_{2\epsilon/3}(N)$, and such that for each $p \in N$, and unit $v \in M_p$, $Z \circ c_v$ is parallel along the geodesic $c_v : (0, \epsilon/3) \to M$, $c_v(t) = \exp(tv)$. If $\pi$ denotes the projection of the unit tangent bundle $\tau^1 M$, then $\pi \circ Z = 1_{M \setminus N}$, and the restriction of $e$ to $M \setminus N$ may be expressed as

$$e = (\pi \circ Z)^* e = Z^* \pi^* e = Z^* d\Omega = dZ^* \Omega,$$

with $\Omega$ denoting the $(2k-1)$-form on $T^1 M$ from Theorem 6.1. Let $p \in N$, and $H : S^{n-1} \times [0, \epsilon] \to M$ as above. Although $Z$ does not extend continuously to $B_\epsilon(p)$, we obtain a differentiable vector field $Y$ along $H$ by setting

$$Y(v, t) = Z \circ H(v, t) \text{ for } t > 0, \text{ and } Y(v, 0) = \lim_{t \to 0^+} Z \circ H(v, t).$$

Stokes' theorem then implies

$$\int_{\overline{B_\epsilon}(p)} e = \int_{H(S^{n-1}\times[0,\epsilon])} dZ^*\Omega = \int_{S^{n-1}\times[0,\epsilon]} H^* dZ^*\Omega = \int_{S^{n-1}\times[0,\epsilon]} dH^* Z^*\Omega$$

$$= \int_{S^{n-1}\times\epsilon} H^* Z^*\Omega - \int_{S^{n-1}\times 0} H^* Z^*\Omega$$

$$= \int_{\partial B_\epsilon(p)} Z^*\Omega \quad - \int_{S^{n-1}\times 0} H^* Z^*\Omega.$$

Since the degree of $Y \circ \imath_0$ equals the index of $X$ at $p$, we have

$$\int_{S^{n-1}\times 0} H^* Z^*\Omega = \int_{S^{n-1}} (Y \circ \imath_0)^*\Omega = \operatorname{ind}_p X \int_{S^{n-1}} \Omega = -\operatorname{ind}_p X.$$

Thus,

$$\int_M e = \int_{M\setminus\cup_{p\in N} B_\epsilon(p)} e + \sum_{p\in N} \int_{\overline{B_\epsilon}(p)} e$$

$$= \int_{M\setminus\cup_{p\in N} B_\epsilon(p)} dZ^*\Omega + \sum_{p\in N} \left( \int_{\partial B_\epsilon(p)} Z^*\Omega + \operatorname{ind}_p X \right)$$

$$= -\sum_{p\in N} \int_{\partial B_\epsilon(p)} Z^*\Omega + \sum_{p\in N} \int_{\partial B_\epsilon(p)} Z^*\Omega + \sum_{p\in N} \operatorname{ind}_p X$$

$$= \chi(M).$$

$\square$

EXAMPLES AND REMARKS 7.1. (i) For an oriented surface $M^2$, the Euler form at $p \in M$ is given by $e(x,y) = \frac{1}{2\pi}\langle R(x,y)v, u\rangle$, where $u, v$ is a positively oriented orthonormal basis of $M_p$. Thus, $e = \frac{1}{2\pi} K\omega$, with $K$ the sectional curvature, and $\omega$ the volume form of $M$. The 2-dimensional case then reduces to the classical Gauss-Bonnet theorem:

$$\int_M K\omega = 2\pi\chi(M).$$

(ii) A Riemannian manifold is said to be *Einstein* if the Ricci curvature $\operatorname{Ric} = \kappa\langle,\rangle$ of $M$ equals a constant multiple $\kappa$ of the metric, $\kappa \in \mathbb{R}$. Any space of constant curvature is Einstein of course, but so is for example $S^2 \times S^2$ with the product metric.

Consider a compact, oriented 4-dimensional Einstein manifold. Given $p \in M$, let $e_1, \ldots, e_4$ denote a positively oriented orthonormal basis of $M_p$. If

$$\alpha_2 = \frac{1}{\sqrt{2}} e_1 \wedge e_2, \qquad \alpha_3 = \frac{1}{\sqrt{2}} e_1 \wedge e_3, \qquad \alpha_4 = \frac{1}{\sqrt{2}} e_1 \wedge e_4,$$

then $\alpha_i \pm \star\alpha_i$ $i = 2, 3, 4$, is an orthonormal basis of $\Lambda_2^{\pm}(M_p)$. We claim that $\langle R\alpha, \beta\rangle = 0$ for all $\alpha \in \Lambda_2^+(M_p)$ and $\beta \in \Lambda_2^-(M_p)$. To see this, notice first that

$$0 = \operatorname{Ric}(e_2, e_3) = \langle R(e_1, e_2)e_3, e_1\rangle + \langle R(e_4, e_2)e_3, e_4\rangle$$

$$= \langle Re_1 \wedge e_2, e_1 \wedge e_3\rangle - \langle Re_4 \wedge e_2, e_3 \wedge e_4\rangle$$

$$= 2\langle R\alpha_2, \alpha_3\rangle - 2\langle R \star \alpha_3, \star\alpha_2\rangle.$$

Similarly,

$$0 = \text{Ric}(e_1, e_4) = \langle R(e_2, e_1)e_4, e_2 \rangle + \langle R(e_3, e_1)e_4, e_3 \rangle$$
$$= 2\langle R\alpha_2, \star\alpha_3 \rangle - 2\langle R\alpha_3, \star\alpha_2 \rangle.$$

Thus,

$$\langle R(\alpha_2 + \star\alpha_2), \alpha_3 - \star\alpha_3 \rangle = \frac{1}{2}(\text{Ric}(e_2, e_3) - \text{Ric}(e_1, e_4)) = 0.$$

A similar computation shows that $\langle R(\alpha_i + \star\alpha_i), \alpha_j - \star\alpha_j \rangle = 0$ whenever $i \neq j$. When $i = j$,

$$\langle R(\alpha_i + \star\alpha_i), \alpha_i - \star\alpha_i) \rangle = \langle R\alpha_i, \alpha_i \rangle - \langle R \star \alpha_i, \star\alpha_i \rangle = K_{\alpha_i} - K_{\star\alpha_i},$$

with $K_{\alpha_i}$ denoting the sectional curvature of the plane spanned by $e_1$ and $e_i$. This expression is always zero: For example,

$$K_{\alpha_2} - K_{\star\alpha_2} = K_{e_1,e_2} - K_{e_3,e_4}$$
$$= \frac{1}{2}(\text{Ric}(e_1, e_1) + \text{Ric}(e_2, e_2) - \text{Ric}(e_3, e_3) - \text{Ric}(e_4, e_4)) = 0.$$

This establishes our claim that the curvature tensor $R$ of an oriented 4-dimensional Einstein manifold leaves the subspaces $\Lambda_2^{\pm}(M_p)$ invariant; i.e., $R_-^+ = R_+^- = 0$ in the notation of Section 4. By Proposition 5.2, the Euler form of $M$ equals

$$e = \frac{1}{8\pi^2}(|R_+^+|^2 + |R_-^-|^2).$$

According to the Gauss-Bonnet theorem, the Euler characteristic of $M$ is then nonnegative, and is zero iff $M$ is flat.

(iii) One large class of Einstein manifolds is the one consisting of so-called semi-simple Lie groups: The *Killing form* $B : \mathfrak{g} \times \mathfrak{g} \to \mathbb{R}$ of a Lie algebra $\mathfrak{g}$ is the symmetric bilinear form given by $B(X, Y) = \text{tr}\,\text{ad}_X \circ \text{ad}_Y$. A Lie group $G$ is said to be *semi-simple* if the Killing form of its Lie algebra is nondegenerate.

It turns out that a compact Lie group is semi-simple iff it has discrete center $Z(G) = \{g \in G \mid gh = hg \text{ for all } h \in G\}$. To see this, assume first that $G$ is compact and semi-simple. By compactness, there exists an inner product on $\mathfrak{g}$ for which $\text{ad}_X : \mathfrak{g} \to \mathfrak{g}$ is skew-adjoint for each $X \in \mathfrak{g}$, cf. Examples and remarks 1.1(ii) in Chapter 5. If $(a_{ij})$ denotes the matrix of $\text{ad}_X$ with respect to some orthonormal basis of $\mathfrak{g}$, then

$$B(X, X) = \text{tr}\,\text{ad}_X^2 = \sum_{i,j} a_{ij} a_{ji} = -\sum_{i,j} a_{ij}^2 \leq 0,$$

and equals zero iff $\text{ad}_X = 0$. Thus, the kernel of $\text{ad} = \text{Ad}_{*e}$ is trivial (see the observation in Section 5 of Chapter 4), so that $Z(G) \subset \ker \text{Ad}$ has trivial Lie algebra, and must be discrete. Notice that in fact, $Z(G) = \ker \text{Ad}$: If $g \in \ker \text{Ad}$, then for any $X \in \mathfrak{g}$, $X(g) = R_{g*}X(e)$. Thus, the curve $c$, where $c(t) = R_g(\exp tX)$, is an integral curve of $X$ which passes through $g$ when $t = 0$. By uniqueness of integral curves, it must equal $t \mapsto L_g(\exp tX)$. Since the exponential map is onto (we are implicitly assuming $G$ is connected), $g$ belongs to the center. This also implies that conversely, if $Z(G)$ is discrete, then its Lie algebra is trivial, and $B$ is nondegenerate.

Suppose then that $G$ is compact and semi-simple, so that its Killing form is nondegenerate. Then $-B$ is an inner product on $\mathfrak{g}$ for which $\mathrm{ad}_Y$ is skew-adjoint, $Y \in \mathfrak{g}$:

$$
\begin{aligned}
-B(\mathrm{ad}_Y X, Z) = B(\mathrm{ad}_X Y, Z) &= \mathrm{tr}(\mathrm{ad}_{[X,Y]} \circ \mathrm{ad}_Z) \\
&= \mathrm{tr}(\mathrm{ad}_X \circ \mathrm{ad}_Y \circ \mathrm{ad}_Z - \mathrm{ad}_Y \circ \mathrm{ad}_X \circ \mathrm{ad}_Z) \\
&= \mathrm{tr}(\mathrm{ad}_X \circ \mathrm{ad}_Y \circ \mathrm{ad}_Z - \mathrm{ad}_X \circ \mathrm{ad}_Z \circ \mathrm{ad}_Y) \\
&= \mathrm{tr}(\mathrm{ad}_X \circ \mathrm{ad}_{[Y,Z]}) = B(X, \mathrm{ad}_Y Z).
\end{aligned}
$$

Thus, $-B$ induces a so-called *canonical bi-invariant metric* on $G$. (2.4) in Chapter 5 implies that $G$ with the canonical metric is an Einstein manifold with $\kappa = \frac{1}{4}$: Given $X, Y \in \mathfrak{g}$, and an orthonormal basis $Z_i$ of $\mathfrak{g}$,

$$
\begin{aligned}
\mathrm{Ric}(X,Y) = \sum_i \langle R(Z_i, X)Y, Z_i \rangle &= -\frac{1}{4} \sum_i \langle [[Z_i, X], Y], Z_i \rangle \\
&= -\frac{1}{4} \mathrm{tr}(\mathrm{ad}_X \circ \mathrm{ad}_Y) = \frac{1}{4} \langle X, Y \rangle.
\end{aligned}
$$

EXERCISE 159. Prove that a compact, oriented 4-dimensional Riemannian manifold $M$ with constant curvature $\kappa$ has Euler characteristic $\chi(M) = \frac{3\kappa^2}{4\pi^2} \mathrm{vol}(M)$.

EXERCISE 160. Let $G$ be a compact, connected, semi-simple Lie group with its canonical metric, $L : \Lambda_2 \mathfrak{g} \to \mathfrak{g}$ the linear map which on decomposable elements is given by $L(X \wedge Y) = [X, Y]$. Show that for any $\alpha \in \Lambda_2 \mathfrak{g}$, $\langle R\alpha, \alpha \rangle = \frac{1}{4} |T\alpha|^2$. Thus, $G$ has nonnegative-definite curvature operator. The Gauss-Bonnet theorem can be used to show that any Riemannian manifold with nonnegative curvature operator has nonnegative Euler characteristic.

## 8. Complex and Symplectic Vector Spaces

There is yet another characteristic class, called the Chern class, that can be defined on certain bundles possessing additional structure. Before introducing it, we review some basic notions from complex linear algebra. The reader familiar with the material may proceed to Theorem 8.1 below without loss of continuity.

A *complex vector space* $(V, +, \cdot)$ is a set $V$ together with two operations $+$, $\cdot$, satisfying the usual vector space axioms, but taking $\mathbb{C}$ as the scalar field instead of $\mathbb{R}$. A (complex) linear transformation $L : V \to W$ between complex spaces $V$, $W$ is a map that satisfies $L(v + w) = Lv + Lw$, $L(\alpha v) = \alpha Lv$, for $v, w \in V$, $\alpha \in \mathbb{C}$. The standard example of a complex vector space is $(\mathbb{C}^n, +, \cdot)$ where for $v = (\alpha_1, \ldots, \alpha_n)$, $w = (\beta_1, \ldots, \beta_n) \in \mathbb{C}^n$, and $\alpha \in \mathbb{C}$, $v + w = (\alpha_1 + \beta_1, \ldots, \alpha_n + \beta_n)$, $\alpha \cdot v = (\alpha \alpha_1, \ldots, \alpha \alpha_n)$. All standard notions from real linear algebra, such as linear independence, bases, etc., carry over. If $\mathbf{e}_j$ denotes the $n$-tuple with 1 in the $j$-th slot and 0 elsewhere, then any $n$-dimensional complex vector space is isomorphic to $\mathbb{C}^n$ by mapping a basis $v_1, \ldots, v_n$ of $V$ pointwise to $\mathbf{e}_1, \ldots, \mathbf{e}_n$ and extending linearly.

DEFINITION 8.1. The *realification* $V_{\mathbb{R}}$ of a complex vector space $(V, +, \cdot)$ is the real vector space $(V, +, \cdot_{|\mathbb{R}})$ with scalar multiplication restricted to the reals.

An endomorphism $J$ of a real vector space $V$ is said to be a *complex structure* on $V$ if $J^2 = -1_V$, cf. Exercise 57 in Chapter 2. The realification $V_{\mathbb{R}}$ of a complex space $V$ admits a canonical complex structure given by $Jv = iv$ for $v \in V$. Conversely, any real space with a complex structure $J$ becomes a complex space when defining $(a + ib)v = av + bJv$.

DEFINITION 8.2. The *complexification* $V_{\mathbb{C}}$ of a real vector space $V$ is the complex space determined by the (real) space $V \oplus V$ together with the complex structure $J$ given by $J(v, w) = (-w, v)$.

Thus, the isomorphism $(V_{\mathbb{C}})_{\mathbb{R}} \cong V \oplus V$ maps $u + iv$ to $(u, v)$. One customarily thinks of $\mathbb{C}^n$ as the complexification of $\mathbb{R}^n$, so that the identification between the underlying real spaces is given by

$$h : (\mathbb{C}^n)_{\mathbb{R}} \longrightarrow (\mathbb{R}^n_{\mathbb{C}})_{\mathbb{R}} = \mathbb{R}^n \times \mathbb{R}^n,$$
$$v \longmapsto (\operatorname{Re} v, \operatorname{Im} v).$$

The isomorphism $h$ induces a linear map $h : M_{n,n}(\mathbb{C}) \to M_{2n,2n}(\mathbb{R})$ from the space of $n \times n$ complex matrices to the space of $2n \times 2n$ real ones determined by $h(Mv) = h(M)h(v)$, for $M \in M_{n,n}(\mathbb{C})$ and $v \in \mathbb{C}^n$. Writing $M = A + iB$ with $A, B \in M_{n,n}(\mathbb{R})$, we have for $v = x + iy \in \mathbb{C}^n$,

$$h(M)(x, y) = h(Mv) = h((A + iB)(x + iy)) = h(Ax - By + i(Bx + Cy))$$
$$= (Ax - By, By + Ax).$$

Thus,

$$(8.1) \qquad h(M) = \begin{pmatrix} \operatorname{Re} M & -\operatorname{Im} M \\ \operatorname{Im} M & \operatorname{Re} M \end{pmatrix} \in M_{2n,2n}(\mathbb{R}).$$

If we denote by $GL(n, \mathbb{C})$ the group of all invertible $n \times n$ complex matrices, then $h : GL(n, \mathbb{C}) \to GL(2n, \mathbb{R})$ is a group homomorphism. Identifying $GL(n, \mathbb{C})$ with its image shows that it is a Lie subgroup of $GL(2n, \mathbb{R})$ of dimension $2n^2$.

DEFINITION 8.3. A *Hermitian inner product* on a complex vector space $V$ is a map $\langle , \rangle : V \times V \to \mathbb{C}$ satisfying

(1) $\langle \alpha v_1 + v_2, v \rangle = \alpha \langle v_1, v \rangle + \langle v_2, v \rangle$,
(2) $\overline{\langle v_1, v_2 \rangle} = \langle v_2, v_1 \rangle$, and
(3) $\langle v, v \rangle > 0$ if $v \neq 0$,

for all $\alpha \in \mathbb{C}$, $v, v_i \in V$.

By (1) and (2), $\langle v, \alpha w \rangle = \overline{\alpha} \langle v, w \rangle$. For example, the *standard Hermitian inner product* on $\mathbb{C}^n$ is given by

$$\langle v, w \rangle = \sum_j \alpha_j \overline{\beta_j}, \qquad v = (\alpha_1, \dots, \alpha_n), \quad w = (\beta_1, \dots, \beta_n).$$

If $J$ denotes a complex structure on a real space $V$, there is always a (real) inner product on $V$ for which $J$ is skew-adjoint: Let $v_1$ be any nonzero vector, and set $v_{n+1} := Jv_1$, $W_1 = \operatorname{span}\{v_1, v_{n+1}\}$. Since $J^2$ equals minus the identity, $W_1$ is invariant under $J$. Arguing inductively, $V$ decomposes as a direct sum $\oplus_k W_k$ of $J$-invariant planes $W_k = \operatorname{span}\{v_k, v_{n+k} = Jv_k\}$, $k = 1, \dots, n$. If $\langle , \rangle$

denotes that inner product for which the basis $v_1, \dots, v_{2n}$ is orthonormal, then the matrix of $J$ with respect to this basis is

$$\begin{pmatrix} 0 & -I_n \\ I_n & 0 \end{pmatrix} \in M_{2n,2n}(\mathbb{R}).$$

This shows that $J$ is skew-adjoint, and in fact isometric: $\langle Jv, Jw \rangle = -\langle J^2 v, w \rangle = \langle v, w \rangle$.

PROPOSITION 8.1. *Let $V$ be a complex vector space, $J$ the induced complex structure on the underlying real space $V_\mathbb{R}$. Given any inner product $\langle , \rangle$ on $V_\mathbb{R}$ for which $J$ is skew-adjoint, the formula*

(8.2) $$\langle v, w \rangle_\mathbb{C} := \langle v, w \rangle + i \langle v, Jw \rangle$$

*defines a Hermitian inner product $\langle , \rangle_\mathbb{C}$ on $V$. Conversely, if $\langle , \rangle_\mathbb{C}$ is a Hermitian inner product on $V$, then the real part of $\langle , \rangle_\mathbb{C}$ is an inner product on $V_\mathbb{R}$ with respect to which $J$ is skew-adjoint, and $\langle , \rangle_\mathbb{C}$ is given by (8.2).*

PROOF. Given a real inner product on $V_\mathbb{R}$, (8.2) defines a complex-valued function on $V \times V$ that is clearly additive in the first variable. Given $\alpha = a + ib \in \mathbb{C}$,

$$\begin{aligned} \langle \alpha v, w \rangle_\mathbb{C} &= \langle (a + ib)v, w \rangle + i \langle (a + ib)v, Jw \rangle \\ &= a \langle v, w \rangle + b \langle iv, w \rangle + ia \langle v, Jw \rangle + ib \langle iv, Jw \rangle \\ &= a \langle v, w \rangle - b \langle v, Jw \rangle + ia \langle v, Jw \rangle + ib \langle v, w \rangle \\ &= (a + ib)(\langle v, w \rangle + i \langle v, Jw \rangle) = \alpha \langle v, w \rangle_\mathbb{C}. \end{aligned}$$

The second axiom follows from $\overline{\langle v, w \rangle_\mathbb{C}} = \langle v, w \rangle - i \langle v, Jw \rangle = \langle w, v \rangle + i \langle w, Jv \rangle = \langle w, v \rangle_\mathbb{C}$ by the skew-adjoint property of $J$. For the same reason $\langle v, Jv \rangle = 0$, so that $\langle v, v \rangle_\mathbb{C} = \langle v, v \rangle > 0$ if $v \neq 0$. Thus, $\langle , \rangle$ is Hermitian. Conversely, if $\langle , \rangle_\mathbb{C}$ is a Hermitian inner product on $V$, it is elementary to check that its real part $\langle , \rangle$ is an inner product on $V_\mathbb{R}$. Furthermore,

$$\mathrm{Im}\langle v, w \rangle_\mathbb{C} = \mathrm{Re}(-i \langle v, w \rangle_\mathbb{C}) = \mathrm{Re}\langle v, iw \rangle_\mathbb{C} = \langle v, Jw \rangle,$$

so that (8.2) holds. Finally,

$$\langle Jv, w \rangle = \mathrm{Re}\langle iv, w \rangle_\mathbb{C} = \mathrm{Re}(i \langle v, w \rangle_\mathbb{C}) = -\mathrm{Im}\langle v, w \rangle_\mathbb{C} = -\langle v, Jw \rangle,$$

where the last equality follows from the previous equation, so that $J$ is skew-adjoint. □

There is an alternative way of describing complex structures:

DEFINITION 8.4. A *symplectic form* on a real vector space $V$ is a nondegenerate, skew-symmetric, bilinear form $\sigma$ on $V$. $(V, \sigma)$ is then called a *symplectic vector space*.

For example, the *canonical symplectic form* on $\mathbb{R}^{2n}$ is $\sigma_0 = \sum_{k=1}^{n} u^k \wedge u^{n+k}$. It is the bilinear form associated to the canonical complex structure $J_0$ on $\mathbb{R}^{2n}$, in the sense that $\sigma_0(v, w) = \langle J_0 v, w \rangle$; this follows for instance from the fact that the matrix of $\sigma_0$ with respect to the standard basis is given by

$$\begin{pmatrix} 0 & I_n \\ -I_n & 0 \end{pmatrix},$$

so that $\sigma_0(\mathbf{e}_i, \mathbf{e}_j) = -\langle J_0\mathbf{e}_j, \mathbf{e}_i \rangle = \langle J_0\mathbf{e}_i, \mathbf{e}_j \rangle$. Notice also that

$$\sigma_0(J_0 v, J_0 w) = \langle J_0^2 v, J_0 w \rangle = -\langle v, J_0 w \rangle = \langle J_0 v, w \rangle = \sigma_0(v, w).$$

Up to isomorphism, $\sigma_0$ is the only symplectic form: For any symplectic vector space $(V, \sigma)$, there exists an isomorphism $L : V \to \mathbb{R}^{2n}$ such that $\sigma_0(Lv, Lw) = \sigma(v, w)$ for all $v$, $w \in V$ (and in particular, $V$ is even-dimensional). This is the essence of the following:

PROPOSITION 8.2. *For any symplectic vector space $(V, \sigma)$, there exists a basis $\alpha_1, \ldots, \alpha_{2n}$ of the dual $V^*$ such that $\sigma = \sum_{k=1}^n \alpha_k \wedge \alpha_{n+k}$.*

PROOF. Let $v_1$ be any nonzero vector in $V$. Since $\sigma$ is nondegenerate, there exists some $w \in V$ with $\sigma(v_1, w) = 1$. Set $v_{n+1} := w$, $W := \text{span}\{v_1, v_{n+1}\}$, and $Z := \{v \in V \mid \sigma(v, v_1) = \sigma(v, v_{n+1}) = 0\}$. Any $v \in V$ can then be written as $v = w + (v - w) \in W + Z$, where $w = \sigma(v, v_{n+1})v_1 - \sigma(v, v_1)v_{n+1}$. If $u \in W \cap Z$, then $u = av_1 + bv_{n+1}$ for some $a$, $b \in \mathbb{R}$, and since $u$ also belongs to $Z$, $0 = \sigma(u, v_1) = b\sigma(v_{n+1}, v_1) = -b$. Similarly, $0 = \sigma(u, v_{n+1}) = a\sigma(v_1, v_{n+1}) = a$. Thus, $W \cap Z = \{0\}$, so that $V = W \oplus Z$, and the restriction of $\sigma$ to $Z$ is symplectic. Arguing inductively, we obtain a basis $v_1, \ldots, v_{2n}$ of $V$, with dual basis $\alpha_1, \ldots, \alpha_{2n}$, such that $\sigma = \sum \alpha_k \wedge \alpha_{n+k}$. $\qquad\square$

A symplectic form $\sigma$ and a complex structure $J$ on $V$ are said to be *compatible* if $\sigma(Jv, Jw) = \sigma(v, w)$ for all $v$, $w \in V$.

PROPOSITION 8.3. *If a real vector space has a complex structure $J$, then it admits a compatible symplectic form $\sigma$. Conversely, any symplectic form $\sigma$ on $V$ induces a compatible complex structure $J$. In each case, there exists an inner product on $V$ such that $\sigma(v, w) = \langle Jv, w \rangle$, and $J$ is an isometry.*

PROOF. Given a complex structure $J$ on $V$, choose an inner product for which $J$ is skew-adjoint, and hence also isometric. Then $\sigma$, where $\sigma(v, w) := \langle Jv, w \rangle$ is symplectic. Furthermore,

$$\sigma(Jv, Jw) = \langle J^2 v, Jw \rangle = -\langle v, Jw \rangle = \langle Jv, w \rangle = \sigma(v, w).$$

Conversely, if $\sigma$ is a symplectic form on $V$, choose a basis $v_k$ such that $\sigma = \sum \alpha_k \wedge \alpha_{n+k}$, where $\alpha_k$ denotes the basis dual to $v_k$, see Proposition 8.2. Consider the inner product on $V$ for which $v_k$ is orthonormal, and define $Jv := (\iota_v \sigma)^\#$; i.e., $\langle Jv, w \rangle = \sigma(v, w)$. The matrix of $J$ with respect to the basis $v_k$ has as $(k, l)$-th entry $\langle Jv_l, v_k \rangle = \sigma(v_l, v_k)$, and is thus given by

$$\begin{pmatrix} 0 & -I_n \\ I_n & 0 \end{pmatrix}.$$

This clearly implies that $J$ is a complex structure and an isometry. Compatibility of $J$ and $\sigma$ follows as before. $\qquad\square$

PROPOSITION 8.4. *If $V$ is a complex vector space, then its realification $V_{\mathbb{R}}$ inherits a canonical orientation.*

PROOF. An arbitrary basis $\{v_1, \ldots, v_n\}$ of $V$ induces an element

$$v_1 \wedge \cdots \wedge v_n \wedge Jv_1 \wedge \cdots \wedge Jv_n \in (\Lambda_{2n} V_{\mathbb{R}}) \setminus \{0\}.$$

The component of $(\Lambda_{2n} V_{\mathbb{R}}) \setminus \{0\}$ containing it is independent of the original basis: If $w_1, \ldots, w_n$ is another basis of $V$, and $\langle , \rangle$ is an inner product on $V_{\mathbb{R}}$ for which $J$ is skew-adjoint, then

$$\langle v_1 \wedge \cdots \wedge v_n \wedge Jv_1 \wedge \cdots \wedge Jv_n, w_1 \wedge \cdots \wedge w_n \wedge Jw_1 \wedge \cdots \wedge Jw_n \rangle = \det \begin{pmatrix} A & -B \\ B & A \end{pmatrix},$$

where the components $a_{ij}$, $b_{ij}$, of $A$, $B$, are given by $a_{ij} = \langle v_i, w_j \rangle = \langle Jv_i, Jw_j \rangle$, and $b_{ij} = \langle Jv_i, w_j \rangle = -\langle v_i, Jw_j \rangle$. The matrix above is the image via $h$ of $M = A + iB \in M_{n,n}(\mathbb{C})$. The claim now follows from: □

LEMMA 8.1. For $M \in M_{n,n}(\mathbb{C})$, $\det h(M) = |\det M|^2$.

PROOF. The claim is easily seen to be true for diagonalizable matrices. But the latter are dense in $M_{n,n}(\mathbb{C})$; in fact, we may assume that $M \in M_{n,n}(\mathbb{C})$ is in Jordan canonical form. If not all the diagonal terms are distinct, then modifying them appropriately yields a matrix arbitrarily close to $M$ with $n$ distinct eigenvalues. The latter is then diagonalizable. □

REMARK 8.1. There is another orientation on $V_{\mathbb{R}}$ that is commonly used, namely the one induced by $v_1 \wedge Jv_1 \wedge \cdots \wedge v_n \wedge Jv_n$, where $v_k$ is a basis of $V$. It coincides with ours only when $[n/2]$ is even. The reason behind our choice is that it makes the isomorphism $h : (\mathbb{C}^n_{\mathbb{R}}, \omega) \to (\mathbb{R}^{2n}, \text{can})$ orientation-preserving, where $\omega$ denotes the orientation of $\mathbb{C}^n_{\mathbb{R}}$ from Proposition 8.4, and can the canonical orientation of $\mathbb{R}^{2n}$.

We next look at isometric automorphisms of a Hermitian inner product space; i.e., automorphisms that preserve the Hermitian inner product. Since such a space is linearly isometric to $\mathbb{C}^n$ with the standard Hermitian inner product, we only need to study linear transformations $L : \mathbb{C}^n \to \mathbb{C}^n$ that satisfy $\langle Lv, Lw \rangle = \langle v, w \rangle$, $v, w \in \mathbb{C}^n$. Recall that the *adjoint* $L^* : \mathbb{C}^n \to \mathbb{C}^n$ of $L$ is defined by $\langle L^*v, w \rangle = \langle v, Lw \rangle$ for $v$, $w$ in $\mathbb{C}^n$. If $v_k$ is an orthonormal basis of $\mathbb{C}^n$, then the matrix $[L^*]$ of $L$ in this basis is the conjugate transpose of the matrix $[L]$ of $L$:

$$[L^*]_{ij} = \langle L^*v_j, v_i \rangle = \langle v_j, Lv_i \rangle = \overline{\langle Lv_i, v_j \rangle} = \overline{[L]}_{ji}.$$

Now, the transformation $L$ preserves the Hermitian inner product iff $\langle Lv, Lw \rangle = \langle v, w \rangle$, iff $\langle L^*Lv, w \rangle = \langle v, w \rangle$ for all $v$ and $w$ in $\mathbb{C}^n$; equivalently, $L^*L = LL^* = 1_{\mathbb{C}^n}$. In terms of matrices, $[L]\overline{[L]}^t = \overline{[L]}^t[L] = I_n$.

DEFINITION 8.5. The *unitary group* $U(n)$ is the subgroup of $GL(n, \mathbb{C})$ that preserves the Hermitian inner product:

$$U(n) = \{M \in GL(n, \mathbb{C}) \mid \overline{M}^t M = M\overline{M}^t = I_n\}.$$

LEMMA 8.2. $h(U(n)) = h(GL(n, \mathbb{C})) \cap SO(2n)$.

PROOF. By Lemma 8.1, it suffices to show that $M \in U(n)$ iff $h(M) \in O(2n)$. But since $h(\overline{M}^t) = h(M)^t$, we have that $A \in U(n)$ iff $A\overline{A}^t = I_n$ iff $h(A)h(A)^t = I_{2n}$ iff $h(A) \in O(2n)$. □

For example, $U(1)$ is the group of all complex numbers $z$ such that $\langle zv, zw \rangle = \langle v, w \rangle$. Since $\langle zv, zw \rangle = zv\overline{zw} = z\overline{z}\langle v, w \rangle$, $U(1)$ is the group $S^1$ of all unit complex numbers, and

$$h(U(1)) = \{ \begin{pmatrix} a & -b \\ b & a \end{pmatrix} \mid a^2 + b^2 = 1 \}.$$

The exponential map $e : M_{n,n}(\mathbb{C}) \to GL(n, \mathbb{C})$ is defined as in the real case by

$$e^M := \sum_{k=0}^{\infty} \frac{M^k}{k!};$$

cf. Examples and Remarks 4.2(iv) in Chapter 4. Since $h(MN) = h(M)h(N)$, we have that $h(e^M) = e^{h(M)}$, and the diagram

$$
\begin{array}{ccc}
M_{n,n}(\mathbb{C}) & \xrightarrow{\ h\ } & M_{2n,2n}(\mathbb{R}) \\
{\scriptstyle e}\downarrow & & \downarrow{\scriptstyle e} \\
GL(n, \mathbb{C}) & \xrightarrow[\ h\ ]{} & GL(2n, \mathbb{R})
\end{array}
$$

commutes. In particular, $e$ is one-to-one on a neighborhood $V$ of $I_n \in M_{n,n}(\mathbb{C})$, so that given $A \in U(n) \cap e^V$, $A = e^M$ for a unique $M$. Then $I = A\overline{A}^t = e^M e^{\overline{M}^t}$, and $e^{\overline{M}^t} = (e^M)^{-1} = e^{-M}$; i.e., $M + \overline{M}^t = 0$. Conversely, if $M + \overline{M}^t = 0$, then $e^M \in U(n)$. This shows that $U(n)$ is an $n^2$-dimensional Lie subgroup of $GL(n, \mathbb{C})$ with Lie algebra $\mathfrak{u}(n)$ canonically isomorphic to

$$\{ M \in M_{n,n}(\mathbb{C}) \mid M + \overline{M}^t = 0 \} \cong \{ \begin{pmatrix} A & -B \\ B & A \end{pmatrix} \mid A + A^t = 0, B = B^t \},$$

a fact that also follows from Lemma 8.2.

PROPOSITION 8.5. *For any $M \in \mathfrak{u}(n)$, there exists $A \in U(n)$ such that*

$$AMA^{-1} = \begin{pmatrix} i\lambda_1 & & \\ & \ddots & \\ & & i\lambda_n \end{pmatrix}, \qquad \lambda_i \in \mathbb{R};$$

*equivalently,*

$$h(A)h(M)h(A)^{-1} = \begin{pmatrix} & & & -\lambda_1 & & \\ & & & & \ddots & \\ & & & & & -\lambda_n \\ \lambda_1 & & & & & \\ & \ddots & & & & \\ & & \lambda_n & & & \end{pmatrix}.$$

PROOF. Recall that an endomorphism $L$ of $\mathbb{C}^n$ is *normal* if $LL^* = L^*L$. The spectral theorem asserts that a normal endomorphism of $\mathbb{C}^n$ has $n$ orthonormal eigenvectors. The endomorphism $v \mapsto Lv := M \cdot v$ is skew-adjoint, hence normal, so that there exists a matrix $A \in GL(n, \mathbb{C})$ such that $AMA^{-1}$ is diagonal. Since the eigenvectors are orthonormal, $A \in U(n)$. Furthermore, the conjugate transpose of

$$AM\overline{A}^t = \begin{pmatrix} \lambda_1 & & \\ & \ddots & \\ & & \lambda_n \end{pmatrix}$$

is $-AM\overline{A}^t$, so that $\lambda_i + \overline{\lambda}_i = 0$, and each $\lambda_i$ is imaginary. $\qquad\square$

A *polynomial on* $\mathfrak{u}(n)$ is a map $p : \mathfrak{u}(n) \to \mathbb{R}$ such that $p \circ h^{-1} : h(\mathfrak{u}(n)) \subset M_{2n,2n}(\mathbb{R}) \to \mathbb{R}$ is a polynomial in the usual sense. $p$ is said to be *invariant* if it is invariant under the action of $U(n)$. For example, $f_{2k} \circ h$ and Pf $\circ h$ are invariant polynomials, because $h(U(n)) \subset SO(2n)$. Notice that $f_k$ can actually be defined on $M_{n,n}(\mathbb{C})$ as in Example 1.1, but is not, in general, real-valued. However, the polynomial $f_k^i$, where $f_k^i(M) = f_k(iM)$, is real-valued on $\mathfrak{u}(n)$:

$$\sum_{k=0}^{n} (-1)^k \overline{f_k^i}(M) x^{n-k} = \sum_{k=0}^{n} (-1)^k \overline{f_k}(iM) x^{n-k} = \overline{\det}(xI_n - iM)$$

$$= \det(xI_n + i\overline{M}^t) = \det(xI_n - iM)$$

$$= \sum_{k=0}^{n} (-1)^k f_k^i(M) x^{n-k}.$$

THEOREM 8.1. *Any invariant polynomial on the Lie algebra* $\mathfrak{u}(n)$ *is a polynomial in* $f_1^i, \ldots, f_n^i$.

PROOF. For $z_1, \ldots, z_n \in \mathbb{C}$, let $(z_1 \ldots z_n)$ denote the matrix

$$\begin{pmatrix} z_1 & & \\ & \ddots & \\ & & z_n \end{pmatrix}.$$

Given an invariant polynomial $f$ on $\mathfrak{u}(n)$, it suffices to show that there exists a polynomial $p$ such that

$$f(i\lambda_1 \ldots i\lambda_n) = p(f_1^i(i\lambda_1 \ldots i\lambda_n), \ldots, f_n^i(i\lambda_1 \ldots i\lambda_n))$$

for all $\lambda_i \in \mathbb{R}$. To see this, denote by $q$ the polynomial given by $q(\lambda_1, \ldots, \lambda_n) = f(i\lambda_1 \ldots i\lambda_n)$. Since any pair $(\lambda_k, \lambda_l)$ can be transposed when conjugating the matrix $(i\lambda_1 \ldots i\lambda_n)$ by an appropriate $A \in U(n)$, $q$ is symmetric, so that

$$q(\lambda_1, \ldots, \lambda_n) = p(s_1(\lambda_1, \ldots, \lambda_n), \ldots, s_n(\lambda_1, \ldots, \lambda_n))$$

for some polynomial $p$. Then

$$f(i\lambda_1 \ldots i\lambda_n) = q(\lambda_1, \ldots, \lambda_n) = p(s_1(\lambda_1, \ldots, \lambda_n), \ldots, s_n(\lambda_1, \ldots, \lambda_n))$$

$$= p(f_1(\lambda_1 \ldots \lambda_n), \ldots, f_n(\lambda_1 \ldots \lambda_n)).$$

But $(\lambda_1 \ldots \lambda_n) = -i(i\lambda_1 \ldots i\lambda_n)$, so that $f_k(\lambda_1 \ldots \lambda_n) = (-1)^k f_k^i(i\lambda_1 \ldots i\lambda_n)$. Thus,

$$f(i\lambda_1 \ldots i\lambda_n) = p(-f_1^i(i\lambda_1 \ldots i\lambda_n), \ldots, (-1)^n f_n^i(i\lambda_1 \ldots i\lambda_n)),$$

as claimed. $\qquad\square$

According to Theorem 8.1, $f_{2k} \circ h$ and $\mathrm{Pf} \circ h$ are polynomials in $f_1^i, \ldots, f_n^i$. These polynomials can be described explicitly:

PROPOSITION 8.6. $f_{2k} \circ h = (-1)^k \sum_{l=0}^{2k} (-1)^l f_l^i f_{2k-l}^i$, $\mathrm{Pf} \circ h = (-1)^{[n/2]} f_n^i$.

PROOF. By Lemma 8.1,

$$|\det(xI_n - M)|^2 = \det(h(xI_n - M)) = \det(xI_{2n} - h(M)) = \sum_{k=0}^{n} x^{2(n-k)} f_{2k} \circ h(M).$$

On the other hand,

$$|\det(xI_n - M)|^2 = |\det(xI_n - i(-iM))|^2 = \left| \sum_{k=0}^{n} (-1)^k (-i)^k f_k^i(M) x^{n-k} \right|^2$$

$$= |x^n + ix^{n-1} f_1^i(M) - x^{n-2} f_2^i(M) - ix^{n-3} f_3^i(M) + \cdots |^2$$

$$= |(x^n - x^{n-2} f_2^i(M) + x^{n-4} f_4^i(M) \cdots)$$

$$+ i(x^{n-1} f_1^i(M) - x^{n-3} f_3^i(M) + x^{n-5} f_5^i(M) \cdots)|^2$$

$$= (x^n - x^{n-2} f_2^i(M) + x^{n-4} f_4^i(M) \cdots)^2$$

$$+ (x^{n-1} f_1^i(M) - x^{n-3} f_3^i(M) + x^{n-5} f_5^i(M) \cdots)^2.$$

The coefficient of $x^{2n-2k}$ in the last equality is

$$\sum_{k-j \text{ even}} (-1)^{(k-j)/2} f_{k-j}^i(M) (-1)^{(k+j)/2} f_{k+j}^i(M)$$

$$+ \sum_{k-j \text{ odd}} (-1)^{(k-j+1)/2} f_{k-j}^i(M) (-1)^{(k+j+1)/2} f_{k+j}^i(M)$$

$$= (-1)^k [ \sum_{k-j \text{ even}} f_{k-j}^i(M) f_{k+j}^i(M) - \sum_{k-j \text{ odd}} f_{k-j}^i(M) f_{k+j}^i(M)]$$

$$= (-1)^k \sum_{k-j} (-1)^{k-j} f_{k-j}^i(M) f_{k+j}^i(M)$$

$$= (-1)^k \sum_{l} (-1)^l f_l^i(M) f_{2k-l}^i(M).$$

This establishes the first identity in the proposition. For the one involving the Pfaffian, it suffices to check the formula in the case when $M = (i\lambda_1 \ldots i\lambda_n)$. Then

$$\mathrm{Pf}(h(M)) = \mathrm{Pf} \begin{pmatrix} & & -\lambda_1 & & \\ & & & \ddots & \\ & & & & -\lambda_n \\ \lambda_1 & & & & \\ & \ddots & & & \\ & & \lambda_n & & \end{pmatrix}$$

$$= \epsilon^{1(n+1)2(n+2)\ldots n(2n)} (-1)^n \lambda_1 \cdots \lambda_n$$

$$= (-1)^{[n/2]} (-1)^n \lambda_1 \cdots \lambda_n = (-1)^{[n/2]} \det iM = (-1)^{[n/2]} f_n^i(M). \qquad \square$$

EXERCISE 161. Show that any Hermitian inner product is determined by its norm function. Specifically,

$$\text{Re}\langle v, w \rangle = \frac{1}{4}(|v + w|^2 - |v - w|^2), \qquad \text{Im}\langle v, w \rangle = \frac{1}{4}(|v + iw|^2 - |v - iw|^2).$$

EXERCISE 162. Show that a symplectic form on $V$ induces a Hermitian inner product on $V$, and that conversely, if $\langle, \rangle$ is a Hermitian inner product on $V$, then $\sigma(v, w) = -\text{Im}\langle v, w \rangle$ defines a symplectic form on $V$.

EXERCISE 163. Fill in the details of the proof of Lemma 8.1.

## 9. Chern Classes

A *complex rank $n$ vector bundle* is a fiber bundle with fiber $\mathbb{C}^n$ and structure group $GL(n, \mathbb{C})$. Thus, the fiber over each point inherits a complex vector space structure. The realification $\xi_\mathbb{R}$ of a complex bundle $\xi$ and the complexification $\xi_\mathbb{C}$ of a real bundle $\xi$ are defined in the same way as for vector spaces. In particular, $\xi_\mathbb{R}$ is orientable, with a canonical orientation.

A *Hermitian metric* on a complex vector bundle $\xi = \pi : E \to M$ is a section of the bundle $\text{Hom}(\xi \otimes \xi, \mathbb{C})$ which is a Hermitian inner product on each fiber. Such a metric always exists, since one can choose a Euclidean metric on $\xi_\mathbb{R}$, and this metric induces, by Exercise 161, a Hermitian one on $\xi$. A *Hermitian connection* $\nabla$ on $\xi$ is one for which the metric is parallel. In this case,

$$X\langle U, V \rangle = \langle \nabla_X U, V \rangle + \langle U, \nabla_X V \rangle, \qquad X \in \mathfrak{X}M, \quad U, V \in \Gamma\xi.$$

Just as in the Riemannian case, the curvature tensor $R$ of a Hermitian connection is skew-adjoint:

$$\langle R(X, Y)U, V \rangle = -\langle U, R(X, Y)V \rangle.$$

Thus, given $p \in M$, and an orthonormal basis $b : \mathbb{C}^n \to E_p$, $b^{-1} \circ R(x, y) \circ b \in \mathfrak{u}(n)$ for any $x, y \in M_p$.

Let $g_k^i$ denote the polarization of the polynomial $f_k^i$ from the previous section. By Proposition 1.1, $g_k^i$ induces a parallel section $\bar{g}_k^i$ of $\text{End}_k(\xi)^*$, and $\bar{g}_k^i(R^k)$ is a $2k$-form on $M$. By Theorem 1.1, this form is closed, and its cohomology class is independent of the choice of connection.

DEFINITION 9.1. The *$k$-th Chern class* $c_k(\xi) \in H^{2k}(M)$ of $\xi$ is the class determined by the $2k$-form

$$c_k = \frac{1}{(2\pi)^k} \bar{g}_k^i(R^k).$$

$c_k$ is called the *$k$-th Chern form* (of the connection). The *total Chern class* of $\xi$ is

$$c(\xi) = c_0(\xi) + c_1(\xi) + \cdots + c_n(\xi),$$

where $c_0(\xi)$ denotes the class containing the constant function 1.

EXAMPLE 9.1. A complex line bundle (or, more accurately, its realification) is equivalent to an oriented real plane bundle: To see this, it suffices to exhibit a complex structure $J$ on an oriented plane bundle $\xi = \pi : E \to M$. Choose a Euclidean metric on $\xi$, and for nonzero $u$ in $E$, define $Ju$ to be the unique vector of norm equal to that of $u$, such that $u, Ju$ is a positively oriented orthogonal basis of $E_{\pi(u)}$. $J$ is then a complex structure on $\xi$, and it makes sense to talk about the first Chern class $c_1(\xi)$ of $\xi$. Given a Hermitian connection on $\xi$, the Chern form $c_1$ at $p \in M$ is given by

$$c_1(x,y) = \frac{1}{2\pi}\operatorname{tr} iR(x,y) = \frac{1}{2\pi}\langle iR(x,y)u, u\rangle = \frac{1}{2\pi}\langle R(x,y)u, -iu\rangle$$
$$= \frac{1}{2\pi}\langle R(x,y)iu, u\rangle \in \mathbb{R},$$

for unit $u$ in $E_p$. In terms of the underlying real plane bundle,

$$c_1(x,y) = \frac{1}{2\pi}\langle R(x,y)Ju, u\rangle,$$

where $\langle,\rangle$ now denotes the Euclidean metric on $\xi_{\mathbb{R}}$ induced by the real part of the Hermitian metric on $\xi$. By Examples and Remarks 3.1(i), the first Chern class of a complex line bundle equals the Euler class of its realification.

More generally, consider a complex rank $n$ bundle $\xi = \pi : E \to M$ with Hermitian connection $\nabla$. The real part of the Hermitian metric is a Euclidean metric which is parallel under $\nabla$. Thus, $\nabla$ induces a Riemannian connection $\tilde{\nabla}$ on $\xi_{\mathbb{R}}$. Since $iU$ is parallel along a curve whenever $U$ is, the complex structure $J$ is parallel.

A Hermitian orthonormal basis $b : \mathbb{C}^n \to E_p$ induces an isomorphism $B : \mathfrak{u}(E_p) \to \mathfrak{u}(n)$. There is a corresponding Euclidean orthonormal basis $b \circ h^{-1} : \mathbb{R}^{2n} \to E_p$ that induces an isomorphism $\tilde{B} : \mathfrak{o}(E_p) \to \mathfrak{o}(2n)$. Denote by $\tilde{h}$ the corresponding homomorphism $\tilde{B}^{-1} \circ h \circ B : \mathfrak{u}(E_p) \to \mathfrak{o}(E_p)$. If $R, \tilde{R}$ denote the curvature tensors of $\nabla$ and $\tilde{\nabla}$, then $\tilde{R} = \tilde{h} \circ R$. Thus,

$$\tilde{B}\tilde{R} = \tilde{B} \circ \tilde{h} \circ R = \tilde{B} \circ \tilde{B}^{-1} \circ h \circ B \circ R = h(BR),$$

and by Proposition 8.6,

$$\bar{g}_{2k}(\tilde{R}^{2k}) = f_{2k}(\tilde{B}\tilde{R}) = f_{2k} \circ h(BR) = (-1)^k \sum_{l=0}^{2k}(-1)^l f_l^i(BR) f_{2k-l}^i(BR)$$

$$= (-1)^k \sum_{l=0}^{2k}(-1)^l \bar{g}_l^i(R^l) \wedge \bar{g}_{2k-l}^i(R^{2k-l}).$$

Similarly,

$$\bar{\mathrm{pf}}(\tilde{R}^n) = \mathrm{Pf}(\tilde{B}\tilde{R}) = \mathrm{Pf} \circ h(BR) = (-1)^{[n/2]} f_n^i(BR) = (-1)^{[n/2]} \bar{g}_n^i(R^n).$$

Summarizing, we have proved the following:

THEOREM 9.1. If $\xi$ is a complex rank $n$ bundle, then

$$p_k(\xi_{\mathbb{R}}) = (-1)^k \sum_{l=0}^{2k}(-1)^l c_l(\xi) \cup c_{2k-l}(\xi), \qquad k = 1, \ldots, n,$$

*and*

$$e(\xi_{\mathbb{R}}) = (-1)^{[n/2]} c_n(\xi).$$

In many references, one finds instead $e(\xi_{\mathbb{R}}) = c_n(\xi)$. The sign in Theorem 9.1 stems from our choice of imbedding $h : \mathbb{C}^n_{\mathbb{R}} \to \mathbb{R}^{2n}$ and the resulting orientation on $\xi_{\mathbb{R}}$.

Instead of looking at the Pontrjagin classes of the realification of a complex bundle $\xi$, one can instead begin with a real bundle $\xi$, and look at the Chern classes of its complexification $\xi_{\mathbb{C}} = (\xi \oplus \xi, J)$, where $J(u, v) = (-v, u)$.

THEOREM 9.2. *If $\xi$ is a real vector bundle, then $c_{2k}(\xi_{\mathbb{C}}) = (-1)^k p_k(\xi)$.*

PROOF. Consider a Euclidean metric $\langle , \rangle$ on $\xi$, and denote by $\langle , \rangle_{\mathbb{C}}$ the Hermitian metric on $\xi_{\mathbb{C}}$ the real part of which is $\langle , \rangle$. Then

$$\langle (U_1, U_2), (V_1, V_2) \rangle_{\mathbb{C}} = \langle U_1, V_1 \rangle + \langle U_2, V_2 \rangle + i(\langle U_2, V_1 \rangle - \langle U_1, V_2 \rangle),$$

for $U_i, V_i \in \Gamma\xi$; cf. Exercise 164. A Riemannian connection $\nabla$ on $\xi$ induces one on $\xi \oplus \xi$, with

$$\tilde{\nabla}_x(U_1, U_2) = (\nabla_x U_1, \nabla_x U_2).$$

If $(U_1, U_2)$ is parallel along a curve, then so is $J(U_1, U_2) = (-U_2, U_1)$, implying that $J$ is parallel. Furthermore, if $(V_1, V_2)$ is also parallel, then the function $\langle (U_1, U_2), (V_1, V_2) \rangle_{\mathbb{C}}$ is constant. Thus, $\tilde{\nabla}$ is a Hermitian connection, with curvature tensor

$$\tilde{R}(x, y)(U_1, U_2)(p) = (R(x, y)U_1(p), R(x, y)U_2(p)), \qquad x, y \in M_p, \quad U_i \in \Gamma\xi.$$

Denoting by $[R] \in \mathfrak{o}(n)$ the matrix of $R(x, y)$ in an orthonormal basis $b : \mathbb{R}^n \to E_p$ of $E_p$, we have that the matrix of $\tilde{R}(x, y)$ in the basis $(b, b)$ of $E_p \oplus E_p$ is given by

$$\begin{pmatrix} [R] & 0 \\ 0 & [R] \end{pmatrix}.$$

In the corresponding Hermitian basis $(b, b) \circ h$, this matrix is just the original $[R]$. In other words, if $B : \mathfrak{o}(E_p) \to \mathfrak{o}(n)$ is the isomorphism induced by $b$, and $\tilde{B} : \mathfrak{u}(E_p \oplus E_p) \to \mathfrak{u}(n)$ the one induced by $(b, b) \circ h$, then $\tilde{B}\tilde{R} = BR$. Thus,

$$\bar{g}^i_{2k}(\tilde{R}^{2k}) = f^i_{2k}(\tilde{B}\tilde{R}) = f^i_{2k}(BR) = (-1)^k f_{2k}(BR) = (-1)^k \bar{g}_{2k}(R^{2k}),$$

which establishes the claim. $\square$

To account for the odd Chern classes that are missing in the above theorem, define the *conjugate bundle* $\bar{\xi}$ of a complex bundle $\xi$ to be the (complex) bundle with the same underlying total space and addition, but with scalar multiplication $\bullet$ given by $\alpha \bullet u = \bar{\alpha} u$, where the right side is the usual scalar multiplication in $\xi$. Although the identity is a *real* bundle equivalence, $\xi$ and its conjugate need not be equivalent as complex bundles; i.e., there may not be an equivalence $h : \xi \to \bar{\xi}$ satisfying $h(\alpha u) = \alpha \bullet h(u) = \bar{\alpha} h(u)$. Such an $h$ does, however, exist when $\xi$ is the complexification $\eta_{\mathbb{C}}$ of a real bundle $\eta$: It is straightforward to verify that the formula $h(u, v) = (u, -v)$ defines such an equivalence.

PROPOSITION 9.1. *If $\xi$ is a complex bundle, then the total Chern class of its conjugate is given by*

$$c(\bar{\xi}) = 1 - c_1(\xi) + c_2(\xi) - c_3(\xi) + \cdots$$

PROOF. A Hermitian inner product $\langle,\rangle$ on $\xi$ induces a Hermitian inner product $\overline{\langle,\rangle}$ on $\overline{\xi}$ given by

$$\overline{\langle U, V\rangle} := \overline{\langle U, V\rangle} = \langle V, U\rangle, \qquad U, V \in \Gamma\xi.$$

A Hermitian connection $\nabla$ on $\xi$ then becomes also a Hermitian connection $\overline{\nabla}$ on the conjugate bundle, and their curvature tensors are related by

$$\overline{\langle \overline{R}(x,y)u, v\rangle} = \langle v, R(x,y)u\rangle = \overline{\langle R(x,y)u, v\rangle}.$$

Since the eigenvalues of $R$ and $\overline{R}$ are imaginary, $\overline{R}(x,y) = -R(x,y)$. Thus,

$$\overline{g}_k^i(\overline{R}) = \overline{g}_k^i(-R) = (-1)^k \overline{g}_k^i(R),$$

which establishes the claim.     $\square$

We have seen that given a real bundle $\xi$, its complexification $\xi_{\mathbb{C}}$ is equivalent, in the complex sense, to the conjugate bundle $\overline{\xi}_{\mathbb{C}}$. Proposition 9.1 then implies the following:

COROLLARY 9.1. *If $\xi$ is a real bundle, then the odd Chern classes of its complexification are zero.*

THEOREM 9.3. *For complex bundles $\xi$ and $\eta$, $c(\xi \oplus \eta) = c(\xi) \cup c(\eta)$.*

PROOF. Notice that for complex matrices $M$, $N$,

$$f_k^i(A \circledast B) = \sum_{l=0}^{k} f_l^i(A) f_{k-l}^i(B).$$

The statement now follows by an argument similar to that in Theorem 4.2.     $\square$

EXAMPLE 9.2. Consider an oriented rank 4 bundle $\xi_0$, and suppose its structure group reduces to $S_+^3 \subset SO(4)$; cf. Section 5. Thus, if $P$ denotes the total space of the corresponding principal $S^3$-bundle, then $\xi_0 = \pi : P \times_{S_+^3} \mathbb{H} \to P/S_+^3$, with $S_+^3$ acting on $\mathbb{H} = \mathbb{R}^4$ by left multiplication. Any quaternion $q = a + bi + cj + dk$ can be written as $(a + bi) + j(c - di) = z_1 + jz_2$ for some complex numbers $z_1$, $z_2$. The map

$$\overline{h} : \mathbb{H} \longrightarrow \mathbb{C}^2,$$

$$z_1 + jz_2 \longmapsto (z_1, z_2), \qquad z_i \in \mathbb{C}$$

becomes a complex isomorphism if we define scalar multiplication in $\mathbb{H}$ by $\alpha(z_1 + jz_2) := (z_1 + jz_2)\alpha = z_1\alpha + jz_2\alpha$.

The map $\overline{h}$ in turn induces a homomorphism $\overline{h} : GL(1, \mathbb{H}) \to GL(2, \mathbb{C})$ determined by

$$\overline{h}(qu) = \overline{h}(q)\overline{h}(u), \qquad q \in \mathbb{H} \setminus \{0\}, \quad u \in \mathbb{H}.$$

Given $q = z_1 + jz_2 \in S^3$ and $u = w_1 + jw_2 \in \mathbb{H}$,

$$qu = (z_1 + jz_2)(w_1 + jw_2) = (z_1w_1 - \overline{z_2}w_2) + j(z_2w_1 + \overline{z_1}w_2).$$

Recalling that $z_1\overline{z_1} + z_2\overline{z_2} = 1$, we conclude that

$$\overline{h}(z_1 + jz_2) = \begin{pmatrix} z_1 & -\overline{z_2} \\ z_2 & \overline{z_1} \end{pmatrix} \in U(2).$$

This exhibits $\xi_0$ as the realification of a complex bundle $\xi$ with group $U(2)$. By Theorem 9.1,

$$p_1(\xi_0) = -(2c_2(\xi) - c_1(\xi) \cup c_1(\xi)) = 2e(\xi_0) + c_1^2(\xi).$$

Consider the map $L : \mathbb{H} \to \mathbb{H}$ that sends $q \in \mathbb{H}$ to $qj$. $L$ preserves addition, and given $\alpha \in \mathbb{C}$, $q = z_1 + jz_2 \in \mathbb{H}$,

$$L(\alpha q) = L(z_1\alpha + jz_2\alpha) = (z_1\alpha + jz_2\alpha)j = z_1j\overline{\alpha} + jz_2j\overline{\alpha} = (z_1 + jz_2)j\overline{\alpha}$$
$$= \overline{\alpha}Lq.$$

Thus, $L$ induces a complex equivalence $\xi \cong \overline{\xi}$, so that $c_1(\xi) = 0$ by Proposition 9.1, and

$$p_1(\xi_0) = 2e(\xi_0),$$

a property already observed earlier in the special case that the base is a 4-sphere, cf. Corollary 5.1.

EXERCISE 164. Let $\xi$ be a real vector bundle with complexification $\xi_{\mathbb{C}} = (\xi \oplus \xi, J)$, $J(u, v) = (-v, u)$. A Euclidean metric on $\xi$ extends naturally to $\xi \oplus \xi$ by setting

$$\langle (U_1, U_2), (V_1, V_2) \rangle = \langle U_1, V_1 \rangle + \langle U_2, V_2 \rangle, \qquad U_i, V_i \in \Gamma\xi.$$

By Exercise 161, there exists a unique Hermitian metric $\langle , \rangle_{\mathbb{C}}$ on $\xi_{\mathbb{C}}$ the norm function of which equals that of the Euclidean metric. Prove that

$$\langle (U_1, U_2), (V_1, V_2) \rangle_{\mathbb{C}} = \langle U_1, V_1 \rangle + \langle U_2, V_2 \rangle + i(\langle U_2, V_1 \rangle - \langle U_1, V_2 \rangle).$$

EXERCISE 165. Determine the total Chern class of $\gamma_{1,1}^{\mathbb{C}}$ (observe that $G_{1,1}^{\mathbb{C}}$ is just $\mathbb{C}P^1 = S^2$).

# Bibliography

[1] W. Ballmann, *Spaces of Nonpositive Curvature*, Birkhäuser, Basel 1995.

[2] W. Ballmann, V. Schroeder, M. Gromov, *Manifolds of Nonpositive Curvature*, Birkhäuser, Boston 1985.

[3] M. Berger, B. Gostiaux, *Géométrie différentielle*, Armand Colin, Paris 1972.

[4] A. Besse, *Einstein Manifolds*, Springer-Verlag, Berlin Heidelberg 1987.

[5] R. Bott, L. W. Tu, *Differential Forms in Algebraic Topology*, Springer-Verlag, New York 1982.

[6] J.-P. Bourguignon, H. B. Lawson, Jr., *Stability and isolation phenomena for Yang-Mills theory*, Comm. Math. Phys. **79** (1982), 189–230.

[7] J.-P. Bourguignon, H. B. Lawson, Jr., J. Simons, *Stability and gap phenomena for Yang-Mills fields*, Proc. Nat. Acad. Sci. U.S.A. **76** (1979), 1550–1553.

[8] G. E. Bredon, *Introduction to Compact Transformation Groups*, Academic Press, New York 1972.

[9] J. Cheeger, *Some examples of manifolds of nonnegative curvature*, J. Diff. Geom. **8** (1973), 623–628.

[10] J. Cheeger, D. Ebin, *Comparison Theorems in Riemannian Geometry*, North Holland, New York 1975.

[11] J. Cheeger, D. Gromoll, *On the structure of complete manifolds of nonnegative curvature*, Ann. of Math. **96** (1972), 413–443.

[12] M. Do Carmo, *Differential Forms and Applications*, Springer-Verlag, Berlin Heidelberg 1994.

[13] ———, *Riemannian Geometry*, Birkhäuser, Boston 1992.

[14] S. Gallot, D. Hulin, J. Lafontaine, *Riemannian Geometry* (2nd edition), Springer-Verlag, Berlin Heidelberg 1990.

[15] D. Gromoll, W. Klingenberg, W. Meyer, *Riemannsche Geometrie im Großen* (2nd edition), Springer-Verlag, Berlin Heidelberg 1975.

[16] M. Gromov, J. Lafontaine, P. Pansu, *Structures métriques pour les variétés Riemanniennes*, Cedic/Fernand Nathan, Paris 1981.

[17] S. Helgason, *Differential Geometry, Lie Groups and Symmetric Spaces*, Academic Press, New York 1962.

[18] D. Husemoller, *Fiber Bundles* (3rd edition), Springer-Verlag, New York 1994.

[19] J. Jost, *Riemannian Geometry and Geometric Analysis*, Springer, Berlin Heidelberg 1995.

[20] S. Kobayashi, *Differential Geometry of Complex Vector Bundles*, Iwanami Shoten and Princeton University Press, Princeton 1987.

[21] S. Kobayashi, K. Nomizu, *Foundations of Differential Geometry*, John Wiley & Sons, New York 1963.

[22] H. B. Lawson, Jr., *The theory of Gauge fields in four dimensions*, Amer. Math. Soc, CBMS **58**, Providence 1985.

[23] H. B. Lawson, Jr., M.-L. Michelson, *Spin Geometry*, Princeton University Press, Princeton 1989.

[24] J. Milnor, *Morse Theory*, Princeton University Press, Princeton 1963.

[25] ———, *Topology from the Differentiable Viewpoint*, University Press of Virginia, Charlottesville 1965.

[26] J. Milnor, J. Stasheff, *Characteristic Classes*, Princeton University Press, Princeton 1974.

[27] B. O'Neill, *The fundamental equations of a submersion*, Michigan Math. J. **13** (1966), 459–469.

[28] _____, *Semi-Riemannian Geometry*, Academic Press, New York 1983.

[29] M. Özaydın, G. Walschap, *Vector bundles with no soul*, Proc. Amer. Math. Soc. **120** (1994), 565–567.

[30] G. Perelman, *Proof of the soul conjecture of Cheeger and Gromoll*, J. Differential Geom. **40** (1994), 209–212.

[31] P. Petersen, *Riemannian Geometry*, Springer, New York 1998.

[32] W. A. Poor, *Differential Geometric Structures*, McGraw-Hill, New York 1981.

[33] A. Rigas, *Some bundles of nonnegative curvature*, Math. Ann. **232** (1978), 187–193.

[34] M. Spivak, *A Comprehensive Introduction to Differential Geometry*, Publish or Perish, Inc., Berkeley 1979.

[35] N. Steenrod, *The Topology of Fiber Bundles*, Princeton University Press, Princeton 1951.

[36] F. W. Warner, *Foundations of Differentiable Manifolds and Lie Groups*, Springer-Verlag, New York 1983.

[37] G. Whitehead, *Elements of Homotopy Theory*, Springer-Verlag, New York 1978.

# Index

# Graduate Texts in Mathematics

*(continued from page ii)*